DIGITAL (IN)JUSTICE IN THE SMART

Edited by Debra Mackinnon, Ryan Burns, and V

In the contemporary moment, smart cities have become the dominant paradigm for urban planning and administration, which involves weaving the urban fabric with digital technologies. Recently, however, the promises of smart cities have been gradually supplanted by recognition of their inherent inequalities, and scholars are increasingly working to envision alternative smart cities.

Informed by these pressing challenges, *Digital (In)Justice in the Smart City* foregrounds discussions of how we should think of and work towards urban digital justice in the smart city. It provides a deep exploration of the sources of injustice that percolate throughout a range of sociotechnical assemblages, and it questions whether working towards more just, sustainable, liveable, and egalitarian cities requires that we look beyond the limitations of "smartness" altogether. The book grapples with how geographies impact smart city visions and roll-outs, on the one hand, and how (unjust) geographies are produced in smart pursuits, on the other. Ultimately, *Digital (In)Justice in the Smart City* envisions alternative cities – smart or merely digital – and outlines the sorts of roles that the commons, utopia, and the law might take on in our conceptions and realizations of better cities.

(Technoscience and Society)

DEBRA MACKINNON is an assistant professor in the Interdisciplinary Studies Department at Lakehead University.

RYAN BURNS is an assistant professor in the Department of Geography at the University of Calgary.

VICTORIA FAST is an associate professor in the Department of Geography at the University of Calgary.

If our world and our futures are technoscientific, then how should we organize this world? And how should we understand these futures? Technoscience and Society seeks to provide new analytical tools to do this, as well as new empirical insights into the changes happening around us. The series encourages shorter, punchier scholarly books providing a crossover forum in which both established researchers and new and emerging scholars can present their investigations into the ever-changing relationship between technoscience and society.

Also in the series:
An Anthropogenic Table of Elements: Explorations in the Fundamental, ed. Timothy Neale, Thao Phan, and Courtney Addison

Digital (In)Justice in the Smart City

EDITED BY DEBRA MACKINNON, RYAN BURNS, AND VICTORIA FAST

UNIVERSITY OF TORONTO PRESS
Toronto Buffalo London

Toronto Buffalo London
utorontopress.com

ISBN 978-1-4875-2715-0 (cloth)
ISBN 978-1-4875-2716-7 (paper)
ISBN 978-1-4875-2718-1 (EPUB)
ISBN 978-1-4875-2717-4 (PDF)

Library and Archives Canada Cataloguing in Publication

Title: Digital (in)justice in the smart city / edited by Debra Mackinnon, Ryan Burns, and Victoria Fast.
Names: Mackinnon, Debra, editor. | Burns, Ryan (Lecturer in geography), editor. | Fast, Victoria, editor.
Description: Series statement: Technoscience and society | Includes bibliographical references and index.
Identifiers: Canadiana (print) 20220398437 | Canadiana (ebook) 2022039847X | ISBN 9781487527150 (cloth) | ISBN 9781487527167 (paper) | ISBN 9781487527181 (EPUB) | ISBN 9781487527174 (PDF)
Subjects: LCSH: Smart cities – Social aspects. | LCSH: Technology – Social aspects. | LCSH: Urban geography – Social aspects. | LCSH: Social justice. | LCSH: City planning.
Classification: LCC TD159.4 .D54 2023 | DDC 307.760285 – dc23

We wish to acknowledge the land on which the University of Toronto Press operates. This land is the traditional territory of the Wendat, the Anishnaabeg, the Haudenosaunee, the Métis, and the Mississaugas of the Credit First Nation.

University of Toronto Press acknowledges the financial support of the Government of Canada, the Canada Council for the Arts, and the Ontario Arts Council, an agency of the Government of Ontario, for its publishing activities.

Contents

Figures

DIGITAL (IN)JUSTICE IN THE SMART CITY

Introduction
Towards Urban Digital Justice: The Smart City as an Empty Signifier

RYAN BURNS, VICTORIA FAST, AND DEBRA MACKINNON

This book is situated at the critical juncture where future cities are imagined, conceptualized, and operationalized. In the contemporary moment, *smart cities* have become the dominant paradigm for urban planning and administration around the world, which in practice involves suffusing the urban fabric with digital technologies. Kitchin, Lauriault, and McArdle (2015, 18) describe smart cities as "the roll-out of new information and communication technologies ... and neoliberal visions of market-led and technocratic solutions to city governance and development, [which] are promoted as pragmatic, nonideological and commonsensical in approach." This is happening at the same time as – and likely because of – federal governments launching national smart city challenges, private firms looking to take over public services, and public conversations invoking new tech utopias. *Smartness* as a signifier applied to cities is not a new idea, combining elements of *smart growth* and *intelligent cities* as far back in the literature as the late 1990s (Vanolo 2014), largely attendant with promises of improved efficiency and optimized city services (Kitchin 2013; Powell 2021).[1]

Recently, however, the promises of smart cities have been gradually supplanted by recognition of their inherent inequalities and even revanchist urban governance, and scholars are thus increasingly working to envision *alternative* smart cities (Marvin, Luque-Ayala, and McFarlane 2015). Such visions are usually premised on models that reclaim long-standing social movements: social justice in the smart city (Masucci, Pearsall, and Wiig 2019; Rosol, Blue, and Fast 2019), gender in the smart city (Datta 2020), sustainable cities (Karvonen, Cugurullo, and Caprotti 2018) and the right to the smart city (Cardullo, Di Feliciantonio, and Kitchin 2019). These conversations have emerged concomitant with growing calls to explore empirically *actually existing* smart cities, rather than ideal-type imaginaries of planned cities and marketing materials (Shelton, Zook, and Wiig 2015).

These efforts, however, present us with a dilemma: *smart* is usually taken for granted, seen as self-explanatory and as an inherently and unquestionably good value to pursue, and yet undesirable outcomes are understood to be aberrations from an otherwise positive movement. This treatment of the term leaves unexplored and underexamined multiple and contradictory meanings that the term assumes in different articulations. This omission constitutes a technicist politics that forecloses exploring the concept's attendant inequalities, political economies, and ideologies, making it particularly amenable to deleterious processes such as capital accumulation, corporatist ideology, and entrenched state power.

Here, we step back to challenge the assumptions that get worked into *smart* when applied to a particular form of urbanism, and to offer broadly deeper insights into how injustice circulates through digital urban systems. Drawing on the thought of Ernesto Laclau and Chantal Mouffe, we argue that the qualifier *smart*, when applied to cities, is an empty signifier – a pointer without a referent, a signifier without a signified. It contains no inherent meaning on its own, but remains powerful in its flexible application potential. The concept of empty signifiers is a useful theoretical and analytical tool to illuminate the uneven, contingent, and contested implications of the ambivalent meanings of terms and ideas that actors use to characterize smart cities. Conceiving of *smart* in smart city as an empty signifier, then, opens new avenues for pursuing digital justice. As we work to make sense of the ways in which *smart* seems to enrol simultaneously powerful, contradictory, normative, and fearful urban discourses and epistemologies, the empty signifier offers a productive way forward for thinking about the injustices of smartness.

Putting social justice front and centre means that implications for justice, equality, access, and inclusion are foregrounded, which together do more conceptual work for urban scholars than existing critiques of smart cities' social, economic, and political implications. We contend that, although *social justice* can mean multiple things to different people, it does not warrant treatment as an empty signifier, since it is grounded in the concrete material and symbolic realities of those experiencing injustice; any conceptual slippage of the term merely emanates from the complex web of relations and processes that constitute social *in*justice.

As smart cities produce, amplify, and mediate digital urban justice, the authors in this volume reveal new critical insights into *smartness* and its variegated implications. They foreground relations between smartness and social justice, and offer ways to think of and work towards urban digital justice in the smart city. This volume creates space to explore the sources of these injustices that percolate throughout a range of sociotechnical assemblages, questioning whether working towards more just, sustainable, liveable, and egalitarian cities requires that we look beyond the limitations of *smartness* altogether. For this reason, this volume anchors on contemporary debates around smart cities, while keeping a broad perspective on how we might think of urban digital (in)justice writ large. Indeed, *smart cities* as a label captures what, with varying levels of overlap, historically has been called the *digital city*, *intelligent cities*, and so on, and we therefore maintain that the lessons this volume presents will outlast the *smart city* label itself.

By foregrounding relations between smartness and social justice specifically, this volume follows a range of inquiries that we think have emerged as profoundly important to contemporary cities: How do the particular assemblages of technologies, data flows, and actors across geographic contexts influence the differential implications of smart cities? In what ways can smart technologies support or detract from conceptions of social justice? How is *smartness* conceived differently in different contexts, and with what effects on urban governance and models of service provision? Is it desirable – or even possible – to pursue justice while retaining the structure and framing of smartness, or must we abandon smartness in favour of other organizing principles and rationales?

To initiate an exegesis of smartness and smart cities, we next describe empty signifiers and their dangers before establishing smartness as just such an empty signifier; we follow this by reviewing the broad contours of the debates around justice, to remind readers of the long tradition within which this book sits, and then consider the horizons for reframing justice for a digital world. Finally, we lay out the broad contours of this book, and the dialogue and chapters within each.

Empty Signifiers and Their Dangers

The promise of the empty signifier concept derives from the way it highlights the indeterminacy of terms, phrases, ideas, symbols, images, and language – collectively called *signifiers*. In this, we draw upon Laclau (2007) to disrupt the notion that signifiers must always contain concrete and overdetermined meanings. Instead, as Saussure (1959) argued, language is a system of differences, and meaning is constructed *relationally*, through contrast, opposition, or proximity to other signifiers. Signifiers invoke all other signification in the entire system in order to differentiate themselves. This relationality is the source of the meaning, rather than the meaning being inherent to the term or idea itself. This is not to say that signifiers are meaningless – this would make language impossible – as much as that their meaning is contingent on context, history, geography, relationality, ideology, and power. They are ontologically unstable, subject to differentiation across multiple axes of social processes.

Consequently, the construction of a signifier's meaning is itself a politics. Think, for example, of the way *freedom* as a term and idea operates in different contexts for very different purposes. In post–September 11 America, *freedom* worked to legitimize colonialist imaginaries of Afghanistan and Iraq, while simultaneously subtending a jingoist nationalism in popular electoral politics. It was often leveraged to signify something that the United States had and was able to – if not obligated to – deliver to other places that did not have it. At the same time, one would be hard pressed to argue that the value *freedom* is not worth pursuing on some level, and indeed has been effectively leveraged in liberal governance models from the Enlightenment through – albeit often not directly – to the various liberation movements of the twentieth and twenty-first centuries. As further evidence of the emptiness of this signifier, it has also been leveraged in maligned social movements such as the so-called freedom convoys in Canada and around the world protesting government restrictions related to COVID-19.

Signifiers' *identity* construction – or how a term acquires meaning – operates through an exclusionary and inclusionary dialectic. For a term to convey meaning, it must indicate that which it is *not* as much as it must draw "chains of equivalence" (Laclau 2007, 42) with other meanings and signifiers. The signifier must draw a boundary that demarcates its identity from signifiers with different identities (Glasze 2007). Laclau calls this the signifier's radical exclusionary limit – the boundary between affinities and the signifier's radical other. On the side of this boundary opposite the mobilized signifier are signs and symbols meant to be irreconcilably different to it, but on the other side the signifier must oscillate between finding similarities with its proximal signifiers – they are, after all, on the same side of the exclusionary limit – and expressing some form of difference from them. Without this difference, a term's identity is void, as it would be equivalent to other signifiers on this side of the boundary. This is the condition of possibility for the empty signifier, the signifier without a signified. The *empty* of this concept refers to the quality of having ambiguous relations with signifiers on either side of the exclusionary limit; it exemplifies "pure cancellation of all difference" (Laclau 2007, 38). It can mean anything, because of the way it defies any particular form of difference.

A simple example could be Arnstein's (1969) demonstration that *participation* for city administrators can entail everything from duplicitous tokenism and manipulation to full citizen control in a decision-making setting. Every rung of the ladder connecting these two extremes constitutes *participation* to some degree, yet because *participation* defies any signification of its radical other, it can take contradictory forms. In some of its forms, *participation* could even entail non-involvement – or rather, active exclusion from involvement – as a tokenist approach to urban planning, for example, is taken precisely to demobilize entire communities. Participation, as a concept, always entails non-participation, both in conceptual terms (i.e., systems of difference) and in practical terms (i.e., as instituted in urban political relations). In sum, *participation* is an empty signifier because it negates difference: it can be articulated to signify a range of contradictory meanings precisely because it has no grounding.

More concretely, we can see the empty signifier at work in the production of political movements and solidarity. The antagonisms at the heart of identity construction function similarly in the constitution of social movements, epitomized in the struggle between the proletariat and owners of the means of production. The collective, political identity of "proletariat" must congeal around a set of common or contingently agreed-upon goals and struggles, which historically has been the negative opposition to a "common enemy" (Laclau 2007, 41). Such negative constitution of a movement's identity exists in dialectic relation to its positive other: for a social movement to attach to collective goals, strategic actors must articulate chains of equivalence to which its adherents might orient.

To be clear, Laclau and Mouffe's precise formulation of empty signifiers should be understood as distinct from some common misuses and related but irreconcilable metaphors. For one, *empty signifier* does not imply that the term is meaningless. Rather, the empty signifier framing suggests that there is no inherent grounding to a concept, and thus a concept can be used in multiple, often contradictory ways. A concept's meaning arises from the way in which it links with chains of equivalence and negates difference; more plainly, concepts are very meaningful, but those meanings arise in their use rather than before. Second and relatedly, to call a concept an empty signifier does not mean that the concept has no need for differentiation from other concepts. This common slippage comes from misconstruing how meaning is formed across the radical exclusionary limit. Rather than simply claiming that concepts have no meaning, empty signifiers, again, *acquire* meaning in the ways they link with other concepts, terms, and ideas. Third, one might be tempted to think of empty signifiers using related metaphors such as a "Trojan horse." This line of argument would draw attention to the fact that the *true* meaning of the horse-shaped gift was disguised from the recipients of the gift – or, in other words, the transaction masked as gift giving actually served the more nefarious purpose of transporting and unleashing combatant forces. While a potent metaphor, it is different from empty signifiers because the horse had an *actual* grounding, a realizable *true* meaning, even if it was not detectable by the gift recipients. To argue for empty signifiers implies that there is no such grounding for the term.

Smartness as Empty Signifier

> Sidewalk Toronto is not a smart city. It is a colonizing experiment in surveillance capitalism attempting to bulldoze important urban, civic and political issues.
>
> – Jim Balsillie, Research in Motion

The epigraph above points to a common way of thinking about the qualifier *smartness* within smart cities discourses. Jim Balsillie, in the opinion piece from which the epigraph is taken (Balsillie 2018), argues that Toronto's (now defunct) Sidewalk Labs project exudes qualities and characteristics that are incommensurate with the ontologically stable, conceptually straightforward, and ultimately desirable idea of *smart*. For Balsillie, *smart* means something clear: it is the opposite of "a colonizing experiment in surveillance capitalism," however that opposite might be construed. One might surmise that, here, Balsillie draws chains of equivalence to intelligent and wise, prudent and discerning, participatory and liberatory. The rest of the commentary from which this comes is scattered with similarly obtuse terms and ideas used to describe Sidewalk Toronto, such as "gaslighting," "flawed," and "valuable"; he situates it as part of the turn towards "data-driven" governance and as counterproductive to a "sovereign" smart cities strategy. What most stands out about the piece, though, is its authorship by a former chief executive officer of Research in Motion, a multinational corporation perhaps most popularly known for having developed the Blackberry mobile phone. Arguably, Balsillie is himself situated within the global class that stands to benefit from *surveillance capitalism*, or at least from the centralization and protection of capital and governance power that a Sidewalk Toronto might entail. For Balsillie to argue against *surveillance capitalism* might seem from this angle to be working against his own interests. But more pertinently, Balsillie invokes the smart cities vulgate that actors as diverse as global capitalists and grassroots organizers mobilize to great political effect: the *smartness* of smart cities has a definable, stable, and *good* meaning that is categorically worth pursuing.

Why do we start with this discussion? The Balsillie quote usefully captures much of the work *smart* does in popular conversations – and in academic conversations as well, but less frequently. Arguably, one key distinction of *smart* from previous digital city conceptual frameworks (e.g., city of bits, cybercities) is how *smart* articulates keenly with the normative values already mentioned. It is difficult to argue that *smart* is not what we want cities to be, without identifying what, precisely, is meant by the term. Indeed, this is the crux of the issue this Introduction seeks to address – that the term *smart* means nothing on its own, but has been signified by a small number of powerful actors in such a way that it seems to be an inherently desirable vision for cities. This is precisely how the term has gained its power: it was first applied to cities by large technology corporations such as Cisco and IBM, and thus to argue against smartness was either to reject the term for some replacement or to reveal the term's absent grounding.

The first strategy – to reject *smart* altogether – would appear to appeal to terms that might be framed as outside its radical exclusionary limit: *dumb*, for instance. A dumb city, given its own inherent emptiness, could be construed broadly as a city that ultimately rejects digital technologies, is unintelligent and poorly advised, or is framed as some opposite of a Cisco-dominated city. We are drawn also to the idea of the *idiotic city* characterized by "moments of misbehaving, recalcitrance and indifference" (Tironi and Valderrama 2018, 166) that are key to all forms of urbanism, but in particular smart cities. Insofar as idiocy can be pursued as a more desirable value than smartness, it would be framed as a rejection of smartness for an alternative set of norms and values.

Elements of rejecting *smartness* can be seen in scattered examples across the globe, manifesting in prescriptive forms of technological development, creation of legislative limits, and community resistance to smart city development and platform capitalism. Prescriptive forms of technological development, often related to engineering, can employ forms of data segregation on device processing and storage. This can be seen with closed-circuit television cameras that limit recording or wearables where data are stored and analysed on the device. These privacy-by-design solutions attempt to place limits on interoperability and transmission; however, with the incentivization of central processing (e.g., data extraction, machine learning), 5G and other technological advances that are aligned with smart development are sure to curtail these weak pre-emptive forms of resistance. We also see the beginnings of the creation of legislative limits. In 2019, in the wake of the Clearview AI revelations and Black Lives Matter protests, San Francisco and other municipalities banned the use of facial recognition. Others in Silicon Valley (including IBM, Microsoft, and Amazon), not wanting to be on the *wrong side* of a political moment, have put a temporary moratorium on selling facial recognition software. Lastly, community resistance has blocked and stopped smart city development and related platform capitalism. In Toronto, the #blocksidewalk campaign, in part, resulted in the exit of Sidewalk Labs from the Toronto waterfront development. Similarly, in Queens, New York, backlash from activists, union leaders, and legislators made Amazon retract its HQ2 plans in Long Island City. In Berlin, #FuckOffGoogle – a community-led antigentrification organization – prevented the opening of a Google campus in the Kreuzberg neighbourhood. In India, Navi Mumbai rejected participating in the 100 Smart Cities Mission.

The second strategy – to reveal smartness as an empty signifier – might allow a city to pursue a range of future visions under the banner of *smart*, so long as the city can articulate why one approach or another *still* counts as *smart* despite not aligning with *smartness* as articulated by IBM, Cisco, and so on. Rearticulating *smart* is famously observed in the way Barcelona has leveraged its smart city program to enrol the democratic participation of its citizens (Charnock, March, and Ribera-Fumaz 2021; Lynch 2020). Relatedly, in Columbus, Ohio, the city leveraged the 2015 national Smart Cities Challenge to increase transit availability to residents, a barrier to justice that had been expressed by community residents and advocacy organizations (Green 2019). This is not to say that either case is unproblematic; only that the cities have pursued social justice within the *smart* framework by rearticulating what smartness can mean. In other words, they have drawn new chains of equivalence that position *smartness* closer to democratizing and pluralizing processes.

Regardless of the political tactics a city uses to resist *smartness*, the term's malleability and tenacity under diverse sociopolitical conditions and towards complex, contradictory ends underscore its inherent emptiness. Laclau clearly shows, however, that empty signifiers retain important political power precisely because of their emptiness. More directly, empty signifiers enable a hegemonic relation to signification that suppresses class struggle: "The presence of empty signifiers … is the very condition of hegemony" (Laclau 2007, 43). Hegemony, for Laclau, is the *imposition* of a signification process on particular social or political groups, rather than "emerging from the political interaction between groups" (44). Empty signifiers enable a politics of signification, such that terms, concepts, and images come to be aligned with particular chains of equivalence and not others; those imposing the hegemonic relation must claim that only *they* have articulated the empty signifier in a way that benefits society broadly. For the political project of smart cities, the lack that smartness seeks to address can be framed or articulated as unoptimized systems (Powell 2021), inefficient use of natural or human resources (Luque-Ayala, McFarlane, and Marvin 2016), lack of channels for participation (Cardullo and Kitchin 2019), and others. Each such framing implicates the political systems and imaginaries that a smart city administrator mobilizes to address those lacks – for example, shaping how we think about citizenship, participation, natural environments, and class struggle in the smart city.

The status of *smart* as an empty signifier goes a long way towards explaining its various uses within this volume. On the one hand, some of the authors take the meaning of *smart* as essentially synonymous with intelligent, prudent, and tech-saturated, and instead seek to understand the justice implications of smart pursuits. Other authors seek to understand the way *smart* came to hold particular meaning in a unique geographic, political, and sociotechnical context. Still others explore the sets of knowledges and logics that a particular articulation of *smart* enables and supports, showing how smart's chain of equivalences frames much more than access to or distribution of resources, but reshapes how we think of urbanism itself.

Reframing Justice in a Digital World

In choosing to rearticulate *smartness*, we think it pertinent to put *smartness* in conversation with social justice. Indeed, a growing chorus of critical voices is increasingly calling for attention to social justice in the smart city (Araya 2015; Cardullo, Di Feliciantonio, and Kitchin 2019; Coletta et al. 2018; Rosol, Blue, and Fast 2018). However, conversations around social justice in the smart city tend to forego positioning their mobilization of justice within the term's longer conceptual history. In many cases, the term *social justice* is taken to be self-explanatory, or at least not worth unpacking, and instead the debate is often how to achieve *more just* smart cities. Of course, working towards justice is difficult without grappling with the complex ways in which we can think of related ideas such as fairness, equal opportunity, equal distribution, dignity, and so on. We are concerned that we cannot achieve more just smart cities – or more just cities writ large – if we do not know what goal we are aiming for.

Justice and *social justice* do not have singular, straightforward meaning; instead, the potentialities of various conceptions have been drawn out in several directions for millennia. Here, we intend to briefly invoke important threads of debates around social justice, while also setting departure points introduced by digital and smart transformations, in order to provide a more nuanced understanding of the concept. Our objective is not to deeply engage with any particular theory, but instead to survey broadly, in order to demonstrate the complexity of ideas tied in with social justice. To be sure, in this discussion we acknowledge great overlap in philosophical engagements with morality, ethics, and justice, but we focus on the intellectual tradition of justice specifically.

Our engagement with theories of justice is inescapably Western, a reflection of the unfortunately persistent Eurocentrism of urban theory (Derickson 2015; Roy 2016). Although a great simplification, we see the canon as largely deliberating over what constitutes *fair* treatment among individuals, collectives, and institutions such as the state or an economy. Following Sen (1999, 55), different justice frameworks use different "informational bases" to evaluate whether a decision, society, or distribution is fair. That is to say, different frameworks take into account different information to evaluate as criteria for justice. Utilitarianism, for example, takes as its evaluative information happiness or pleasure: a decision or society is just if it maximizes happiness and satisfaction to the highest intensity across the highest number of people (Mill 1906). Thus, to evaluate whether or not a phenomenon is just requires information regarding how much happiness (among other attributes) it cultivates and sustains. In contrast, Nozick's (1974) libertarian framework necessitates prioritization of liberty and protection of rights and property; therefore, its informational basis relates to the degree to which liberties are withheld or threatened.

Social justice frameworks often differ on several axes. First, a framework might consider whether justice entails equal receipt of goods or values, or differential receipt in order to correct for past injustices. If two citizens petition their government for assistance, should they each be given the same amount or a different amount determined from their relative need, the relative impact, or the historic bases of individuals' needs? For Rawls (2001, 43), a principle of justice should "be to the greatest benefit of the least-advantaged members of society" – meaning that, provided all other claims to justice and liberties are satisfied, a decision, society, or distribution is just if it has a disproportionately strong advantage for marginalized individuals and communities. Second, frameworks differ on whether justice should be ensured merely in procedures or in the outcomes of a particular procedure. Stuntz (2011), for example, argues that racial disparities in the US criminal justice system result from justice's being tied to its procedures – how laws are written, enacted, and enforced – rather than to its outcomes; for example, the "War on Drugs." Others, such as Rawls, frame this consideration as "fair equality of *opportunity*" attached to "pure procedural justice" (1971, 73, emphasis added). Here, Rawls's emphasis is on opportunity rather than on outcomes: a system is just if it gives everyone equal opportunity of success, regardless of whether those opportunities are taken on an individual level or whether the opportunities produce equal effects.

Of course, these considerations play out differently across different arenas. For Fraser (1995, 2009), justice can be conceived in a tripartite relation of redistribution, representation, and recognition. Historically, different social movements and social justice frameworks have pursued each of these separately, albeit shifting between them at particular moments. Redistribution historically has been the materialist, political-economic, Marxian approach to social justice: it involves organizing wealth, resources, and material goods in a more just distribution across societies. Social movements that seek redistributive justice typically focus on what is construed as an *economic* sphere. Recognition, in contrast, pertains more to the symbolic and immaterial: it involves working to value different forms of work, bodies, sexualities, and expressions of difference. Movements pursuing this form of justice typically have worked within the realm often thought of as *cultural*; a key example for Fraser is the struggle to recognize women's domestic labour as valuable labour, and the work of antiracist activists to secure the legitimacy of people of colour within the public sphere. The last element of justice, representation, Fraser conceives as the frames within which political claims may be made, and who may make them. Fraser envisions this as analogous to Arendt's (1979, 296) "right to have rights," insofar as representational injustices preclude some individuals and communities from being able to affect political changes and make claims of the state. Recently, Blue, Rosol, and Fast (2019) have argued that this is not merely a dichotomous (in)justice, but rather a modulation of justice that can be understood in relation to Arnstein's ladder of participation, where even having a seat at the table (so to speak) is not inherently enough to remedy misrecognition or maldistribution (see also Blue and Rosol 2022; Fraser 2007).

Iris Marion Young (1990) has provided another important conceptualization of justice that largely concurs with Fraser's framework. For Young, redistributive justice insufficiently tackles relations of exploitation, marginalization, powerlessness, cultural imperialism, and structural violence that make up oppression and injustice. Although, on an important level, distributive justice both causes and stems from these injustices, redistributive practices would fail to remedy their root causes.

The disciplines of geography and urban studies have also fostered some noteworthy conceptions of social justice. David Harvey's (2009) paradigm-shifting "Social Justice and the City" stands out for analytically bringing together capitalist accumulation processes and spatial-urban forms. In the Marxian tradition, Harvey calls for a radical reformulation of power and processes of production in cities, largely orienting his critique around a redistributive approach to justice. A just city would safeguard the ability of its denizens to specify what, precisely, constitutes a just distribution of resources: "We are seeking, in short, a specification of a just distribution justly arrived at" (Harvey 2009, 98). Thus, Fraser's dimension of representation is not absent, but is instead a means to arrive at redistributive justice. In a similar vein, Soja (2010) argues that justice has a geography, and that instances of injustice are tied directly to the inequitable and uneven distribution of resources and services. Fainstein's (2009b, 5) "three hallmarks of urban justice – material equality, diversity, and democracy" differ from the previous two by harkening back to Fraser's conception. In fact, Fainstein complicates the analytically distinct but empirically overlapping dimensions of social justice mentioned above, not by rejecting them but by seeing how they come together in important ways: "When we think about planning for cities, therefore, we must realize that *substance* and *procedure* are inseparable. Open processes do not necessarily produce just outcomes" (2009a, 34, emphasis added).

In what ways should this plethora of frameworks inform how we think of social justice in a digital world? Many answers to this question are available, but here we focus on the question of who is responsible for *making* justice claims, and who is responsible for *satisfying* justice claims. Paralleling Fraser's (2005, 69) assertion that "[g]lobalization is changing the way we argue about justice," we assert that our increasingly digital cities also change the way we argue about justice. For Fraser, globalization has undermined the assumed Keynesian-Westphalian state bases for justice: rather than individuals having at their recourse only claims made of the nation-state, many now recognize that their lives are increasingly framed by transnational agencies such as the United Nations or World Trade Organization, and that justice claims legitimately can be made beyond the nation-state boundary. Similarly, technologies that shape material, symbolic, and political aspects of urban life are not locatable only within the jurisdiction of the state, and much less so within individual municipalities. Private, often multinational, corporations – e.g., Alphabet, IBM, Microsoft – are developing and licensing hardware, software, and platforms that run the smart city. In so doing, they frame claims of what is *important* to know about urbanism and the ways in which we *should* know it, and even provide us imaginaries of "the good city" (Amin 2006). To contest those framings, epistemologies, and visions is, according to many frameworks we discuss above, an issue of *justice*, and claims of *injustice* require appeals often located at multiple scales not reducible to the state or a municipality. Instead, a real case might be made that to redress injustices in the smart city requires demands at the supranational scale.

Still, smart cities have germinated new relations between urban denizens and various scales of the state. Nation-states all over the world have deployed smart initiatives, with notable examples being the US Department of Transportation's Smart City Challenge, Infrastructure Canada's Smart Cities Challenge, and India's 100 Smart Cities Mission. In these and other cases, the way in which the state functions as a conduit for corporate interests complicates efforts to redress injustice: to whom should such claims be made? Elsewhere, we have noted that smart cities presume a tech-savvy individual who is expected to perform tasks in "loving service" of the smart city (Burns and Welker 2022a; Burns and Andrucki 2021). This, of course, is an often-overlooked suite of responsibilities entailed in new conceptualizations of smart *citizens*. If the smart city put the responsibility of ensuring just and equitable outcomes of its initiatives onto ordinary citizens, is this not a new source of injustice, particularly against those who sit askew to the presumed subject of the smart city (see also Burns and Welker 2022b)?

In short, digital urbanism introduces a range of complexities for how individuals and collectives should work towards social justice. Rather than presuming *social justice* has a singular, self-explanatory meaning, we contend that it is deeply important to consider what frameworks we mobilize to evaluate fairness, oppression, and justice. Fraser (2012, 43) states that "only by pondering the character of what we consider unjust do we begin to get a sense of what would count as an alternative." To date, these remain holistically and systematically underexamined in smart cities literature.

The Structure of the Book

This book features twenty-four original chapters that offer novel, complementary, and contrasting interrogations of digital urban injustice. The volume is divided into five thematic sections, each contributing a topically focused set of contributions to the volume's central questions. The organization and classification of chapters into this set of themes arise *both* from what we see as the most dynamic debates in smart cities research pertaining to social justice, *and* from inductively observing the themes that cohere the chapters that each author has submitted. After the first section establishes important conceptual foundations for questioning and challenging the origins, assumptions, and implications of *smart* as a signifier, the second section opens an important dialogue between smart cities and burgeoning data justice research. The next three sections explore a particular dimension of (in)justice in smart cities: infrastructures, digital disparities, and citizenship. With few exceptions, primarily in the first section, each chapter mobilizes empirically grounded studies towards developing theoretical material. Taken in whole, this organization both advances extant debates and offers new empirical and theoretical resources where debate is actually *needed* in consideration of fresh attention to long-standing social justice scholarship.

Each section begins with a dialogue between the editors and a prominent scholar. These conversations provide keyholes to gain insight into long-standing and contemporary debates. Addressing themes and theories raised by authors throughout the book, the editors and dialoguers discuss questions, examples, and provocations at greater depth, identify points of importance and friction between and across the sections, and illuminate future directions for urban digital justice. To a degree, the dialogues normatively scope out the current spaces of social justice in the smart city and where such research should be headed. Especially, the dialogues grapple with the question of the theoretical perspective(s) through which this research should be conducted.

The first section, *Challenging the Foundations of Smart*, explores the ideologies and assumptions that predate and underpin smart city development and establishes conceptual foundations for a new smart city research agenda focused on feminist, queer, and more than human diffractions of digital justice. In the first dialogue, Stephen Graham offers a genealogy of the smart city, placing its roots in 1960s North American counterculture and tracing its many mutations through to Silicon Valley entrepreneurialism, early cyber libertarianism, and current waves of fascism. We go on to discuss enduring characteristics of smart cities as well as the renewed urban challenges they pose. This dialogue problematizes the way a seeming "newness" of smart cities drives planning adoption, research funders, and public imaginaries, while sweeping a plethora of consequent urban problems under the rug of a thinly veiled modernist desire for control in the city. These perspectives preface a collection of chapters that likewise see smart cities within a broader historical lineage, and develop the conceptual resources that we can use to think about (in)justice in the smart city. In Chapter 1, Carina Listerborn examines what smart city strategies can tell us about contemporary urban visions, conditions, and gendered social relations. Listerborn extends feminist urban critique to analyse power relations that are being reproduced through technocapitalist urban developments in Canada and Sweden. In Chapter 2, Ryan Burns offers a novel framework for theorizing digital urban justice, reclaiming space to consider the gendered, racialized, and more-than-human actors of smart cities. Burns leverages relational-geographic thought to challenge anthropocentric and enlightenment framings of justice, and calls for more feminist, queer, and hybrid approaches to digital justice. In Chapter 3, Maroš Krivý examines the nexus of smart cities and the professionalization of urban systems. Krivý contends the resulting "epistemology of complexity" and its affinity with neoliberalism compounds elite power, making injustice and inequality appear inevitable. Tracing a similar genealogy of the smart city and its connections to contemporary neoliberal theories, policies, and discourses, in Chapter 4 Joe Daniels, Micah Hilt, and Elvin Wyly explore how Hayekian epistemologies have been encoded into postmodern media infrastructures of post-truth American digital pop culture and virulent conspiracy theories. In Chapter 5, Güneş Tavmen offers a historical account that traces the development of smart city logics. Tavmen highlights the convergence of neoliberal subjectivities, cybernetics, and emergence of the smart citizen and connections to data behaviourism. Tavmen's intervention sets up the final section on citizenship, discussed in detail below, quite well.

The second section, *Data Decisioning and Data Justice*, analyses the creation, logics, rationalities, governance, and experimentation of urban data. Building on critical data studies and data justice, the authors in this section identify (in) justice within data-based sociotechnical arrangements. In our second dialogue, with Rob Kitchin, we unpack approaches and discuss disjunctures between dominant conceptions of rights and the data-based assemblages, rationalities, and relationalities of smart urbanism. This dialogue establishes that data are central to any notion of digital (in)justice in smart cities, a point that is elucidated and substantiated in each of the section's chapters. Opening with a broad discussion of the forms and functions of urban data in Chapter 6, Jonathan Gray and Noortje Marres question how data, and data culture, play a role in reshaping city life, for whom, and to what end. In Chapter 7, Torin Monahan interrogates how platform companies have actively established themselves as obligatory passage points for individuals seeking city information. Monahan draws upon a two-year study of large-scale digital platforms in US cities to show that potentials for justice are eroded through the platformization of cities, particularly as rights are recast as consumer choices and collective resources are depleted. In Chapter 8, Miguel Valdez, Matthew Cook, and Helen Roby extend understanding of epistemic justice to document shifting (dis)engagements and power relations, as data infrastructures rendered different versions of the United Kingdom's Milton Keynes visible, actionable, and even possible. Valdez, Cook, and Roby contend that the smart city's pretence of "seeing everything from nowhere" introduces "missing body problems," as bodies are extracted, absent, uncounted, and rendered invisible by data infrastructures presented as rational and just. Responding to these silences and absences, in Chapter 9 Nathan Olmstead and Zachary Spicer propose a more deliberate "politics of re-membering" that balances accountability to citizens, including the previously unaccounted for, with the need to protect individual persons. Drawing on posthuman notions of "biodegradability," Olmsted and Spicer position data as a communal resource and prioritizes their role in government accountability and responsivity.

The third section, *Infrastructures of Injustice,* offers empirical cases of smart cities ranging from prototypes to those that are *actually existing*, from both the global North and global South. Each contribution foregrounds the uneven, granular, and scalar contexts of smart cities and their (un)just potentials and capacities. This section features a dialogue with Vincent Mosco, in which we explore the co-constitution of the smart city from the infrastructures and mythologies that power it. Revisiting what *smart* signifies, we unpack Mosco's manifesto for smart cities and its implications for environmental, social, and digital justice. Mosco suggests always keeping the vision of a "just smart city" within the broader goals of reaching a "just city" – that is to say, although the digital infrastructures of smart cities reproduce many injustices and likewise can be reappropriated towards just ends, neither possibility should detract from the broader political objectives beyond the *smart* label. This, of course, speaks to the broader objectives of this edited volume to theorize the infrastructures of digital (in)justice that will inform urban thinking after *smart cities* are replaced by a new planning and urban studies paradigm. With the passage of smart to autonomous urbanism fuelled by artificial intelligence, in Chapter 10 Federico Cugurullo returns to questions of data decision-making and data ethics and the implications for multiple spheres of urban sustainability. Cugurullo critically examines how diverse artificial intelligences process issues of ethics in the city. In Chapter 11, Alberto Vanolo reflects on the legitimation of digital corporations and smart interventions during the COVID-19 pandemic. Vanolo considers the case of *Immuni*, the Italian digital app for tracking potential infections, and speculates about the transformation and diffusion of smart city rationales during the crisis. Further interrogating the layer of material infrastructures in Chapter 12 Liam Heaphy traces the evolution of smart transport technologies in Dublin. Heaphy provides a framework for interrelating these technologies, and identifies phases of experimentation and opportunities for shaping the future of data. In Chapter 13, Karol Kurnicki analyses the digitization of car parking in London, Warsaw, and Utrecht. Kurnicki challenges the solutionism of these tools, and contends that the design and implementation of digital solutions should be informed by a comprehensive understanding of rights and justice – including accessibility, sustainability, and equitability. Also exploring the limitations of this solutionism, in Chapter 14 Lorena Melgaço and Lígia Milagres mobilize a decolonial lens to examine the logics of technological determinism and developmentalism that undergird the "climate smart city." Melgaço and Milagres explicate the connections between global climate protection targets, the adoption of smart sustainable agendas by national and local governments, and the reproduction of unequal relations between the global North and South.

The fourth section, *Complicated and Complicating Digital Divides,* challenges and updates long-standing discussions of the digital divide in order to examine landscapes of inequality and (in)justice. Chapters in this section refuse to take for granted what work the digital divide concept can do for us, and instead use the lens of the *digital* to think about social, political, and economic inequalities across a spectrum of foci. In our fourth dialogue, we discuss with Ayona Datta the complex variegations of smart cities, and interrogate what constitutes the digital, as well as barriers to access. Using a postcolonial frame, Datta draws on examples from India to explicate the complicated intersections of technology, urbanism, and inequality to reconsider justice and the emancipatory potentials of technology in the global South and North. This dialogue brings to light the way digital divides (re)produced by smartness occur across a range of intersecting scales, from the body (e.g., different *individual* access to and experiences of smartness) to the global (e.g., lingering effects of colonialism and neocolonialism). Datta encourages us to think through Chakrabarty's (2000) notion of *provincializing* to "chip away at the cracks" of smart cities as a normative epistemological and urban-political framework. In Chapter 15, Martin Tironi and Camila Albornoz analyse the invisible ecology of practices, abilities, and forms of knowledge that "delivery partners" have adopted in order to contend with the precarity imposed by the Uber Eats platform in Santiago de Chile. Charting forms of surveillance, resistance, adaptation, and injustice within these socio-technical arrangements, Tironi and Albornoz propose a decolonized vision of the smart city, showing how this paradigm reproduces a productivist, anthropocentric, and universalist vision. In Chapter 16, Alan Smart examines the intersections of smart city strategies and the formalization of informality: "smart formalization." Rather than using technology to eradicate forms of informality, Smart contends that we must better acknowledge and appreciate the power of informality to provide for citizens. In Chapter 17, Alan Wiig, Hamil Pearsall, and Michele Masucci explore how youth in Philadelphia have leveraged and used digital technologies in the COVID-19 context. The authors argue that a deepening inaccessibility to computers and the Internet, combined with meagre prospects for employment due to physical distancing requirements, represent a new digital divide. Unable to solve pressing social issues, smart initiatives create a paradox for the youth who produce digital smart city knowledge while remaining on the margins of smart urbanization itself. In Chapter 18, Teresa Abbruzzese and Brandon Hillier turn a critical intersectional lens on the role of public libraries in producing the smart city narrative in Toronto. Although the library is a key site of innovative partnerships with the neoliberal state, Abbruzzese and Hillier examine the contradictory discursive strategies centred around its role in facilitating inclusion and democratic interventions in the smart city. In Chapter 19, Yonn Dierwechter reads smart city discourses through three small regional Canadian cities to draw in stark relief the way smart city knowledge typically is generated about large cities. Dierwechter forwards the notion of *just renewal*, to reconsider not only how wider smartness discourses shape local policy dynamics, but also how less-favoured cities in a Canadian province still territorialize and rework wider smartness discourses into local space economies and social policies.

The fifth and final section, *Urban Citizenship and Participation,* explores the political geographies of digital injustice, focusing on the classic themes of citizenship, belonging, community, and the public sphere. Contributors in this section examine what constitutes *participation*, and whether such a concept is a worthwhile goal in struggles for justice. In our fifth and concluding dialogue, Alison Powell returns to promises of smart cities and their value today by juxtaposing them with countervailing tendencies of idiocy, banality, and ferality. We discuss what is being optimized through these forms of development and science, and the type of participation, citizenship, and community they might entail and curtail.

Exploring another facet of "smart citizenship," in Chapter 20 Marikken Wathne, Andrew Karvonen, and Håvard Haarstad examine how citizenship is framed through participatory processes in three EU Lighthouse smart cities developments. The authors contend that this citizen focus is an attempt to move away from technocentric modes of urban development towards urban futures that are socially inclusive and environmentally sustainable; however, it is unclear how citizens actually participate in these smart city agendas. Challenging typical forms of "participation washing," in Chapter 21 Nina David provides a grounded model for how to structure and combine digital and non-digital participation to achieve inclusivity in the planning process. Focused on smart participation in the pre-planning, implementation, and post-planning stages, David highlights the importance of capacity building and of expanding participation opportunities in order to secure a public role in plan implementation. In Chapter 22, I-Chun Catherine Chang and Ming-Kuang Chung consider how Taipei's citizen-centric initiatives have played a key role in unequally redistributing municipal resources, and the implications of this shift for digital justice. Rather than focus on digital technologies that improve municipal functions, Chang and Chung argue that the Ko administration's development of "perceptible" initiatives that affect everyday citizens has led to widespread populist support at the cost of underresourcing disadvantaged parts of the city. Returning to questions of data veracity, infrastructures, and relationality, in Chapter 23 Orlando Woods contrasts open data policies – often championed in city developments – with Singapore's recently passed Protection from Online Falsehoods and Manipulation Act. By reconciling these two distinct treatments and understandings of the (im)partiality of "truth," Woods considers the ways in which (in)justice can be reimagined through the treatment of data and (mis) information within the smart city. In Chapter 24, Inka Santala and Pauline McGuirk explore "sharing platforms" as an alternative political-economic ordering mechanism in Sydney, Australia. Although sharing platform initiatives often presents alternative, more equitable, and inclusive modes of organizing, they still might be exclusionary and limited by time and resource constraints. Santala and McGuirk argue that "the sharing city" needs support and action on multiple sociopolitical scales to foster social justice in the city.

Conclusion

This volume enters long-standing and diverse discussions concerning the imbrication of technologies into the urban fabric. With an appreciation of the antecedents and genealogies of the smart city, the volume brings together discussions across a range of disciplines and geographic contexts to foreground justice, equality, access, and inclusion. Drawing attention to spaces of (in)justice, the contributions highlight the need not only to centre justice in these sociotechnical entanglements, but also to envision and create spaces for alterity. By conceptualizing the smart city as an empty signifier, on the one hand we acknowledge the current composition of smart cities; with top-down imposition of these systems, the concept is filled with forms of data extractivism, surveillance capitalism, and more, which reinforce social sorting, datafied prejudice, and discrimination. On the other hand, however, contributors push back against such technological determinism, highlighting resistance, co-constitution, and the conditions of possibility that would allow it to be otherwise. What if the smart city were one that promoted human and non-human flourishing?

Yet this is far from the city we have today. Discussing our (dystopian) realities and the inability of technology "to save us"[2] with novelist and tech writer Tim Maughan, we reflect on the role of critique:

> When I'm criticizing this technology, I am not meaning to be dismissive of how useful and powerful it is, and how many people depend on it. But we need to ask who is the smart city primarily built for? They are these generic, extractive, off-the-shelf solutions – that narrowly conceive of cities as problems to be solved. It is that terrible Silicon Valley philosophy, "find a problem you can technologically fix," whether that problem needs fixing or not. It is often the most marginalized people and communities that have the most to lose. This technology is aimed at the model citizen – a bourgeois middle-class tech-savvy kind of person – rather than those in marginalized communities. As we steamroll this technology in, we do not talk to the community that it's been dropped on top of, and we do not consider social or digital justice.
>
> In fact, I am not even always sure what the term social justice means. I have been thinking a lot about Astra Taylor's distinction between "activism" and "organizing." She argues there is a lot of talk about activism these days, but it is a short-term, sexy, and often limited to social media. Whereas, actual political change is done through organizing, and that is a much longer process, and often a much smaller process. It starts at a community level and involves a lot of planning and dedicating, if not all your life, maybe a certain chunk of your life, towards a political cause.
>
> To do this – talk to communities, foreground justice, organize, and make that space – we need to slow things down. We need to get people to pause and think

> about technological adoption before it is inevitable. Because once it has happened, it is very hard to come back. How do you rachet backwards? I worry it might be too late for some of these things. This is the biggest ideological problem that I run into constantly – we just do not stop and think about it. I think it's up to artists, activists, journalists, and academics to try and encourage that – to try and create a space between the emergence of technology and its acceptance, where we can have these conversations. But it is incredibly hard to do. (Tim Maughan, personal communication, 2021)

Sharing Maughan's sentiment, we begin from the perspective that cities are often made into better places not by those in power, but by activists, academics, everyday citizens, journalists, and others, and we hope that this book, alongside many other ongoing conversations, serves as one means of carving out this space towards digital urban justice.

NOTES

1 To be sure, *smart* has been applied to many non-urban cases, such as *smart technology* and *smart phones* in particular. While these earlier uses initiated a lineage of the term's use, here we are most interested in its application to the urban, and so we forego putting it in a longer non-urban lineage, even though such a consideration could be useful to explore.

2 See Paris Marx's podcast, "Tech Won't Save Us," online at https://podcasts.apple.com/us/podcast/tech-wont-save-us/id1507621076

REFERENCES

Amin, A. 2006. "The Good City." *Urban Studies* 43 (5/6): 1009–23. https://doi.org/10.1080/00420980600676717

Araya, D., ed. 2015. *Smart Cities as Democratic Ecologies*. London: Palgrave Macmillan.

Arendt, H. 1979. *The Origins of Totalitarianism*. San Diego: Harcourt Brace.

Arnstein, S.R. 1969. "A Ladder of Citizen Participation." *Journal of the American Planning Association* 35 (4): 216–24. https://doi.org/10.1080/01944366908977225

Balsillie, J. 2018. "Sidewalk Toronto has only one beneficiary, and it is not Toronto." *Globe and Mail*, 5 October. Online at https://www.theglobeandmail.com/opinion/article-sidewalk-toronto-is-not-a-smart-city/

Blue, G., M. Rosol, and V. Fast. 2019. "Justice as Parity of Participation." *Journal of the American Planning Association* 85 (3): 363–76. https://doi.org/10.1080/01944363.2019.1619476

Burns, R., and M.J. Andrucki. 2021. "Smart Cities: Who Cares?" *Environment & Planning A* 53 (1): 12–30.

Burns, R., and P. Welker. 2022a. "'Make Our Communities Better through Data': The Moral Economy of Smart City Labor." *Big Data & Society* 9 (1): 1–13. https://doi.org/10.1177/20539517221106381

Burns, R., and P. Welker. 2022b. "Interstitiality in the Smart City: More than Top-Down and Bottom-Up Smartness." *Urban Studies* OnlineFirst: 1–17. https://doi.org/10.1177/00420980221097590

Cardullo, P., C. Di Feliciantonio, and R. Kitchin, eds. 2019. *The Right to the Smart City.* Bingley, UK: Emerald Publishing.

Cardullo, P., and R. Kitchin. 2019. "Being a 'Citizen' in the Smart City: Up and Down the Scaffold of Smart Citizen Participation." *GeoJournal* 84 (1): 1–13. https://doi.org/10.1007/s10708-018-9845-8

Chakrabarty, D. 2000. *Provincializing Europe: Postcolonial Thought and Historical Difference.* Princeton, NJ: Princeton University Press.

Charnock, G., H. March, and R. Ribera-Fumaz. 2021. "From Smart to Rebel City? Worlding, Provincialising and the Barcelona Model." *Urban Studies* 58 (3): 581–600. https://doi.org/10.1177/0042098019872119

Coletta, C., L. Evans, L. Heaphy, and R. Kitchin, eds. 2018. *Creating Smart Cities.* London: Routledge.

Datta, A. 2020. "The 'Smart Safe City': Gendered Time, Speed, and Violence in the Margins of India's Urban Age." *Annals of the American Association of Geographers* 20 (5): 1318–34. https://doi.org/10.1080/24694452.2019.1687279

Derickson, K. 2015. "Urban Geography I: Locating Urban Theory in the 'Urban age.'" *Progress in Human Geography* 39 (5): 647–57. https://doi.org/10.1177/0309132514560961

Fainstein, S. 2009a. "Planning and the Just City." In *Searching for the Just City: Debates in Urban Theory and Practice*, ed. P. Marcuse, J. Connolly, J. Novy, I. Olivo, C. Potter, and J. Steil, 19–39. New York: Routledge.

Fainstein, S. 2009b. "Spatial Justice and Planning." *justice spatiale | spatial justice* 1: 1–13.

Fraser, N. 1995. "From Redistribution to Recognition? Dilemmas of Justice in a 'Post-Socialist' Age." *New Left Review* 212: 68–93.

Fraser, N. 2005. "Reframing Justice in a Globalized World." *New Left Review* 36: 69–88.

Fraser, N. 2007. *Scales of Justice: Re-framing Justice in a Globalizing World.* New York: Columbia University Press.

Fraser, N. 2009. "Social Justice in the Age of Identity Politics: Redistribution, Recognition, and Participation." In *Geographic Thought: A Praxis Perspective*, ed. G. Henderson and M. Waterstone, 72–90. New York: Routledge.

Fraser, N. 2012. "On Justice." *New Left Review* 74: 41–51.

Glasze, G. 2007. "The Discursive Constitution of a World-Spanning Region and the Role of Empty Signifiers: The Case of Francophonia." *Geopolitics* 12 (4): 656–79. https://doi.org/10.1080/14650040701546103

Green, B. 2019. *The Smart Enough City: Putting Technology in Its Place to Reclaim Our Urban Future.* Cambridge, MA: MIT Press.

Harvey, D. 2009. *Social Justice and the City*, rev. ed. Atlanta: University of Georgia Press.
Karvonen, A., F. Cugurullo, and F. Caprotti, eds. 2018. *Inside Smart Cities: Place, Politics and Urban Innovation*. London: Routledge.
Kitchin, R. 2013. "The Real-Time City? Big Data and Smart Urbanism." *GeoJournal* 79 (1): 1–14. https://doi.org/10.1007/s10708-013-9516-8
Kitchin, R., T. Lauriault, and G. McArdle. 2015. "Smart Cities and the Politics of Urban Data." In *Smart Urbanism: Utopian Vision or False Dawn?*, ed. S. Marvin, A. Luque-Ayala, and C. McFarlane, 16–33. New York: Routledge.
Laclau, Ernesto. 2007. *Emancipation(s)*. Brooklyn: Verso.
Luque-Ayala, A., C. McFarlane, and S. Marvin. 2016. "Introduction." In *Smart Urbanism: Utopian Vision or False Dawn?*, ed. S. Marvin, A. Luque-Ayala, and C. McFarlane, 1–15. New York: Routledge.
Lynch, C.R. 2020. "Contesting Digital Futures: Urban Politics, Alternative Economies, and the Movement for Technological Sovereignty in Barcelona." *Antipode* 52: 660–80. https://doi.org/10.1111/anti.12522
Marvin, S., A. Luque-Ayala, and C. McFarlane, eds. 2015. *Smart Urbanism: Utopian Vision or False Dawn?* London: Routledge.
Masucci, M., H. Pearsall, A. Wiig. 2019. "The Smart City Conundrum for Social Justice: Youth Perspectives on Digital Technologies and Urban Transformations." Annals of the American Association of Geographers 110 (2): 476–84. https://doi.org/10.1080/24694452.2019.1617101
Mill, J.S. 1906. *Utilitarianism*, 13th London ed.. Chicago: University of Chicago Press.
Nozick, R. 1974. *Anarchy, State, and Utopia*. Oxford: Blackwell Publishers.
Powell, A. 2021. *Undoing Optimization*. New Haven, CT: Yale University Press.
Rawls, J. 1971. *A Theory of Justice*. Cambridge, MA: Harvard University Press.
Rawls, J. 2001. *Justice as Fairness: A Restatement*. Cambridge, MA: Belknap Press.
Rosol, M., G. Blue, and V. Fast. 2018. "'Smart,' aber ungerecht? Die Smart City-Kritik mit Nancy Fraser denken" [Smart but injust? thinking the smart city critique with Nancy Fraser]. In *Smart City – Kritische Perspektiven auf die Digitalisierung in Städten* [Smart city – critical perspectives on digitization in cities], ed. S. Baruiedl and A. Strüver, 87–98. Transcript Verlag.
Rosol, M., G. Blue, and V. Fast. 2019. "Social Justice in the Digital Age: Re-thinking the Smart City with Nancy Fraser." UCCities – Global Urban Research at the University of Calgary Working Paper 1.
Rosol, M., and G. Blue. 2022. "From the Smart City to Urban Justice in a Digital Age." *City* 26 (4): 684–705. https://doi.org/10.1080/13604813.2022.2079881
Roy, A. 2016. "Who's Afraid of Postcolonial Theory?" *International Journal of Urban and Regional Research* 40 (1): 200–9. https://doi.org/10.1111/1468-2427.12274
Saussure, F. 1959. *Course in General Linguistics*. New York: Columbia University Press.
Sen, A. 1999. *Development as Freedom*. New York: Anchor Books.
Shelton, T., M. Zook, A. Wiig. 2015. "The 'Actually Existing Smart City.'" *Cambridge Journal of Regions, Economy, and Society* 8 (1): 13–25. https://doi.org/10.1093/cjres/rsu026

Soja, E. 2010. *Seeking Spatial Justice*. Minneapolis: University of Minnesota Press.
Stuntz, W. 2011. *The Collapse of American Criminal Justice*. Cambridge, MA: Belknap Press.
Tironi, M., and M. Valderrama. 2018. Acknowledging the Idiot in the Smart City: Experimentation and Citizenship in the Making of a Low-Carbon District in Santiago De Chile. In *Inside Smart Cities: Place, Politics and Urban Innovation*, ed. A. Karvonen, F. Cugurullo, and F. Caprotti, 163–81. London: Routledge.
Vanolo, A. 2014. "Smartmentality: The Smart City as Disciplinary Strategy." *Urban Studies* 51 (5): 883–98. https://doi.org/10.1177/0042098013494427
Young, I.M. 1990. *Justice and the Politics of Difference*. Princeton, NJ: Princeton University Press.

PART ONE

Challenging the Foundations of Smart

A Dialogue with Stephen Graham

Part One explores the ideological and theoretical underpinnings of digital urbanism and smart cities in order to establish the conceptual foundations for a new smart city research agenda focused on digital justice. Tracing antecedents and kernels of the smart city, authors in this section challenge the foundations of this development, exploring the political, economic, gender, racial, and anthropocentric logics of *smartness*, complexity, cybernetic urbanism, and urban systems. This part provides alternative – and competing – conceptions of what sort of justice smart cities scholars and activists should pursue. Each contribution contributes towards the question of whether a rights-based conception of social justice is ideal for the smart city. In doing so, authors across this volume question whether working towards more just, sustainable, liveable, and egalitarian cities requires that we look beyond the epistemological limitations of *smartness* altogether.

Discussing these questions and exploring the genealogy of the smart city, as well as its antecedents and mutations, here we dialogue with Dr Stephen Graham, Professor of Cities and Society, Newcastle University. Graham's long-standing work focuses on the transformation of cities through infrastructure, digital media, surveillance, security, and verticality.

Editors: In approaching this volume, we have had many conversations about the work "smart" does as a qualifier. In our Introduction, we contend that "smart" is an empty signifier and discuss the implications this has for justice; others in the volume, such as Alan Smart and Vincent Mosco, call for different interpretations of what makes a city or its inhabitants "smart." To further interrogate this concept, what, if anything, is different about smart cities from previous incarnations of "cyber cities," "city of bits," "intelligent cities," and so on?

SG: I am not sure it is that different. There was an influential website in the mid-1990s that identified the continually expanding list of technological place-marketing labels then in circulation. They put together hyperlinks where words denoting a place, such as *highway, mountain, city, district, neighbourhood*, were connected with those denoting some evocation of high technology, such as *intelligent, smart, knowledge-based, digital, techno, silicon*, and so on. The marketing discourses and bubbles of that period have continually evolved since then and have a life of their own. The *smart* city phenomenon is only one of the latest and most influential urban technology policy bandwagons in a long and complex – and continuing – genealogy. The technological paradigms are not that different, and often the underlying sociotechnical materialities are not that different, either.

I would concur with the critical perceptions that the smart city paradigm is similar and related to previous corporate technological utopias. This is especially so in the way that big tech are selling a cookie-cutter vision of technological optimization: a highly technocratic and technophiliac vision, where, *a priori*, complex urban social problems are deemed to be solvable – and *perfectible* – through new media or computational technologies. What strikes me is the startling and endless similarity that these visions have. Granted, it is not that surprising, since it is the same big IT companies pushing their urban operating systems to try and fix their share in the burgeoning market.

Smart has more political and economic momentum than the earlier bandwagons of cyber cities, virtual cities, and digital communities. However, the qualitative difference is the sheer scale of the corporate information technology push, which has packaged and sold *smart* as the latest thing. The big IT corporations, like Cisco, Google, IBM, Hitachi, Siemens, and so on, have all converged and colonized this business model. They saw the city and the urban not only as a market to colonize, but also as the place where their technological fortunes were going to be decided, and realized they could package-up their products and services using these discourses. The core idea of the smart city is that *optimization would solve your problems*. Urban, governmental, and social problems are cast as being solely the result of the accretion of multiple, previous, technological systems that have accumulated haphazardly in ways that do not intersect, that are siloed and incompatible. The panacea or technological utopia is thus attainable if only technologies are made to intersect to provide that seamless sense of pervasive knowledge that allows for a deeply technophiliac (and often utopian) vision of urban optimization and perfection.

One of the things that has changed is the political economy of technological urbanism. Data harvesting has become one of the biggest industries of our times. As an economic model, Sidewalk Labs' hugely contested project on

Toronto's lakefront is a great example of this. The projected real estate is profitable, the projected Sidewalk-based services are profitable, but what would be most profitable is the continuous harvest of data generated by that community and its inhabitants as they are going about their daily lives.

As with the previous generations, going right back to the telegraph, telephone, videotext, cable, and community telematics, and so on, the discourse is always, "we have got all of these messy, crazy, dysfunctional, chaotic, non-optimized technological landscapes! This creates huge inefficiencies and costs and prevents the realization of the technological potential! *This time* we will provide optimized perfect real-time control." It is always *this time*, which is part of the ahistorical forgetting. That is exactly how cable was sold and all the rest of it. The forgetting is convenient and probably deliberate.

Editors: This forgetting seems so common, especially in the corporate and policy communities. In many ways, not seeing connections to antecedents helps camouflage processes of uneven development and neoliberalization. To politicize practices of remembering, Maroš Krivý and Güneş Tavmen highlight the revolving door of elites responsible for stabilizing the smart city movement. Similarly in Part Two, Nathan Olmstead and Zachary Spicer call for a politics of "re-membering" that would foreground the biodegradability of data and perhaps new understandings of responsibility. What are other implications of "forgetting" and what we can do to move forward?

SG: Because it is so fetishistic, the entire discourse on smart cities is profoundly ahistorical and inhabits a permanent paradigm of the present problem, followed, simplistically, by the future (technological) solution. The discourse thus fails almost completely to engage with the rich and complex history of urban communication, mediation, visualization, and calculation. This problem was a key motivation for putting together my *Cyber Cities Reader* in 2004 (Graham 2004). The book was an attempt to take original sources and package them together with lots of editorial interpretations in a way that could get beyond the superficial nature of these debates and respond more critically, and more analytically, to those usual, superficial, claims.

The world has changed dramatically since then. What has *not* changed is the profound set of stories and histories here, and that we still fail to learn from them. We just forget them in our ahistorical obsession with the allegedly *smart* world of contemporary urban mediation, communication, and visualization. The crucial and ignored histories of those worlds just go into the abyss, as we continually invent some new bandwagon that will, once again, be a panacea for all urban ills. I am jaded about the endless succession of urban, digital, policy bandwagons, I just wait for the next round …

This is an amnesia that Shannon Mattern (2017) writes so brilliantly about: What happens when visions of the future become archaeology, waste, history. She is thoroughly embedded in a deep historical understanding of these things in a way that very few scholars are (see also Parikka 2013). We are living in and amongst the debris of fifteen generations of utopian technological thinking: the often-abandoned material legacies of those periods are around us, between us, above us, below us, sometimes *within* us – they *are* the *city*. In Swedish cities, these material legacies are being *mined* for the metals whose natural ores are so increasingly scarce and expensive in the wider world of non-urban peripheries.

The problem with such ahistoricity, at the policy level, is the way it tends to plunge policymakers – who all too easily end up believing that new digital media have some magical quality in and of themselves, and can thus be merely unleashed into the world to do their stuff – into a cycle of making the same mistakes again and again. Policymakers rarely have a background in critical understandings of histories and dynamics of urban media. They are susceptible to smart city marketing – that a wholesale investment in one smart city package will necessarily and inevitably augur in some optimum and omniscient platform that will miraculously solve deep-seated social, institutional, governmental, or environmental problems.

The reality is that hyped up expectations of *smart city* promoters are, at best, dramatically overegged; however, the realizations of this are often long, embarrassing, and painful. There might be some improvements in certain aspects of knowledge, of substantial financial, institutional, and personal investments to maintain often fragile technosocial systems in some form of functioning order over time. There will, inevitably in complex social, institutional, and governance contexts, be a lot of messiness, politics, and infighting. Most likely, the technology will be revealed not to work as promised, at least against the often wildly utopian and technophiliac promises used to sell and implant it. It will emerge that the utopian, totalizing omniscience promised by the sales*men* – as it is almost always men – is a chimera, and that the technological and media systems that sustain contemporary urbanism will always and inevitably be a *kludge*: a messy assemblage of bits and pieces that sometimes work together some of the time and sometimes most powerfully do not. In the situational reality of people and machines, strung out across distance, the simplistic and utopian promises will jar against the (re)emerging axiom that the rich, messy, and socially embedded nature of these things is – and will always be – messy and complex and resistant to totalization.

This, of course, is inevitably oversimplistic. Some policy communities are much more reflective and are able to be critical and nuanced. Too often, *Big Tech* and its boys come to town and say "you are going to be X," "this is your great place marketing moniker," and "you will be a burgeoning hotbed of digital this and technological that as long as you give us all your dollars for our smart city platform," and make the sale. Of course, it never works as promised. But the *boys* are long gone by then. On to their next sales pitch.

We have seen these bandwagons many times. For example, in the 1980s and 1990s, there were the global Technology, Science Park, and Technopole movements, based on the idea that new *clusters* of technological innovation could be initiated in a distributed way. Experience, not surprisingly, did not replicate a mini *Silicon Valley* in every locality. The glitzy new campuses which studded rust-belt and peripheral cities too often ended up being filled with routine call centre workers and data clerks. The economic geographies of digital capitalism are very resilient to change. One cannot just build some entrepreneurial, innovative, digital economy from scratch; in many peripheral locations, at best, you might be able to bring in the odd routine workers that propel the digital economy's backstage.

Editors: Early predictions and rhetorics of cyberspace were grounded in forms of libertarianism and touted an inherently democratic transformation. However, as we have seen with online platforms like 4chan, 8chan, Twitter, YouTube, and Facebook, something quite different from an Athenian or Jeffersonian age of democracy has been forged in the global information infrastructure. Many contributions in our volume, and especially in this part, trace the genealogy of smart city thinking – the ideologies, rationalities, and logics that underpin these urban forms. For instance, Joe Daniels, Micah Hilt, and Elvin Wyly contend that "[a]ll who are committed to social justice must remember that, in the cognitive hierarchies of the surveillance capitalist smart city bequeathed by Hayek and Thiel, intersectionality is complex, multidimensional, non-linear, and evolutionary." Could you speak a bit more to the intended and unintended consequences of the "cyber" or "smart"? Or more directly, the politics that underpin it, and alternatives?

SG: We are currently facing the enormous power of populist-right and alt-right mobilizations around the world. Looking back at the quaint cyber-libertarian and utopian discourses of the early 1990s, it feels like a very long time ago! Reflect on Michael Benedikt's 1991 book, *Cyberspace: First Steps*. Like many at the time, he gushed about "casting away the ballast of materiality" – of abandoning the very idea of travel as the world moved into the utopia of a limitless, infinitely extendible, and inherently democratic electronic world. As digital computing, telecommunications, and visual media converged and fused, the resulting domain of *cyberspace* was assumed to be a world of inherently libertarian potential. It was a spectacularly naive perspective to have, but we need to position this again in a historical genealogy.

There is a direct connection between those discourses and the mutation of the alt-right. This starts with the 1960s Californian, anti–Vietnam War, anti-capitalist counterculture, and their embrace of 1970s decentralized communications technologies like cable TV and community computing to confront the massive and centralized complexes of *Fordist* media, newspapers, and TV stations. These ideologies were articulated through publications like the *Whole Earth Catalogue*, which envisioned new appropriate and alternative technologies organized at the local scale to facilitate utopian communities.

Such thinking mutated and re-emerged in the world of electronic culture in the 1980s and early 1990s on the West Coast. Some of the key 1960s figures, like Timothy Leary, imbued Sixties counterculture into their increasingly libertarian 1980s and 1990s *cyberspace* rhetoric. Then, with magazines like *Wired* and *Mondo 2000*, it became a glossy, yuppie thing, inflected with oh-so-fashionable cyberpunk dystopia. A lot of these same people were involved in the establishment of the Big Tech startups that became hegemonic within platform capitalism, which have metastasized to become some of the big players in global platform urbanism and global capitalism (see Nagle 2017).

Their politics have mutated from left-libertarianism to a right- or even quasi-fascist libertarianism. Peter Thiel, an early investor in Facebook and a founder of PayPal, is particularly powerful (Burrows 2018). Thiel believes that progress and democracy are antithetical. He is helping to fund things like the Seasteading Institute that are pushing a hyper-libertarian and sometimes alt-right fantasy of building island city-states floating around the Earth as some putative solution to the world's problems. Such *Seasteads* are projected to be libertarian bolt-holes for people who want to escape from the perceived problems of Western liberal democracy and the messiness of liberal, cosmopolitan, terrestrial cities.

There is a direct genealogy here which leads straight back to the West Coast, antistate, localism of the hippie 1960s, but which mutated into a rabidly libertarian and rightist obsession with the idea of escape – for those who can. Whilst the Seastead idea is one of *horizontal* secession, the obsession of *vertical* escape through space travel is also a powerful trope amongst the tiny number of multibillionaires who sit on the evermore obscene mountains of personal wealth that accrue when you are fortunate enough to be located at the very centripetal heart of centralizing logics of monopolistic, platform capitalism.

Many of the big-name players and CEOs in Silicon Valley are or have been playing with this alt-right world. It is a very dangerous thing because they are clearly now in a position to have launched, sold, and supported the social media landscapes around the world that allowed them to add huge power to the selling of alt-right, antidemocratic, profoundly anti-urban, sometimes outright fascist, sometimes white supremacist, visions. There exists here a direct and complex, but very poorly studied, genealogy.

The new media ecology enables alt-right activists to sustain and inhabit a closed world based entirely on disinformation and conspiratorial messaging. There is an even weirder twist to this, which my colleague Roger Burrows is writing wonderfully about, and that is how some of the leftist cultural theorists of the early days of *cyber* studies have mutated to become key philosophers supporting the alt-right. An important example is Nick Land, who used to hang out with Sadie Plant in the Cybernetic Culture Research Unit at the University of Warwick. Far from poststructuralist and postmodern theorizations, deep feminist theory, and neo-Marxist critiques of cyber capitalism, Nick Land has mutated into a sort of demigod of alt-right philosophy. He lives in Shanghai, and through blogs, video interviews, and his own e-book publishing house, he is selling a vision based explicitly on the alleged imperative to dismantle liberal, Western democracy entirely, and create a myriad of independent mini-states. He, quite openly, wants to do that violently through a radical acceleration of digitally mediated processes that, allegedly, are already bringing about such states' demise. Land is arguing for a digital-media-based accelerationism which would fragment the world into a myriad of different enclaves. It is splintering urbanism on steroids.

There is nothing new here, of course: cities have generated anti-urban hatred since they have existed, especially amongst political elites. Now we are seeing the alt-right and the mutations of ethnonationalist violence right across the West, and it is much more complicated elsewhere. The heartland argument of these paradigms (especially of the populist far right), that is sustaining so many powerful ethnonationalist political movements is that the city is now no longer *of the nation* – the city is somehow not authentic. It is a mutation of both the Christian and Islamist right that often justifies and sustains violence against the messy cosmopolitanism of the contemporary city. There is a repulsive website called "Crush the Urbanite" which summarizes this. It invokes the rural or exurban white resident as authentically pure and national – to be projected against the impure, amoral, cosmopolitan, materialist urbanite in some supposedly essential and violent fight to save some ethnonationalist purity for the nation. Such perspectives have a looser relationship with the alt-right, and the sorts of things Nick Land is arguing for.

There is a big, powerful connection to these men, and the things that give the likes of Trump – and other violently anti-urban populists – their power. However, because so many discussions about place and digitality are ahistorical, we miss these really vital genealogies, and I think they are so important. I mean, people who learn about *smart cities* will never be taught about Nick Land and the *Whole Earth Catalogue*, or the way the sort of hippie ideologies mutated into neofascist ones. Nor will they learn about the interconnected histories of military technoscience, cybernetic theory, the military industrial complex, and urban cable networks or urban renewal programs. Yet such genealogies are crucial.

Editors: As you, Joe Daniels, Micah Hilt and Elvin Wyly, and others note, the mutation of cyberlibertarianism into fascism in the technology sector is pronounced. Can you draw out those connections to the urban? For instance, Güneş Tavmen offers a genealogy of urban cybernetics and the implications for citizenship, concluding that [d]ecentralization, horizontalization, and self-organization do not inexorably pave the way to an equitable society in a world of inequalities" and that open-data-driven smart cities must learn from these lessons. Similarly, in Part Five, Orlando Woods traces connections between Singapore's Protection from Online Falsehoods and Manipulation Act and its implications for smart citizenship. How have these dehumanizing ideologies mutated, been obscured, and or repackaged?

SG: I think equally important is the deep genealogy of anti-urbanism within discussions of electronic communication and media. Going back to the origins of electrical communications technologies (see Graham 2004), there has been a profound and occasionally dominant strand of thinking which sees such innovations as somehow miraculously *ungluing* the city, or sees dense urbanism as an inherently problematic phenomenon that needs to be *uninvented* through new electronic communication systems. Relying on basic technological determinism, it is a short step to assume that, just because it is now possible to send data instantaneously around the world, the city is somehow doomed.

Many renderings of the smart city, whilst discursively packaged as essentially urban, are actually fundamentally underpinned by profoundly *anti*-urban ideas. They are sold as visions of Splintering Urbanism (see Graham and Marvin 2001) – it is the *attraction* of *smart* capsules, secured enclaves, bourgeois environmentalists' paradises that are closed off from the wider, messy urban world.

The *urban* and the *city* are complete afterthoughts. If you have ever worked with technologists who proffer these models of layered, top-down, corporate smart city policymaking, the urban is most definitely not something they think about: it is merely a burgeoning domain within which market share can be built. Such specialists usually inhabit a profoundly geographical world of coding, business plans, economic projections, and methodologies where the inevitable and fundamental spatiality of human life is usually absent or denied. There is no room for actual constitutive spatial or place-based thinking – let alone urban history – in their educational backgrounds or policy worlds. That comes across powerfully in their visions of urbanism, as understandings of people in places are almost completely absent. When I go to trade shows or see their marketing, often there *are* no people. There are just CGI cookie-cutter landscapes of buildings, technologies, infrastructure, and vehicles clumsily thrown around apparently *green* landscapes to produce the clichéd signifiers of the *smart city*.

A huge amount of greenwashing underpins the *smart* city movement. The *smart city* and the *ecocity* are two worlds that have evolved in parallel, often overlapping in the same places, through the same discourses and initiatives. Visualized new urban landscapes are ridiculously verdant: every surface hosts dense vegetation as a simple signifier of *greenness*. However, this is a *bourgeois* environmentalism: an apparently *green* landscape configured for the aesthetic pleasure of powerful and elite residents does little to address – and even camouflages – the underlying ecological or environmental problems, effects, and *footprints* of the development.

When we look beyond the shiny monikers and seductive, glitzy marketing, many smart city visualizations are violent in the way they *denude* the urban of its social and human qualities. Their mobilization and implementation often camouflage the problematic processes of privatization, dispossession, and the forcible imposition of technocratic, domineering, sometimes authoritarian ideologies. They are pushing technological worlds which are at their core about microsurveillance and encroaching biometric systems of tracking and control. As I demonstrate in my 2010 book *Cities Under Siege: The New Military Urbanism* (Graham 2010), they also overlap in hugely problematic ways with the widespread encroachment of militarized notions of targeting, tracking, and control into social and civilian life.

Many new smart city or ecocity initiatives can best be understood as what Mike Davis and Daniel Monk (2011) called "dream worlds of neoliberalism," because their fundamental point of origin lies deep within the political economies of urban neoliberalism. You see this powerfully in some of the most (in) famous initiatives. Sidewalk Labs on Toronto's waterfront was paradigmatic of the ways in which the smart city rhetoric, and the logics of "running a city like a Big Tech company," are used to allow a corporate body to build and control an entirely privatized new urban district based on the specious, utopian promises of ICTs. It was hugely problematic in that respect, and the robust and widespread mobilizations against it thankfully have seen the project's withdrawal. Other cities have not been so lucky …

Editors: Your work has long highlighted the implications and epistemologies that undergird ways of thinking about technology and space. Authors throughout this volume, return to fundamental theories of space, place, and the implication of the digital for (in)justice, and promote new theoretical interventions, theoretically and conceptually. What do we need to advance, and what, if anything, needs to be left behind?

SG: As an intellectual architecture, the analytical frameworks provided by large technical systems theory, perspectives emphasizing the social constructions of technology, the cyborgian nature of urbanization, and political economies of neoliberal – and post-neoliberal – capitalism are all helpful in different ways (see Graham and Marvin 2001). Technological determinism is clearly less helpful, but it is hugely prevalent in the discourses surrounding the proliferation of smart cities for all the reasons already discussed.

The broader perspectives of *political economies of digital urbanism* or *digital capitalism* are still open to analysis using these frameworks. But one of the things that has changed is the intensification of digitization and datafication, and the ways in which platform capitalism and platform urbanism have become the new mantra for the relationships between digital media, space, and place. These things are articulated through apparently simple interfaces that hide these complex, socionatural, sociotechnical, logistical worlds strung out across so many places and times. Benjamin Bratton's (2016) idea of the *Stack* is an insightful engagement with the intensity and depth of these *hyperobjects* or megastructures that blur the social into the technical, the technical into the *natural*, and so on, in the strung-out logistical and technological assemblages that sustain so-called *smart* cities.

Editors: Amid dominant ideologies that position the experience of technology as apparent inevitability, your work has detailed the complex, hybrid, sociotechnical relationships between ICTs, cities, and urban life, and cautioned against technological determinism. To this point, Ryan Burns calls for more relational approaches – specifically, lessons to be learnt from queer and non-human theorizing. Could you speak more to the intersections of technological determinism, resistance, and justice?

SG: The technological bandwagons surrounding *smart* cities have fairly predictable trajectories. As we have discussed, they often start with this top-down, quite authoritarian, technocratic, technophilic vision, which generates a complex and diverse backlash as it widens its application around the world. The dominant, technophilic paradigm generates a backlash from various aforementioned critical scholarly discussions, but also from citizens, activists, social movements, NGOs, and others. These often grassroots-led movements will often mobilize digital technologies to contest the notion that the *smart* city has to be shaped by a single, top-down, dictatorial initiative.

While technologically deterministic visions are used by the dominant and powerful to sell ICT-based urban change, in reality digital media are immensely flexible and open to innovation, jamming, and repurposing. There is a blossoming array of grassroots-driven or global network-driven movements, where these technologies are repurposed and appropriated in radically different ways to challenge the hegemonic power of globalized neoliberal elites. Often you will see with citizen-led insurgent media, hacktivism, NGO-based efforts, countercartography, countervisualization, antisurveillance, and antimilitarization movements that subvert the logics of centralizing, militarized, algorithmic control. The anti-eviction mapping programs around the Bay Area are an impressive example.

Another great example, which I raise in my 2016 book *Vertical: The City from Satellites to Bunkers* (Graham 2016), is the 2011 Arab Spring, and in particular the Bahrain uprising. In Bahrain, the Sunni elite is repurposing and reorganizing Bahraini territory through the reclamation and terraforming of coasts for real estate projects. The circulation of Google Earth images of these transformation projects triggered uprisings. The Shia majority became politicized and mobilized because they could see their world of densely packed poor neighbourhoods and how it was being cordoned off away from the advancing, privatized, coastline with its superluxury resorts owned by the Sunni elite and its friends. Importantly, this revelatory, vertical gaze was facilitated through a set of systems that are effectively legacies of the US Air Force built for military control and domination.

These counterhegemonic strategies expose the limitations of the top-down corporate vision powerfully. These forms of resistance are incredibly powerful. They are not about considering a world without digital media – this is not the question. Instead, we are looking at a flourishing of insurgent and grassroots-led innovations, where the technologies do and can add power to counterhegemonic movements for environmental and social justice.

Editors: In this part of the book, both Ryan Burns and Carina Listerborn highlight the silences and exclusions found in smart city development and discourses. Is there a disjuncture between dominant conceptions of social justice and sociotechnical rationality, assemblages and relationalities of a smart urbanism? In what ways should social justice be reconceptualized, or is justice even possible when it qualifies the city?

SG: There is a disjuncture. Traditional urban sociology and urban geography inform understandings of social justice in the city. Such perspectives rely on sociospatial and intercorporeal paradigms of thinking about who lives where and how their livelihoods pan out in terms of life chances, health, access to goods and services, freedom from risk and violence, and all those important things. Until very recently, neither subdiscipline has spent much time dealing with how pervasive, electronically mediated communication impinges on all aspects of urban life.

The politics and spatialities of place and the intercorporeal life of things do not cease to matter in a world of pervasive, linked, digital communication and computation – they still shape questions of justice and welfare in cities. But the paradigms of urban sociology and urban geography often fail to consider how every aspect of those other worlds is now mediated and continually remediated through these platform capitalist, platform urbanist, heterogenous media complexes. Those paradigms fail because they do not explore the ways in which algorithmic agency – the automation of massively complex relations between data capture, data storage, and the mediation of life chances strung out over various social landscapes and geographical scales – are profoundly opaque and resistant to analysis, critique, regulation, and challenge. The new powerful transnational worlds of platform capitalism are not open to traditional geographic or traditional sociological thinking. We now have to get into the world of data ethics, data activism, data sovereignty, and algorithmic justice within powerful, complex, space-transcending assemblages, which fundamentally transcend traditional geographical ideas of scale, control, and jurisdiction. What happens to a city when life chances are continually performed and mediated by opaque and unknowable worlds of automated software strung out across planetary media systems where the agency, ethics, and responsibilities of such power are almost impossible to know fully?

In the contemporary world of digitally mediated urbanism, one of the things that always strikes me is the essential invisibility of the processes that produce social injustice – that those favoured and those disfavoured by such processes might be completely unaware that they actually have such a status. An old but early example of algorithmic injustice remains powerful: the algorithmic-queuing of incoming calls within call centres using call line ID. This connects the phone numbers of incoming callers with databases, allowing payment records and assessments of potential profitability to be used automatically to queue incoming calls differentially. Crucially, neither the person who is answered straight away nor the person left hanging on forever is aware of this sorting. When the experience of urban injustice is so individualized and chimerical, it becomes immensely more difficult to build up the vital solidarities and collectivities necessary to start and sustain the social movements which can successfully counter them.

There is some progress. However, urban scholars and activists need to make alliances with critical data technologists, the people who can penetrate the complex black boxes of sociotechnical systems, sociotechnical injustice, and often sociotechnical violence. From the complexes that extend the power of exclusionary geodemographic profiles, to those at the heart of algorithmic exclusion and biometric profiling, analytical expertise is required that can penetrate these algorithmic performances and establish how they can be challenged, exposed, and changed.

Another example I found early on hammered home how the shift to machine-driven urban surveillance worked to try and encode basic and fundamental questions of urban rights, ethics, and justice into esoteric, "black boxed" technological systems that are hard to understand and even harder to challenge. In Boston there was a trial of a then-new "video analytics" system, and the system designers were discussing the use of algorithmic systems to analyse digital video feeds from an urban CCTV to detect *abnormal* or *threatening* people or behaviours. The video surveillance system, in this case in a car park, was programmed to identify a passing cyclist as a *threat* and a driver next to his vehicle as the *target* to be protected. Such a prosaic example demonstrates powerfully that, at all stages where these software platforms are used and operated, there are political and philosophical judgments about worth, fitness, and goodness, within a normative, hugely regressive notion of the city.

Crucially, of course, within contexts of intensifying inequality and the terrifyingly unequal logics of neoliberalized urbanism, pejorative and racialized norms and judgments inevitably creep insidiously into the code itself, and we are only starting to begin to understand (a) how that happens, and (b) how that can be contested and changed. It is a huge challenge, and cuts to the core of the ethics of the so-called *smart* city. Can a politics of such continual technologized agency, screening, and power ever be made understandable or visible? Might it be politicized, rendered visible (or at least *visualizable*), challenged, and reversed? Might it even be remade and reimagined in ways that make it ethically sound or even emancipatory or cosmopolitan?

What is certain is that the rudiments of liberal state power tasked with overseeing and regulating these vast and burgeoning worlds are ill-equipped, poorly informed, and often toothless. Here we confront the ultimate in David and Goliath battles. Traditionally, the data protection and professional regulators who deal with data questions are often just as ill-informed about how these systems work as anybody else, and they, too, are poorly equipped to grapple with the transnational nature of platform capitalism and algorithmically defined urban life. Of course, such regulators are most exposed to the fundamental tension between the inherited system of nested geographical, legal, and political jurisdictions – district-city-nation-supranational bloc – and the near-instantaneous systems of transnational and even near-planetary data circuits that fundamentally underpin contemporary capitalism.

Editors: It sounds like you do not see anything antithetical between smartness and justice. In working on this book, we have been considering whether activists, scholars, and other socially minded people should play into smartness given that it seems it is here to stay, or the "reality" we face, versus the approach that social justice might never be achievable through the lens of the smart city. Instead, meaningfully working towards justice requires putting smartness aside and just pursuing "justice in the city" as conceived by Lefebvre, Harvey, and others.

SG: The political economies of urban development don't work in the same ways as they did in the 1960s and 1970s. Whilst the theoretical arguments of David Harvey's analyses about capitalist urbanization remain powerful and helpful, to understand and challenge the complex dynamics of platform urbanism, a political economy of digital urbanism must be at the centre of contemporary urban thinking. Platform urbanism – that is, an urban universe dominated by centralized worlds of consumption and distribution that combine *front stage* software platforms that operate through the continuous agency of hidden algorithms and function through huge logistical *back stages* extending across global scales – is the urban world now.

These days, the place traditionally labelled the city is a complex and superimposed amalgam of algorithmically organized digital media systems which are layered over and facilitate everything else. For the sheer *materiality* of this point, I always go back to Jay David Bolter and Richard Grusin's superb 2000 book *Remediation: Understanding New Media*, which is all too often missed in debates about digitality and the urban. To them, the *digital* is not a question of some stampede into an immaterial world of *cyberspace*, as in the absurd cyberspace fantasies of the 1980s and 1990s. Rather, it is a complex remediation of the body, the book, the newspaper, the street, the house, the neighbourhood, the electricity system, the transport system, and so on, as the new potentialities of digital media and computing technologies are layered over, through and within all the spaces and circuits of society, facilitating new dynamics and forms in the process.

I was just reading Anna Lauren Hoffman's (2020) work on data ethics in *New Media and Society*, which is absolutely fascinating. Here she is looking at the fact that so many contemporary technoscientific worlds are basically organized around discursive violence. That means that users are being brought into a world that is inherently violent, at least at the discursive level. Sometimes maybe it is better not to want to be included in these systems.

My caveat would be that, sometimes, not to be included in these systems is to be not allowed to live, to exist; it can be a future of being which is ontologically erased. Given my point about the ways in which all of the bases of life, the material basis of life anyway, are often mediated by these systems, it is difficult to survive in the contemporary city without engaging with digital platforms, even unintentionally.

FURTHER READING

Benedikt, M., ed. 1991. *Cyberspace: First Steps*. Cambridge, MA: MIT Press.

Bolter, J.D., and R. Grusin. 2000. *Remediation: Understanding New Media*. Cambridge, MA: MIT Press.

Bratton, B.H. 2016. *The Stack: On Software and Sovereignty*. Cambridge, MA: MIT Press.

Burrows, R. 2018. "Urban Futures and the Dark Enlightenment: A Brief Guide for the Perplexed." In *Towards a Philosophy of the City: Interdisciplinary and Transcultural Perspective*, ed. K. Jacobs and J. Malpas, 245–58. Lanham, MD: Rowman and Littlefield.

Davis, M., and D.B. Monk, eds. 2011. *Evil Paradises: Dreamworlds of Neoliberalism*. New York: New Press.

Graham, S., ed. 2004. *The Cybercities Reader*. London: Psychology Press.

Graham, S. 2011. *Cities Under Siege: The New Military Urbanism*. New York: Verso Books.

Graham, S. 2016. *Vertical: The City from Satellites to Bunkers*. New York: Verso Books.

Hoffmann, L.A. 2020. "Terms of Inclusion: Data, Discourse, Violence." *New Media & Society* 23 (12): 3539–56. https://doi.org/10.1177%2F1461444820958725

Mattern, S. 2017. *Code and Clay, Data and Dirt: Five Thousand Years of Urban Media*. Minneapolis: University of Minnesota Press.

Nagle, A. 2017. *Kill All Normies: Online Culture Wars from 4chan and Tumblr to Trump and the Alt-Right*. Old Alresford, UK: John Hunt Publishing.

Parikka, J. 2013. *What Is Media Archaeology?* Cambridge: John Wiley & Sons.

Wiig, A., A. Karvonen, C. McFarlane, and J. Rutherford. 2022. "*Splintering Urbanism* at 20: Mapping Trajectories of Research on Urban Infrastructures." *Journal of Urban Technology* 29 (1): 1–11. https://doi.org/10.1080/10630732.2021.2005930

1 Who Is Telling the Smart City Story? Feminist Diffractions of Smart Cities

CARINA LISTERBORN

Planning and urban development have fundamentally changed with new technology and new private urban tech actors. Data are needed to plan a city, but how much data are needed, who should analyse them, and with which algorithms? The gathering of data has gained new possibilities as urban life is increasingly digitally mediated. This raises privacy concerns. News outlets report daily on the intrusive conduct of private digital companies and the neglect of the public states. At the same time, the needs and everyday struggles of citizens largely remain the same. Since the 1980s, feminist urban critique has been shedding light on everyday mundane practices and bringing in the embodied and material aspects of urban life into planning, geography, and urban studies. This chapter explores emerging smart city idioms through the lens of feminist urban critique to raise questions about what is new and what remains the same with regard to power relations in the smart future ahead. In particular, it highlights the potential in feminist critique and feminist everyday utopias.

Drawing on both earlier feminist urban critique (Fainstein and Servon 2005; Wajcman 1991; Weisman 1994) and contemporary work (Datta et al. 2020; Parker 2017; Peak and Rieker 2013), this chapter discusses what smart city strategies and controversies can tell us about contemporary urban visions, conditions, and social justice. Two different examples of smart city strategies and conflicts expose how power relations and social differentiation are being produced or reproduced when tech companies enter planning and manifest digital technology in the city: The Sidewalk Toronto project was meant to be an original plan from scratch where new technology could be tested out (see also Krivý, in this volume). The Grow Smarter project in Stockholm is a retrofitted smart city project partly funded by the European Union (see also Wullf-Wathne, Karvonen, and Haarstad, in this volume). Both Canada and Sweden score high on the DiGiX Index (0.80 and 0.88, respectively; BBVA Research 2019), which calculates levels of digitization. In both Stockholm and Toronto, the plans were celebrated by the authorities, at least initially. In Toronto the project was stopped along the way, but in Stockholm it was partly executed. In both cities the critique from the public concerns the people's right to the city and, accordingly, raises questions of who is invited to the storytelling of the smart city future. Whose reality is being manifested through smart city projects, and what can be learnt from the emerging civil critique that follows smart city investments?

Smart cities consist of a complex of corporate targets, urban policy, and branding (Batty 2016; Marvin, Luque-Ayala, and Macfarlane 2016; Schindler and Marvin 2018), steered through technocratic governance (Cardullo and Kitchin 2019). With the emergence of platform capitalism (Srnicek 2017), the influence of private companies has gone beyond gathering data for smart cities (Rose et al. 2020). Digital devices surround and surveil our everyday urban life. Rose et al. (2020) make a distinction, however, between platform urbanism and smart cities, where the latter is systematic and the former is "characterized by being more directly connected to consumers and interactive with users, more intent on rapid scaling-up via network effects and venture capital, and more antagonistic to government regulations and incumbent industries" (see also Sadowski 2020, 2). A smart city strategy may encourage platform urbanism as an innovative part of the urban, but it needs different management and organization to pursue public responsibilities. Smart cities and platform urbanism alternatively could end up in conflictual relations. According to Engelbert, Van Zoonen, and Hirzalla (2019), there is a strong hierarchy among different smart city projects, and they are often driven by supranational institutions and under the influence of global corporations, which questions the role of the city scale. This chapter focuses mainly on smart cities, even though platform urbanism is a large part of the everyday to which feminist scholars have paid attention (Elwood and Leszczynski 2018). These cities, however, are clearly dependent on both supranational (the EU in the case of Stockholm) and global corporations (Alphabet Inc. in the case of Toronto and IBM in the case of Stockholm).

In the next section I present critical feminist perspectives on smart cities, followed by sections discussing the two examples of smart city strategies and contestations seen through the lens of feminist urban theory. I conclude with a future-oriented musing that takes feminist everyday utopias into account.

New Feminist Materialism

There is a need for an intersectional perspective on the dialectics between society justice and technology to understand how the production and use of technology is closely related to gendered power relations. There is nothing intrinsically liberating about new technology, as exemplified by the wife-tracking apps for Saudi Arabian men produced by Apple and Google (Bennett 2019) or the "stalkerware" that facilitates intimate partner violence, abuse, and harassment (Parsons et al. 2019). In 1994, Leslie Kanes Weisman wrote, "We should not be fooled by the perfidious visions of the liberated future filled with technological gadgetry. Such visions are nothing more than patriarchal sophistry. Certainly, technology has eliminated much of the arduous physical strain of household and industrial labour. But it has never freed women or men from the confinement of gender roles" (Weisman 1994, 162).

Smart cities are often associated with public spaces and urban infrastructures, while the home remains a private enclosed space. Feminist scholars have often pointed out the relational aspects of the public and private spheres (Pateman 1989) to overcome this constructed divide. Smart urban systems are articulated and enter everyday life within the home, and the private usage of digital tools and social media invites tech companies to follow our daily habits, both at home and in public. While household work is often neglected in smart homes' visions (Richardson 2008), smart homes promise to make domestic life more efficient and to create time for leisure and pleasure. While modern technology has eased some household work, new tasks and needs have emerged along the way, and new domestic tech-work risks becoming just another chore to keep the home in (technical) function or to increase dependency on external expert support (Strengers and Nicholls 2018). Home automation could free up time for women, who commonly do more household work than men, and simultaneously function as a "wife replacement" for men, performing the stereotypical tasks of a 1950s housewife (Strengers and Nicholls 2018). Richardson (2008, 607) argues that the "gender digital divide is not bridged merely because more women are using ICTs [information and communications technologies]." To further the importance of social justice, a feminist approach is particularly important to point out these contexts, contradictions, and the partiality of the tech boom.

Sexism within the tech industry is an acknowledged problem. In the 1960s and 1970s, programming was mainly a female job, and there were several prominent and influential women in the industry. The tech boom in the 1980s changed gender roles, however, and when stereotypes flourish in an enclosed social environment, it becomes difficult to gain support for ideas that challenge gender norms. Nowadays, in the United States, only 25 per cent of the computing jobs are held by women, black women hold 3 per cent of computing jobs, and Latina women hold 1 per cent (Chang 2018). As the tech industry is a major economic force, this affects the possibilities for women or ethnic minorities to form startup companies, which further reinforces the economic and digital gap between men and women and between white men and ethnic minorities. This downward spiral is reinforced when there are few women in this field at universities and in the business as role models (Chang 2018; Misa 2010). This "gender gap" (Misa 2010) creates shortcomings in representation, biases in knowledge production, and "algorithms of oppression" (Noble 2018). Recent studies illustrate such shortcomings. An example is Google's speech recognition software, reported to be 70 per cent more likely to recognize male speech accurately than female speech (Criado Perez 2019). Face recognition algorithms have proved to be significantly more likely to mix up black women's faces than those of white women, black men's faces than those of white men, and to be most accurate with white men (Simonite 2019). Further, as Slupska (2019) discusses, there is the usage of digital devices in relation to domestic violence, while Elwood and Leszczynski (2018) find that crowd-sourced mapping pays more attention to drinking venues than, for example, to various forms of childcare.

The digital hegemony is skewed not only concerning gender, but also spatially, where the global North dominates in terms of making, and being the subject of, most digital data (McLean, Maalsen, and McNamara 2020). The digital colonialism of the global South is often covered up as economic development, creating new (precarious) job opportunities (Kwet 2019). Exploiting labour in colonized parts of the world or using women as immaterial labour outside factories (Jarrett 2018) is hardly new to the digital economy; however, what was previously driven by nation-states or local industries is now driven by global tech companies, which can act quickly spatially and are always looking out for the next most profitable option. As Kwet (2019, 16) puts it, "colonialism was not just a physical act of aggression, it was an ideology formed to justify conquest and pacify resistance." The digitalization of the everyday gives further emphasis to this statement. The question of who defines what is smart, and useful, technology is then relevant to investigate further – as well as whose experiences are being told within the smart city discourse.

The digitalization of everyday life does not act in a no-(wo-)man's-land. Even though the technology is new, its ideas, norms, and practices are a continuation of previous social and material structures. The neoliberal urban business-oriented agenda of the 1990s and 2000s that focused on attracting the creative class opened up the market for smart city agents. The idealized agent in a creative city is a mobile, autonomous, flexible, and hypercapitalist worker, and the "highest paying industries remained most easily occupied by white, male, elite heterosexual subjects with limited social reproduction responsibilities," concludes Brenda Parker (2017, 168) in her study of neoliberalism in Milwaukee. This echoes what has been observed in the tech industries (Chang 2018). The flip side of the creative city discourse has been the growing urban divides that affect women, youth, elderly, migrants, and single parents through increased costs of living, narrow housing choices, more costly social reproduction, and more limited scope for democratic influence (Curran 2018) – issues that smart city initiatives have hardly approached. The increased privatization of public spaces and resources through smart cities risks reinforcing these patterns of exclusion. Understanding masculinist power as broader than power exercised by men, feminist urban theory can shed light on the "construction of the idealized subject-citizen, a regulatory fiction whose presence delimits the field and agenda of politics" (Roy 2003, 108, quoted in Parker 2017, 3). As discussed above, this masculinist regulatory ideal dominates the smart city discourse.

Going back in time, feminist movements and feminist research have made a significant contribution to broadening the understanding of the urban and the possibility of a "non-sexist city" (Hayden 1980), as well as to revealing material constraints (Listerborn 2020). Nevertheless, as Parker (2016, 1338) states, "the voluminous literature on creative class politics includes only a handful articles that address gender and race," and gender is often left out of planning debates. The rich literature on gender and planning and the feminist historical rewritings of cities are essential in order not to "accept unthinkingly the man-made landscape as a neutral background" (Weisman 1994, 2). This school of thought has theoretically and empirically grounded an understanding of space as socially constructed, and has focused on differences, marginalization, and equity (Fainstein and Servon 2005). Thinking of "feminism as a problem space" (Peake and Rieker 2013, 9) questions and contests contemporary gendered geographies, and is based in an urban feminist materialism (Parker 2016). It adds a different perspective to the ongoing smart city research, and helps to develop a more gender-inclusive vision of *smart* domestic life (Strengers and Nicholls 2018). Therefore, thinking differently about the future when it comes to gender norms and social divides is crucial for actual changes, something that I discuss at the end of the chapter.

The Tech-Lash in Toronto

What initially appeared as a new formula for smart city development from scratch proved to be too difficult to pursue. Waterfront Toronto (a federal, provincial, and municipal partnership) set up a framework agreement in 2017 with Sidewalk Labs (a sister company to Google, part of Alphabet Inc.) to develop the waterfront area of Toronto known as Quayside (Flynn and Valverde 2019). The project, called Sidewalk Toronto, was meant to be a pilot case of how future smart cities could look: "a sensor-laden community that would collect data from the citizens living and moving within it to make the city living easier, in large part by developing new technologies from that data" (*Globe and Mail* 2018).

In 2020, Sidewalk Labs withdrew from the project. According to Sidewalk Labs' CEO, Don Doctoroff, the delays due to COVID-19 and economic uncertainties globally and in the Toronto real estate market led Sidewalk Labs to the conclusion that it would be "too difficult to make the project financially viable without sacrificing core parts of its plan" (*CityNews* 2020). If the project had proceeded, it could have changed our understanding of what a city is and how planning is done. The built environment was to serve as a data mine: "Arguably, this is not a city, as we know it, but rather a physical replica of Google's digital search engine" (Baeten 2020, 27). Although it is difficult to prove, the critique of the project likely played a major role in the decision to withdraw, even though the proponents were several.

Advocates for the plans, such as Richard Florida, argued that Toronto needed to compete with "the best of the best" and create space for the next generation of startup companies. Further, a report commissioned by Sidewalk Toronto predicted that, by 2040, Sidewalk Toronto could lead to 90,000 new jobs in property technology and mobility innovation (*Toronto Life* 2019). The response to critics was that innovation will always bring elements of disruption, risk, and uncertainty with it, but that the digital world is the future. "Without risks, Toronto will continue to be a footnote in the age of urbanism," said architect Alexander Josephson, and Progressive Conservative politician Robert Prichard stated that no more obstacles, such as critiques of the project, were needed because they would sort out the problems along the way: "let's get on with it" (*Toronto Life* 2019).

On the other side of the debate, concerns over lack of transparency and democracy drove a group of thirty Torontonians to set up Block Sidewalk (www.blocksidewalk.ca). Their agenda was to promote urban planning based on public interest, for the benefits and needs of the people of Toronto, and to ensure that the development would prioritize the city's needs first, not the needs and interests of a private corporation. Block Sidewalk raised issues of responsibilities over the installations, the increased surveillance, and the lack of public insight. Bianca Wylie – an open government advocate, the co-founder of the advocacy group Tech Reset Canada, and an activist in Block Sidewalk – referred to previous experiences of Uber, where control over pricing, congestion, climate, and labour was given to the company, and to Google's and Facebook's tendency to control distribution of news, which influences public trust. These were not foreseen consequences of the aforementioned innovations (*Toronto Life* 2019). Wylie (2019a) found Sidewalk Toronto a "Hubristic, Insulting, Incoherent Civic Tragedy." She stated: "they're pushing full-steam ahead on their work to make sure cities, Toronto and elsewhere, are compliant with their quantified worldview. Sidewalk Labs has done several things in the pursuit of this goal. And it's done in the Google way. Cheerful, young, faux-progressive, hip, making this all seem fun and harmless. And attracting a range of stakeholders to make complicit in its work and ideology" (Wylie 2019b).

Similar critiques led several public representatives to resign during the consultation process – claiming that there was a lack of transparency, that decisions were being made without the approval of the board, and that the privacy issue was never dealt with, risking creating a "Smart City of Surveillance" and not prioritizing the interests of the public (Vincent 2018). The story of risk taking, of prioritizing the creative city discourse above public interests, of believing in technology as *the* future force, and of focusing only on one kind of progress, where social equality is not the prime goal, is exactly what the feminist lens reveals. The focus in this project was evidently on new investments and technology-skilled labour, with the aim of reaching the top of urban competition rankings.

The outcome of the critique of the Sidewalk Labs endeavour in Toronto was celebrated by many. For instance, Jim Balsillie, co-founder of Research In Motion, mentioned it as a "'major victory' for those who fought to protect Canada's democracy, civil and digital rights as well as the economic development opportunity" (*CityNews* 2020). But only the future will tell whether this was just a bump on the road or actually a turning point. After all, Sidewalk Labs' investments were large, and their ambition was high: "Sidewalk's mission is not to create a city of the future at all. It is to create the future of cities" (Sidewalk Labs 2017, 12).

Continuous critique is important, not least the feminist one. According to Baeten (2020, 28), "[e]ven though the attempt to implement Sidewalk Labs' plans in Toronto has been aborted, it is of crucial importance that the planning profession carefully considers its explicit and implicit claims and aims related to urban planning and everyday life in the city. Only then we can begin to decide ourselves whether we want a technological construct to 'decide the future of cities' – in Toronto or anywhere else in the world."

Stockholm, the Smartest City in the World

Stockholm, the capital of Sweden, is partaking in the global competition to become the smartest city in the world by 2040. Aiming to turn Stockholm into a city with the highest quality of life for its citizens and the best environment for business, the city council adopted a smart city program in 2017 with the three lead words, "connected, open, and innovative" (Stockholm 2017). The master plan from 2019 required Stockholm to focus particularly on climate-smart implementations. In many ways, Stockholm already has come far in establishing information and communications technology. Since 1994, Stokab, a city-owned company, has been delivering an open fibre-optic communication network with 100 per cent broadband coverage within the Stockholm Region. Moreover, Kista Science Centre was established in 1986 as a site for ICT innovation in a triple helix – that is, a model in which the private sector, academia, and public actors collaborate. Today Stockholm is the largest innovation area in Europe for ICT, and it has large-scale, real environment testbeds, such as the new housing area Royal Seaport. The City of Stockholm has a close collaboration with IBM, which in 2011 registered the trademark "smarter cities" as belonging to it (Söderström, Paasche, and Klauser 2014).

Between 2015 and 2019, Stockholm was the coordinator of an EU Horizon project called Grow Smarter, which received €25 million in funding. Together with Barcelona and Cologne, Stockholm became a so-called Lighthouse city (Grow Smarter n.d.), and these three Lighthouse cities have five Follower cities (Valetta, Suceava, Porto, Cork, and Graz) that are meant to learn from their experiences and adapt their plans. The overall aim of the project was to create 1,500 new jobs in Europe and decrease energy use and transport emissions by 60 per cent by 2020 through twelve smart solutions in energy use, integrated infrastructure, and sustainable transport. As with similar EU-funded projects, innovation in Stockholm tends to build on pre-existing or already planned projects, which then get rebranded as *smart* (Haarstad and Wathne 2019). In Stockholm, three areas were chosen for the Grow Smarter project: one existing housing area, one mixed-use area with small businesses that will be developed into a housing area, and a logistic centre and transport hub.

Of particular interest here is the existing housing area around Valla Torg in South Stockholm. This rental housing block consists of eight hundred apartments in six high-rise buildings and a few lower buildings. It was built in the 1960s and owned by the municipality's housing company, Stockholmshem. The Grow Smarter project leader approached Stockholmshem with the request to use this area to test new technology, which the housing company approved. The renovation included new pipes, new insulation with quadruple-glazed windows, and new bathrooms and balconies, as well as a new waste system, solar systems, carpools, and cargo-bike rental systems. In each apartment, a display was installed to allow tenants to regulate and monitor their energy and water use. Courses were organized to teach tenants how to handle this new technology.

A number of people had lived in the area for decades, about half of them seniors living on pension income. Rents had been low over the years, but with the renovation rents were due to increase radically, by 40–60 per cent over three years. During the renovation, people had to move out; around 20 per cents did not move back in afterwards. One tenant did not agree to the renovation plan and took the case to the National Tenants Court (*Hyresnämnden*). The case was lost, but it did delay the project by eight months as the renovation could not begin before the case was settled. In a local paper (*SöderortDirekt* 2017), critical voices were raised about the rent increase, about how the investment did not improve the quality of living, and about how the smart technology (which the heating system was dependent on) was not working. In the final report about the project, Grow Smarter was regarded as successful because the investments succeeded in decreasing energy use and greenhouse gas emissions by 60 per cent. On the project's website, three content seniors describe the new look of the area, giving the impression that it is unproblematic. The project report, however, left a few unsettled issues – for example, why tenants had to pay the costs of the renovation when the long-term cost benefits of energy saving would profit the housing company.

In a recent smart city index, Stockholm was ranked as low as twenty-fifth among the world's smart cities, far away from the city's goal. This was mainly due to the index being based on surveys among the city's inhabitants and how they prioritized different subject areas. In Stockholm, the need for affordable housing was ranked highest by 67 per cent of the respondents (IMD World Competitiveness Center 2019). In fact, affordable housing does not belong among Stockholm's success stories, and the Grow Smarter project did not address this issue either. Certainly, Stockholm's infrastructural challenges and the ambition to be greenhouse gas emission free in 2040 are urgent matters. Using smart technology to gain these goals might release resources from different stakeholders, but it risks ignoring other social consequences.

The examples from both Toronto and Stockholm illustrate how the belief in smart technology risks setting aside fundamental urban justice issues – for example, low-income households, often single-parent or female-led households, are not considered. When no city wants to become a "footnote in the age of urbanism," the interpretation of what that means is highly normative – following a competitive- and "creative class"-induced urban idiom. In the examples given here, a feminist perspective can help foreground the lack of alternative thinking and critique the norms and values that exclude the importance of the everyday. Smart cities, where the technocratic understanding of the city dominates, risk becoming a continuation and reinforcement of existing power relations. As Strengers and Nicholls (2018, 78) posit, a feminist approach allows us to include issues such as social care, health care, and sociology in the field of design and technology; further, they argue that "such approaches could prove fruitful in deviating from the dominant tech-centric trajectory permeating the smart home, to develop more gender-inclusive visions of 'smart' domestic life."

Time to Refocus

That the future is digital seems difficult to refute, and tech companies are likely to continue to influence urban planning and the everyday life of citizens around the globe (Townsend 2013). The (feminist) question is, then, how can we use the new technology in a way that does not reinforce existing power relations? How can technology support a future that is more liveable, green, and just? And *when* is technology the solution to the problem? According to Elwood and Leszczynski (2018), there is, of course, a feminist potential within digital tools, but there is also a risk of "cracking a nut with a sledgehammer." Engineer Shoshanna Saxe summarizes the digital challenges as follows: smart solutions will be exceedingly more complex to manage; tech products have a short life span and will need to be replaced/updated in a few years' time, leading to disruption and costs; and managing data will require a brand-new municipal bureaucracy – and there is no guarantee that the new technology will solve urban issues more efficiently. Data mines in themselves do not empty public garbage cans when sensors report them full (Saxe 2019). Moreover, what if the voice detector of the home alarm only works for male voices? New technology can create new problems and make everyday life more complicated, stressful, and constrained.

As Stephen Graham states in the introductory dialogue to this part of the book, "the entire discourse on smart cities is profoundly ahistorical and inhabits a permanent paradigm of the present problem, followed, simplistically, by the future (technological) solution." Feminist urban critique precisely highlights the historical injustices connected to technology when it comes to production, access, usages, and applications. For increased technological justice, the shortcomings of including everyday struggles needs to be addressed within a smart future. In *Feminism Confronts Technology*, Wajcman (1991, 166) concludes that "technologies reveal the societies that invent and use them, their notions of social status and distributive justice. In so far as technology currently reflects a man's world, the struggle to transform it demands transformation of gender relations." The development of smart city solutions must be based on an idea of what kind of urban future we desire. This is highly political and cannot be buried by technological determinism or fatalism. Wajcman (1991, 166) further states: "we need to go beyond masculinity and femininity to construct technology according to a completely different set of socially desirable values." This suggests diverting from just reflecting and using reflexive methodology as a critical tool, as it seems to have limitations. It could risk calling for a technology based on specific feminine values instead of changing the underlying power relations. Lykke (2017) draws on Haraway (1997) when she suggests using the concept of diffraction as a metaphor for critical thinking techniques to make a difference. Mirroring, as with a reflection, does not bring us beyond the static of the same. On the other hand, diffraction is a dynamic process urging a "production of different patterns in the world, not just of the same reflected – displaced – elsewhere" (Haraway 1997, 268, quoted in Lykke 2017, 34).

One such source for change could be a turn to feminist utopias and feminist futures (Schalk, Kristiansson, and Mazé 2017). Utopian ideas or projects have been criticized from a feminist perspective for being sexist, colonial, and universalizing hegemonic norms about the "good society" (Bingaman, Sanders, and Zorach 2002). Instead, Cooper's (2014) *Everyday Utopianism* describes utopia as actualized in everyday practice. An everyday utopia is a lived experience in which people challenge oppression and hegemonic discourses. Utopias are about the here and now, not a systematic design or format. Utopianism is thus a response to contemporary societal mechanisms such as neoliberalism, sexism, racism, and colonialism. Everyday utopianism could push the direction of change. Everyday utopian thinking might not be a digital paradise but, rather, affordable housing, collective housing, shared household work, supportive care work, and other urban feminist purposes and urgencies (Hayden 1980). This could include digital tools, but they do not guide utopianism.

A sign of the future can perhaps be seen in the post-COVID era. Returning to the critique of Sidewalk Labs, Waterfront Toronto suddenly took a new turn after Sidewalk Labs dropped its proposal for Quayside in May 2020 in the midst of the pandemic. Leaving the ideas of efficient, high-tech, and urban living, Waterfront Toronto announced that its new plans would instead prioritize "housing affordability and long-term-care housing needs for seniors" (*Toronto Star* 2020). Whether this will be realized remains to be seen, and whether this is an indication of where smart cities are heading is, of course, too early to tell. Nevertheless, words have been spoken about care and affordability instead of smart efficiency. Steven Diamond, chair of Waterfront Toronto, is quoted in the *Toronto Star* newspaper as explaining, "I'm not saying the project won't be innovative, it just won't be data-driven" (*Toronto Star* 2020). The critique of Sidewalk Labs probably played a role in this turnaround.

REFERENCES

Baeten, G. 2020. "Sidewalk Labs' Plans for Toronto Shake the Foundations of Planning as We Know It." *Plan Canada*: 26–8.

Batty, M. 2016. "How Disruptive Is the Smart Cities Movement?" *Environment and Planning B: Planning and Design* 43 (3): 441–3. https://doi.org/10.1177/0265813516645965

BBVA Research. 2019. "Which Countries Are the Most Digitally Advanced?" Online at https://www.bbva.com/en/which-countries-are-the-most-digitally-advanced/, accessed 18 August 2020.

Bennett, C. 2019. "Wife-tracking apps are one sign of Saudi Arabia's vile regime." Others include crucifixion …" *Guardian*, 28 April 2019. Online at https://www.theguardian.com/commentisfree/2019/apr/28/wife-tracking-apps-saudi-arabias-vile-regime-crucifixion, accessed 28 August 2020.

Bingaman, A., L. Sanders, and R. Zorach, eds. 2002. *Embodied Utopias. Gender, Social Change and the Modern Metropolis*, London: Routledge.

Cardullo, P., and R. Kitchin. 2019. "Smart Urbanism and Smart Citizenship: The Neoliberal Logic of 'Citizen-Focused' Smart Cities in Europe." *Environment and Planning C: Politics and Space* 37 (5): 813–83. https://doi.org/10.1177/0263774X18806508

Chang, E. 2018. *Brotopia: Breaking Up the Boy's Club of Silicon Valley*. New York: Penguin Random House.

CityNews. 2020. "Sidewalk Labs pulling out of Quayside Waterfront project." 7 May. Online at https://toronto.citynews.ca/2020/05/07/sidewalk-labs-pulling-out-of-quayside-waterfront-project/, accessed 18 August 2020.

Cooper, D. 2014. *Everyday Utopias: The Conceptual Life of Promising Spaces*. Durham, NC: Duke University Press.

Criado Perez, C. 2019. *Invisible Women: Exposing Data Bias in a World Designed for Men*. New York: Penguin Random House.

Curran, W. 2018. *Gender and Gentrification*. London: Routledge.

Datta, A., P. Hopkins, L. Johnston, E. Olson, and J.M. Silva. 2020. *Routledge Handbook of Gender and Feminist Geographies*. London: Routledge.

Elwood, S., and A. Leszczynski. 2018. "Feminist Digital Geographies." *Gender, Place & Culture* 25 (5): 629–44. https://doi.org/10.1080/0966369X.2018.1465396

Engelbert, J., L. Van Zoonen, and F. Hirzalla. 2019. "Excluding Citizens from the European Smart City: The Discourse Practices of Pursuing and Granting Smartness." *Technological Forecasting and Social Change* 142 (May): 347–53. https://doi.org/10.1016/j.techfore.2018.08.020

Fainstein, S., and L.J. Servon, eds. 2005. *Gender and Planning: A Reader*. New Brunswick, NJ: Rutgers University Press.

Flynn, A., and M. Valverde. 2019. "Planning on the Waterfront: Setting the Agenda for Toronto's 'smart city' Project." Planning Theory & Practice 20 (5): 769–75. https://doi.org/10.1080/14649357.2019.1676566

Globe and Mail. 2018. "Sidewalk Labs' Toronto deal sparks data, innovation concerns." 1 August. Online at https://www.theglobeandmail.com/business/technology/article-new-development-agreement-between-sidewalk-labs-and-waterfront-toronto/, accessed 11 September 2020.

Grow Smarter. n.d. "Lighthouse City: Stockholm." Online at https://grow-smarter.eu/lighthouse-cities/stockholm/, accessed 16 September 2020.

Haarstad, H., and M.W. Wathne. 2019. "Lighthouse Projects: Stavanger, Stockholm, Nottingham." In *Inside Smart Cities: Place, Politics and Urban Innovation*, ed. A. Karvonen, P. Cugurullo, and F. Caprotti, 102–16. London: Routledge.

Haraway, D. 1997. *Modest_Witness@Second_Millenium. FemaleMan©_ Meets_ OncoMouse™: Feminism and Technoscience*. London: Routledge.

Hayden, D. 1980. "What Would a Non-Sexist City Be Like? Speculations on Housing, Urban Design, and Human Work." *Signs* 5 (3): 170–87. https://doi.org/10.1086/495718

IMD World Competitiveness Center. 2019. "Smart City Index." Online at https://www.imd.org/research-knowledge/reports/imd-smart-city-index-2019/, accessed 5 January 2021.

Jarret, K. 2018. "Laundering Women's History: A Feminist Critique of the Social Factory." *First Monday* 23 (3). https://doi.org/10.5210/fm.v23i3.8280

Kwet, M. 2019. "Digital Colonialism: US Empire and the New Imperialism in the Global South." *Race and Class* 60 (4): 3–26. https://doi.org/10.1177/0306396818823172

Listerborn, C. 2020. "Gender and Urban Neoliberalization." In *The Routledge International Handbook of Gender and Feminist Geographies*, ed. A. Datta, P. Hopkins, L. Johnston, E. Olson, and J.M. Silva, 184–93. London: Routledge.

Lykke, N. 2017. "Prologue: Anticipating Feminist Futures while Playing with Materialisms." In *Feminist Futures of Spatial Practice: Materialisms, Activisms, Dialogues, Pedagogies, Projections*, ed. M. Schalk, T. Kristiansson, and R. Mazé, 27–32. AADR – Art Architecture Design Research. https://www.diva-portal.org/smash/get/diva2:1340312/FULLTEXT01.pdf

Marvin, S., A. Luque-Ayala, and C. Macfarlane, eds. 2015. *Smart Urbanism: Utopian Vision or False Dawn?* London: Routledge.

McLean, J., S. Maalsen, and N. McNamara. 2020. "Doing Gender in the Digital: Feminist Geographic Methods Changing Research?" In *The Routledge International Handbook of Gender and Feminist Geographies*, ed. A. Datta, P. Hopkins, L. Johnston, E. Olson, and J.M. Silva, 467–75. London: Routledge.

Misa, T.J., ed. 2010. *Gender Codes: Why Women Are Leaving Computing*. Hoboken, NJ: Wiley.

Noble, S. 2018. *Algorithms of Oppression: How Search Engines Reinforce Racism*. New York: New York University Press.

Parker, B. 2016. "Feminist Forays in the City: Imbalance and Interventions in Urban Research Methods." *Antipode* 48 (5): 1337–58. https://doi.org/10.1111/anti.12241

Parker, B. 2017. *Masculinities and Markets: Raced and Gendered Urban Politics in Milwaukee*. Athens: University of Georgia Press.

Parsons, C., A. Molnar, J. Dalek, J. Knockel, M. Kenyon, B. Haselton, C. Khoo, and R. Deibert. 2019. "The Predator in Your Pocket: A Multidisciplinary Assessment of the Stalkerware Application Industry," Citizen Lab Research Report 119. Toronto: University of Toronto.

Pateman, C. 1989. *The Disorder of Women. Democracy, Feminism and Political Theory*. Cambridge, UK: Polity Press.

Peak, L., and M. Rieker, eds. 2013. *Rethinking Feminist Interventions into the Urban*. London: Routledge.

Richardson, H.J. 2008. "A 'Smart House' Is Not a Home: The Domestication of ICTs." *Information Systems Frontiers* 11 (5): 599–608. https://doi.org/10.1007/s10796-008-9137-9

Rose, G., P. Raghuram, S. Watson, and E. Wigley. 2020. "Platform Urbanism, Smartphone Applications and Valuing Data in a Smart City." *Transactions of the Institute of British Geographers* 46 (1): 59–72. https://doi.org/10.1111/tran.12400

Roy, A. 2003. *City Requiem: Calcutta*. Minneapolis: University of Minnesota Press.

Sadowski, J. 2020. "The Internet of Landlords: Digital Platforms and New Mechanisms of Rentier Capitalism." *Antipode* 52 (2): 562–80. https://doi.org/10.1111/anti.12595

Saxe, S. 2019. "I'm an engineer, and I'm not buying into 'smart' cities." *New York Times*, 16 July. Online at https://www.nytimes.com/2019/07/16/opinion/smart-cities.html.

Schalk, M., T. Kristiansson, and R. Mazé, eds. 2017. *Feminist Futures of Spatial Practice: Materialisms, Activisms, Dialogues, Pedagogies, Projections*, 27–32. AADR – Art Architecture Design Research. https://www.diva-portal.org/smash/get/diva2:1340312/FULLTEXT01.pdf

Schindler, S., and S. Marvin. 2018. "Constructing a Universal Logic of Urban Control?" *City* 22 (2): 298–307. https://doi.org/10.1080/13604813.2018.1451021
Sidewalk Labs. 2017. "Proposal Quayside, Appendix." Toronto.
Simonite, T. 2019. "The best algorithms struggle to recognize black faces equally." *Wired*, 22 July. Online at https://www.wired.com/story/best-algorithms-struggle-recognize-black-faces-equally/, accessed 5 January 2021.
Slupska, J. 2019. "Safe at Home: Towards a Feminist Critique of Cybersecurity." *St Antony's International Review* 15 (1): 83–100. https://ssrn.com/abstract=3429851
SöderortDirekt. 2017. "Ilska över usel renovering vid Valla Torg." 9 November. Online at https://www.mitti.se/nyheter/ilska-over-usel-renovering-vid-valla-torg/repqkg!9XGJTVJdLjNUBSJzCr6hxA/, accessed 16 September 2020.
Söderström, O., T. Paasche, and F. Klauser. 2014. "Smart Cities as Corporate Storytelling." *City* 18 (3): 307–20. https://doi.org/10.1080/13604813.2014.906716
Srnicek, N. 2017. "The Challenges of Platform Capitalism: Understanding the Logic of a New Business Model." *Juncture* 23 (4): 254–7. https://doi.org/10.1111/newe.12023
Stockholm. 2017. "Smart and Connected City." Stockholm: City of Stockholm. Online at https://international.stockholm.se/governance/smart-and-connected-city/, accessed 11 September 2020.
Strengers, Y., and L. Nicholls. 2018. "Aesthetic Pleasures and Gendered Tech-Work in the 21st-Century Smart Home." *Media International Australia* 166 (1): 70–80. https://doi.org/10.1177/1329878X17737661
Toronto Life. 2019. "The sidewalk wars." 4 September. Online at https://torontolife.com/city/18-big-thinkers-take-a-critical-look-at-the-sidewalk-labs-plan/, accessed 5 January 2021.
Toronto Star. 2020. "Waterfront Toronto ditches Sidewalk Labs' vision of high-tech, sensor-driven smart district at Quayside." 30 July. Online at https://www.thestar.com/news/gta/2020/06/30/waterfront-toronto-ditches-sidewalk-labs-vision-of-high-tech-sensor-driven-smart-district-at-quayside.html, accessed 10 August 2020.
Townsend, A. 2013. *Smart Cities: Big Data, Civic Hackers, and the Quest for a New Utopia*. London: Norton.
Vincent, D. 2018. "Tech expert resigns from advisory panel on Sidewalk Toronto over data ownership concerns." *Toronto Star*, 4 October. Online at https://www.thestar.com/news/gta/2018/10/04/tech-expert-resigns-from-sidewalk-toronto-advisory-panel-over-data-ownership-concerns.html, accessed 11 September 2020.
Wajcman, J. 1991. *Feminism Confronts Technology*. Cambridge: Polity Press.
Weisman, L.K. 1994. *Discrimination by Design: A Feminist Critique of the Man-Made Environment*. Chicago: University of Illinois Press.
Wylie, B. 2019a. "Sidewalk Toronto: A Hubristic, Insulting, Incoherent Civic Tragedy – Part II." Online at https://biancawylie.medium.com/sidewalk-toronto-a-hubristic-insulting-incoherent-civic-tragedy-part-ii-334129560cb1, accessed 5 January 2021.
Wylie, B. 2019b. "Sidewalk Toronto: Violating Democracy, Entrenching the Status Quo, Making Markets of the Commons." Online at https://biancawylie.medium.com/sidewalk-toronto-violating-democracy-entrenching-the-status-quo-making-markets-of-the-commons-8a71404d4809, accessed 5 January 2021.

2 More Queer, More than Human: Challenges for Thinking Digital Justice in the Smart City

RYAN BURNS

In recent years, the purported benefits of smart cities have rightly been taken to task. This now robust critical research agenda has been approached from a number of angles, and it is clear that smartness is not straightforwardly effective in fostering empowerment, environmental sustainability, efficiency, or increased access to urban spaces. Put colloquially, in many ways the smart city is not necessarily *smart*, if smart were ever to connect with such connotations. Of course, in many other ways, it was very smart strategy: for capital accumulation (Hollands 2015; Sadowski 2020; Zuboff 2019), governance (Gabrys 2014), worlding (Datta 2016), epistemological claims (Burns and Andrucki 2020; Burns and Wark 2020; Kitchin, Lauriault, and McArdle 2015), reiterating the exclusionary nature of citizenship (Cardullo and Kitchin 2019; Datta 2018; Shelton and Lodato 2019), and related sociopolitical processes. In other words, across the field, there is sufficient evidence that smart cities cause many unevenly distributed harms. Some unsuccessful developments and community resistance notwithstanding (e.g., Alphabet's Sidewalk Toronto, Masdar City, Songdo), the smart city seems likely to remain a dominant urban planning paradigm. Reminiscent of critiques that capitalism absorbs its own contradictions and critiques such that it survives apparent collapses, smart cities might even absorb their critiques into evolved forms of smartness (Levenda and Tretter 2020). Importantly, smart cities are in many ways the latest development of a long tradition of digital urbanism such as "city of bits" and "intelligent cities" (de Jong et al. 2015; Mitchell 1996); however, "smartness" distinctly brings along with it many normative judgments, colloquial connotations, and folkways that did not characterize its predecessors.[1]

Within this context, recent research has begun to ask important questions about whether and how to pursue a conception of the smart city that works harder at ameliorating its unevenly distributed harms. This volume approaches the question from the well-worn tradition of "social justice" within geography, urban studies, and elsewhere, and while a powerful one, it is one among many avenues for such inquiries. And social justice *itself* has long been debated, contested, and (re)worked – it does not have a singular meaning, and is not self-evident (Heynen et al. 2018; Hopkins 2021; Young 1990). This chapter, then, is concerned with the ways in which social justice as a value system, philosophical framework, and political strategy has been taken up in these conversations.

Specifically, this chapter converses with the plethora of recent work emphasizing "the right to the smart city" as a framework for pursuing more just smart cities. While recognizing the value of this approach, I argue here that the limitations of this framework mean that scholars should look elsewhere for pursuing social justice within smart cities. Here, I suggest more-than-human thinking and queer theory as productive ways to address the limitations of the "right to the city" framework. What I present is not a fully mature reconceptualization of justice: to do so would require engaging with moral philosophy and the history of social justice theories to a degree that I am unable to do within these few pages. Rather, I present here three provocations from queer theory and more-than-human thinking that raise important new questions for the field and identify potential new avenues for rethinking justice within the smart context.

The Right to the City

The right to the city has been taken up in diverse ways by spatially minded scholars, but is most notably associated with Henri Lefebvre. Following Purcell (2002, 101), Lefebvre's thinking involved not "piecemeal" reforms, but instead "a radical restructuring of social, political, and economic relations, both in the city and beyond." Lefebvre's vision entails, at its basic core, democratic control over the decision-making processes that affect cities: all denizens of the city have equal say regarding how a city develops and self-administers. Space is *relational* for Lefebvre, produced by, through, and for sociopolitical relations (Lefebvre 1991b, 2003), and thus the production of space is critical for understanding how the right to the city would be achieved (Schmid 2012). Insofar as space crystallizes these relations, decisions that produce urban space – including everything from firms' investment strategies to federal immigration policies – are thus within the purview of one's right to the city. While it has been taken up in different ways since its early formulation, Lefebvre defines the right to the city as:

> The right to the city, complemented by the right to difference and the right to information, should modify, concretize and make more practical the rights of the citizen as an urban dweller (*citadin*) and user of multiple services. It would affirm, on the one hand, the right of users to make known their ideas on the space and time of their activities in the urban area; it would also cover the right to the use of

the centre, a privileged place, instead of being dispersed and stuck into ghettos (for workers, immigrants, the 'marginal' and even for the "privileged." (Lefebvre 1991a; quoted in Kofman and Lebas 1996, emphasis in original)

Importantly in this quote, Lefebvre mobilizes the neologism "citadin" to denote that the right to the city should belong not simply to formal citizens of the nation-state or other jurisdictions, but instead to all who inhabit the spaces of the city. Elsewhere, Lefebvre elaborates on these ideas to argue that such democratic decision-making ultimately would lead to the withering away of the state, aligning with his long-standing commitment to Marxian urban horizons (Purcell 2014).

In this way, Lefebvre understood the right to the city to be not something that an individual or a collective *holds*, but instead as an ongoing political project to assert control over urban space. Purcell (2014) argues that this radical vision of the city should not be seen as a simple addition to liberal-democratic *rights*, but instead as a fundamental revolution in urban politics. This contrasts with many other adaptations of the idea, such as that of Marcuse (2012, 34), who characterizes it as "a claim not only to a right or a set of rights to justice within the existing legal system, but a right on a higher moral plane that demands a better system in which the potential benefits of an urban life can be fully and entirely realized." Marcuse's read combines a structural predilection with an acknowledgment that *rights* on some level can and should be enshrined with "the existing legal system," or in other words, the legal system should be preserved and augmented with further protections that reflect the "higher moral plane." In this way, Marcuse sees the right to the city as emerging within existing legal-juridical frameworks anchored by normative moral philosophy, which reads as less revolutionary than Lefebvre's original writing (Lefebvre 2003).

Since these early conceptual elaborations, the phrase "right to the city" has been taken up by many as "an immediately understandable and intuitively compelling slogan" (Marcuse 2012, 29) somewhat separated from the particularities of Lefebvre's conceptualization. Indeed, Mitchell and Heynen (2009, 616) argue that "the value of the concept of a right to the city is precisely its capaciousness." Many of these later pieces aim to formulate the more general spirit of access to and control over urban spaces, services, and politics; they approximate the timbre of Lefebvre's writing, but often make no explicit references to it. They vary in their application and focus area, from public space writ large (Mitchell 2003) to capitalist modes of production (Harvey 2008).

Limits to the Right to the City

While the right to the city as conceptualized by Lefebvre offers promising avenues for pursuing social justice, it suffers important limitations, both those inherent to the concept itself and in the ways it has been picked up by urban scholars. Here I present these limitations together, and show their implications for the ways scholars have conceived of the right to the smart city. After laying out these limitations and implications, I make a case for addressing them through queer theory and more-than-human thinking.

First, the terminological and conceptual flexibility mentioned earlier necessitates that we ask *what sorts of rights* are one's rights to the city. Attoh (2011) comes closest to directly asking this question and considering its implications, noting that *rights* can be construed in ways that are often in direct contradiction to one another. Rights might entail ability to exclude oneself or others, entitlements claimed of the state, power to make legal or political change, and more. Thus, for Attoh (2011, 676), the flexibility of the term leads *not* to "approaches to the right to the city [that] are commensurable," but rather, "the capaciousness of the right to the city appears less as beneficial and instead as simply confused." A closer investigation is needed to elucidate what sorts of rights constitute the right to the smart city.

Following this enquiry, urban scholars would do well to remember the strong intellectual lineage from rights-talk to Enlightenment visions of the liberal subject (Rorty 1979) that continues to inform legal cultures and human rights (Nelken 2016; Simon 1999).[2] This lineage epistemologically privileges the individual as the holder of rights, as the fundamental social unit (instead of, for example, the community or the planetary), and as in Cartesian duality with nature and environment. The individual is always human and is always the active agent within a passive world (Rose 1992).

This version of the rights-holding liberal subject has been rightly criticized for its linkages with a number of social and political ills. Because in its classical formulations it was always left unspecified, it came to be associated with what Rose (2017, 783) calls "an apparently unmarked cipher: the site of undifferentiated ideas, experience, and resistance … coded as masculine, white, and straight." One might add to this list cisgendered, middle or upper class, and fully rational (Haraway 1991; Rose 1993). Indeed, Nancy Fraser's (1990) critique of Habermas, upon which I expand below, was precisely that his conception of "the public" ignored the exclusionary politics present in the incipient demos itself: that to be present in the public sphere required the individual to be a land-owning male. The liberal Enlightenment subjectivity further informed the human rights tradition dominant through the twentieth and early twenty-first centuries, underwriting new forms of imperialism and domination (Douzinas 2000, 2007).

This intellectual heritage always privileges human subjects independent from their surrounding environments and other non-human actors. In other words, put bluntly, humans may make rights claims, but *rights as such* may not be invoked by an ecological system, and *human* rights do not include livestock and other non-human animals. This orientation fundamentally assumes a pure and coherent *human* who sits in dichotomous relation to animality and diverse natures. For several decades such anthropocentric frameworks have been taken to task in areas as diverse as feminism (Harding 1996), Science and Technology Studies (Law 1991), and relational theory (Whatmore 1999). The principle is becoming increasingly untenable in the wake of the deepening climate crisis, exploitation and eradication of non-human animals, and the growing recognition that human-ness is intricately connected with nature, animal, and machine (Haraway 2015; Wainwright and Mann 2018).

One might rightly point out that this rights framework was not Lefebvre's intention when formulating his right to the city. Indeed, Lefebvre's intention was not necessarily to privilege the individual's held rights, but instead *democratic* control over structural and spatial processes and decision-making practices. While still mobilizing undifferentiated subjects, Lefebvre's conception sought radical, structural change, rather than an individual's claim to justice. However, two critiques remain: first, that Lefebvre was, in fact, decidedly anthropocentric in his conceptualizing the right to the city; and second, that it has since been taken up in ways that depart from Lefebvre's original thought. In other words, one critique is inherent to the concept itself and the other is merely in how it has been theorized by others.

The consequence is that these critiques frame the way we think of social justice in smart cities, leading to inherently exclusionary politics. First, rights-talk has already led to an anthropocentric pursuit of social justice in smart cities. From early critiques of smart cities' inequalities (Greenfield 2013; Hollands 2015) to later explicit engagements with social justice (McFarlane and Söderström 2017), the human is usually front-and-centre as the *subject* of enquiry and intervention. This is, of course, not inevitable, as much of the smart city literature more broadly focuses on ecological concerns and questions (Gabrys 2014; Levenda and Tretter 2020). Second, most research presumes not simply a human subject, but an explicit "smart *citizen*" (Shelton and Lodato 2019; Willis 2019). For example, Kitchin implores scholars to pursue a "genuinely humanizing smart urbanism" (2019, 195) that would "be oriented toward reflecting and serving the interests of *citizens*" (196, emphasis added). Of course, an emphasis on citizens per se neglects many undocumented residents, temporary and permanent residents. It also epistemologically centres the nation-state as the conferrer of citizenship within discussions of social justice. Third, rights-talk discourages attention to the ways in which subjects are coded out of smart justice by virtue of markers of social segmentation. Marginalized communities of colour – in particular, women of colour – are brought into discourses of smart social justice differently than for the normative white heterosexual male (Burns and Andrucki 2021). This difference is meaningful for the specific tactics with which social justice must be pursued, as I explain further below.

More Queer, More-than-Human

To reconstruct a framework of social justice for smart cities, I argue that we can look to queer theory and more-than-human thinking. Working towards this goal, I offer three provocations from these fields of study, which are meant to illustrate "alternative" ways to approach social justice in the smart city. Before that, however, it is important to further unpack those subjects that are included and excluded from public deliberations of social justice. Nancy Fraser (1990) has rightly noted that the liberal-bourgeois model of the public sphere that is central to Habermasian formulation was implicitly imbued with its own exclusionary politics. For Fraser, bourgeois models of the public sphere rest on four assumptions, of which I focus on two. First, they assume that those who enter the public sphere do so on equal grounds; social and political differences are ignored, and thus such sociopolitical equality is not a *prerequisite* of a functioning public sphere, but instead is the *result* of entering a public sphere. This assumption very prominently ignores the fact that, in its early conceptions, those who were not land-owning males were prohibited from entering the public sphere to begin with. Of course, once able to deliberate within the public sphere, people's other linguistic, behavioural, and physical markers of social class retained uneven relations of influence. Fraser notes, however, that this historiography itself omits the strong radical presence of women in the public sphere since its early conception: despite being written out of the public sphere and accounts of it, women were present.

Second, bourgeois models of the public sphere assume that a single hegemonic public sphere is the ideal mode of democracy, rather than "a multiplicity of competing publics" (Fraser 1990, 62). In this formulation, the bourgeois model sees the proliferation of multiple publics as a detraction from the ideal public sphere organization: it is *disorder*, in contrast to a single public sphere's *order*. Proposing the term "subaltern counterpublics" (67), Fraser points out that, historically, less powerful groups such as labourers, women, and people of colour have found it useful to form their own publics as a means by which to agitate oppressive power structures. More recently, Travers (2016) has shown that these counterpublics were not precluded by the move to the digital, but in many ways are reproduced within it. It is precisely these publics' relation to oppressive legal, political, and social structures that necessitates a cohesive, collective voice – they are *public*, the opposite of an enclave – to instigate sociopolitical change.

Michael Warner (2002) picks up this point in his queer theorization of counterpublics. For Warner, a public should be conceived of less as a coherent audience, and much less as a venue for the deliberation of ideas and values. Instead, "[p]ublics are queer creatures" (2002, 7) that are created in the event of being addressed. Publics are *relations* that straddle the boundary of personal and impersonal, "space created by the reflexive circulation of discourse" (90). Publics do not pre-exist the texts and practices that address them *as a public*: their invoked coherence and cohesion is performative and ideological. *Counter*publics are publics for which the dominant modes of publicness are not afforded – in other words, those publics that "are defined by their tension with a larger public" (56). Counterpublics develop unique discourses, codes, and modes of address that distinguish them from common invocations of "the public." By developing these characteristics and cultivating relations to broader publics, counterpublics shape large-scale politics, often cohering as social movements.

This, then, is the first provocation I offer towards a reconceptualization of social justice within the smart city: to draw attention to counterpublics as a way of highlighting the power relations constituting body-subjects within the smart city's public spheres. It would differentiate the *human* and, in particular, the *citizen* that often guides such discussions. This discussion informs smart cities theorization by illuminating the terms on which various publics are "admitted" into dialogue in the smart city public sphere. Far from being egalitarian, or even based on citizenship, venues for political debate have always worked to marginalize and exclude counterpublics and the bodies that compose them. Warner rightly notes that, in addition to Fraser's list of subaltern subjects – labourers, women, and people of colour – queer communities remain omitted from many contemporary publics. According to Warner, "[t]he challenge facing this project in transgender activism, feminism, and *queer theory* is to understand how world making unfolds in publics that are, after all, not just natural collections of people, not just 'communities,' but mediated publics" (2002, 61, emphasis added).

While Warner's particular flavour of queer theory remains largely oriented around the humans who constitute and conjure publics, broader debates within queer theory have problematized the seeming naturalness of dominant categories upon which comfortable epistemologies of the human rest. At varying times, queer theory questions the supposed mutual exclusivity of categories such as "male" and "female," "straight" and "gay" (Butler 2006), being occluded and spectacularized (Sedgwick 1990), or inclusivity and exclusivity (Puar 2017). It asks how we come to see certain epistemologies and ontologies as natural and beyond critique, and then seeks to reveal their origins and disrupt their illusory universality (Brown and Knopp 2008). In this, it often draws on Foucault's archaeological and genealogical methods of taking largely unquestioned concepts such as sexuality, madness, and knowledge, and illustrating their contested historic and geographic origins (Foucault 1978).

This tendency to question supposedly natural categories opens important new avenues for thinking beyond the human as the subject of justice in the smart city, and is my second provocation. Most pertinently, many in this area have begun to theorize the ways in which humans are intertwined in complex relations with non-human animals. In queering the figure of the "crazy cat lady," McKeithen (2017) shows the layers of natural categories that act upon the bodies of women who perform a particular domesticity – that bring them into public discourse in already ostracized ways. Drawing on Haraway's (2003) posthuman ontology to show the deep intimacies between humans and non-human animals, McKeithen argues that "multispecies homemaking unfolds through more-than-human agencies" (2017, 122), renormalizing a queer way of being in the home. Similarly, Brown and Rasmussen (2010) look for encounters that challenge the nature-culture binary through relations between humans and non-humans and the legal structures that emerge to govern such relationships. In this and other related work, we see the ways in which the human is always already intertwined with non-human actors. As Haraway (1991) has shown, the purity of the human as a concept has always been questionable to begin with.

Along these lines, my third provocation turns to recent more-than-human thinking that shows productive ways to reposition humans as actors alongside other important aspects of contemporary urban life. Here, research emphasizes the interdependence of humans and non-humans, with a view to destabilizing the ontological priority given to humans within discussions of justice (Brown, Flemsæter, and Rønningen 2019). With an eye towards decolonial and postcolonial ways of thinking, Collard, Dempsey, and Sundberg (2015, 322) call for a "multispecies abundance … [comprised of] more diverse and autonomous forms of life and ways of living together." This multispecies abundance would recognize the inherent value of non-human lives outside of their relationships with humans (e.g., raising animals for food or for scientific experiments). Blue and Rock (2014) similarly argue for the need to move beyond the rational human at the centre of conceptions of the public sphere, and recognize the interspecies relations and entanglements that characterize contemporary urban life. These questions of whether non-human animals can be legitimate claimants of justice are, of course, not new, but in recent years interest in them has increased in the context of shifting sociopolitical norms and socionatural imperatives such as climate change and deteriorating environments (Deckha 2007).

The promise of this third provocation is in the way it opens conversations of justice to much broader actors relevant for diverse forms of being in smart cities. This is becoming increasingly important in the face of the deepening climate crisis, and raises for us the question of whether non-living beings such as environments should figure into debates about justice in the smart city. For instance, recent research in epigenetics, a field exploring how environments modify the human genetic code, provides new material for ontologically placing the human alongside socionatural environments (Guthman and Mansfield 2013; Pickersgill et al. 2013). Here, humans are deeply connected with their surroundings, and are affected by issues as diverse as pollution, political-economic stressors, and green space. Along similar lines, Pugliese (2020, 4) has recently argued that non-human entities as diverse as "animal, vegetal, and mineral entities," such as soil, trees, rocks, and animal species be included in debates about justice. Should our conceptions of social justice within the smart city be broadened to contend with these non-human actors? Is Lefebvre's right to the city "capacious" enough to bend to these considerations?

I want to ask whether these frameworks go far enough. In an age of deepening algorithmic governance, software-sorted geographies, automated inequality, and data-driven decision-making, machines increasingly are tasked with being the purveyors of injustice. As machines become progressively sophisticated and embedded within human bodies and socionatural environments, urban scholars need to recognize the intertwined-ness of machines, humans, and non-humans, and begin to work towards structures and imperatives that do not task machines with carrying out injustices. The world towards which social justice works is one that seeks justice both for those *receiving* injustice and those actors that are delegated the lamentable task of *enacting* it. In other words, whereas current literature on social justice in the smart city typically focuses on those (human) actors who are unjustly acted upon, I argue that the literature should also consider actors that are forced to carry out the injustices. We need to raise the possibility of justice for machines, as well as for humans and other non-humans alike. We might think of this as "machinic justice," not as a contrast to social justice, but as a necessary component of it.

Can we consider machines to be legitimate claimants of justice? What would that look like? First, it would entail dismantling the digital infrastructures that subtend and enact injustice in the smart city. Doing so would shift attention from the outcomes of digital infrastructures to the sensors, algorithms, databases, robotics, and software themselves, under the principle that forcing machines to carry out injustice is itself an injustice. Second, it would entail mobilizing machines towards more just societies, processes, and relations. By this I mean anything from a digitally mediated community resource allocation system, to constructing machinic assemblages that support anticapitalist labour structures inside firms, to enforcing opt-out parameters to all data-collection mechanisms. In other words, this positive move pertains both to proactive mobilizations and normalized protections for humans and non-humans. Third, it would entail developing new conceptual apparatuses through which to think of machinic justice. Most pressingly, this involves situating machinic justice within theories of moral philosophy and long-standing debates about social justice (to which I have already alluded), and against the rights-based anthropocentric language of the Enlightenment.

Conclusion

In this chapter I have argued that recent discussions of social justice within the smart city need to step back and consider the range of theoretical frameworks that might inform how we conceive and pursue justice. While Henri Lefebvre's *right to the city* framework provides important and productive conceptual grounding for this pursuit, it suffers from a number of shortcomings with which smart cities research must contend for its assumptions and unintended unequal implications. Namely, the right to the city remains primarily concerned with human actors, and risks reiterating the dominance of Enlightenment-based rights-talk that privileges individual legal protections over collective revolutionary action. As well, the right to the city has been taken up in ways that privilege subjects uncoded aside from their status as citizens – that is, within the field of smart cities research, scholars have tended to discuss the right to the city through the lens of citizenship. In this chapter, I have shown ways that queer theory and more-than-human thinking provide productive avenues for expanding justice regimes within smart cities. In particular, I propose the incipient idea of "machinic justice," in which machines, software, algorithms, platforms, and other computational systems are brought into discussions of justice. With this idea, we might reconceptualize machines as legitimate claimants to justice, as they are increasingly tasked with carrying out injustices.

NOTES

1 Debra Mackinnon has helped me with some of this language in personal conversation.
2 Of course, this tradition can be seen in early documents such as the Declaration of the Rights of Man and of the Citizen in France and the Declaration of Independence in the United States, as well as in the philosophy of Jean-Jacques Rousseau, John Locke, and others. Although important differences exist across thinkers and geographic contexts, social scientists today still often speak of an "Enlightenment project" in the way I do here.

REFERENCES

Attoh, K.A. 2011. "What Kind of Right Is the Right to the City?" *Progress in Human Geography* 35 (5): 669–85. https://doi.org/10.1177/0309132510394706
Blue, G., and M. Rock. 2014. "Animal Publics: Accounting for Heterogeneity in Political Life." *Society & Animals* 22 (5): 503–19. https://doi.org/10.1163/15685306-12341350
Brown, K.M., F. Flemsæter, and K. Rønningen. 2019. More-than-Human Geographies of Property: Moving towards Spatial Justice with Response-ability." *Geoforum* 99: 54–62. https://doi.org/10.1016/j.geoforum.2018.12.012

Brown, M., and L. Knopp. 2008. "Queering the Map: The Productive Tensions of Colliding Epistemologies." *Annals of the Association of American Geographers* 98 (1): 40–58. https://doi.org/10.1080/00045600701734042

Brown, M., and C. Rasmussen. 2010. "Bestiality and the Queering of the Human Animal." *Environment and Planning D: Society and Space* 28 (1): 158–77. https://doi.org/10.1068/d5807

Burns, R., and M.J. Andrucki. 2021. "Smart Cities: Who Cares?" *Environment and Planning A* 53 (1): 12–30. https://doi.org/10.1177%2F0308518X20941516

Burns, R., and G. Wark. 2020. "Where's the Database in Digital Ethnography? Exploring Database Ethnography for Open Data Research." *Qualitative Research* 20 (5): 598–616. https://doi.org/10.1177/1468794119885040

Butler, J. 2006. *Gender Trouble: Feminism and the Subversion of Identity*. New York: Routledge.

Cardullo, P., and R. Kitchin. 2019. "Smart Urbanism and Smart Citizenship: The Neoliberal Logic of 'Citizen-Focused' Smart Cities in Europe." *Environment and Planning C: Politics and Space* 37 (5): 813–30. https://doi.org/10.1177/0263774X18806508

Collard, R.-C., J. Dempsey, and J. Sundberg. 2015. "A Manifesto for Abundant Futures." *Annals of the Association of American Geographers* 105 (2): 322–30. https://doi.org/10.1080/00045608.2014.973007

Datta, A. 2016. "The Smart Entrepreneurial City: Dholera and 100 other Utopias in India." In *Smart Urbanism: Utopian Vision or False Dawn?* ed. S. Marvin, A. Luque-Ayala, and C. McFarlane, 52–70. New York: Routledge.

Datta, A. 2018. "The Digital Turn in Postcolonial Urbanism: Smart Citizenship in the Making of India's 100 Smart Cities." *Transactions of the Institute of British Geographers* 43 (3): 405–19. https://doi.org/10.1111/tran.12225

Deckha, M. 2007. "Animal Justice, Cultural Justice: A Posthumanist Response to Cultural Rights in Animals." *Journal of Animal Law and Ethics* 2: 189–229.

de Jong, M., S. Joss, D. Schraven, C. Zhan, and M. Weijnen. 2015. "Sustainable–Smart–Resilient–Low Carbon–Eco–Knowledge Cities; Making Sense of a Multitude of Concepts Promoting Sustainable Urbanization." *Journal of Cleaner Production* 109: 25–38. https://doi.org/10.1016/j.jclepro.2015.02.004

Douzinas, C. 2000. *The End of Human Rights: Critical Legal Thought at the Turn of the Century*. Portland, OR: Hart Publishing.

Douzinas, C. 2007. *Human Rights and Empire: The Political Philosophy of Cosmopolitanism*. New York: Routledge-Cavendish.

Foucault, M. 1978. *The History of Sexuality: An Introduction*. New York: Vintage Books.

Fraser, N. 1990. "Rethinking the Public Sphere: A Contribution to the Critique of Actually Existing Democracy." *Social Text* (25/26): 56–80. https://doi.org/10.2307/466240

Gabrys, J. 2014. "Programming Environments: Environmentality and Citizen Sensing in the Smart City." *Environment and Planning D: Society and Space* 32 (1): 30–48. https://doi.org/10.1068/d16812

Greenfield, A. 2013. *Against the Smart City*. New York: Do Publications.
Guthman, J., and B. Mansfield. 2013. The Implications of Environmental Epigenetics: A New Direction for Geographic Inquiry on Health, Space, and Nature-Society Relations." *Progress in Human Geography* 37 (4): 486–504. https://doi.org/10.1177/0309132512463258
Haraway, D. 1991. *Simians, Cyborgs, and Women*. New York: Routledge.
Haraway, D. 2003. *The Companion Species Manifesto: Dogs, People, and Significant Otherness*. Chicago: Prickly Paradigm Press.
Haraway, D. 2015. Anthropocene, Capitalocene, Plantationocene, Chthulucene: Making Kin." *Environmental Humanities* 6 (1): 159–65. https://doi.org/10.1215/22011919-3615934
Harding, S. 1996. "Feminism, Science, and the Anti-Englightenment Critiques." In *Women, Knowledge, and Reality: Explorations in Feminist Philosophy*, ed. A. Garry and M. Pearsall, 298–320. New York: Routledge.
Harvey, D. 2008. "The Right to the City." *New Left Review* 53: 23–40.
Heynen, N., D. Aiello, C. Keegan, and N. Luke. 2018. "The Enduring Struggle for Social Justice and the City." *Annals of the American Association of Geographers* 108 (2): 301–16. https://doi.org/10.1080/24694452.2017.1419414
Hollands, R.G. 2015. "Critical Interventions into the Corporate Smart City." *Cambridge Journal of Regions, Economy and Society* 8 (1): 61–77. https://doi.org/10.1093/cjres/rsu011
Hopkins, P. 2021. "Social Geography III: Committing to Social Justice." *Progress in Human Geography* 45 (2): 382–93. https://doi.org/10.1177/0309132520913612
Kitchin, R. 2019. "Toward a Genuinely Humanizing Smart Urbanism." In *The Right to the Smart City*, ed. P. Cardullo, C. di Feliciantonio, and R. Kitchin, 193–204. Bingley, UK: Emerald Publishing. https://doi.org/10.1108/978-1-78769-139-120191014
Kitchin, R., T. Lauriault, and G. McArdle. 2015. "Knowing and Governing Cities through Urban Indicators, City Benchmarking and Real-Time Dashboards." *Regional Studies, Regional Science* 2 (1): 6–28. https://doi.org/10.1080/21681376.2014.983149
Kofman, E., and E. Lebas. 1996. "Lost in Transposition – Time, Space and the City." In *Writings on Cities/Henri Lefebvre*, ed. E. Kofman and E. Lebas, 3–60. Malden, MA: Blackwell Publishers.
Law, J. 1991. "Introduction: Monsters, Machines and Sociotechnical Relations." In *A Sociology of Monsters: Essays on Power, Technology and Domination*, ed. J. Law, 1–25. London: Routledge.
Lefebvre, H. 1991a. "Les illusions de la modernité." *Manières de voir* 13: 14–17.
Lefebvre, H. 1991b. *The Production of Space*. Cambridge, MA: Blackwell Publishing.
Lefebvre, H. 2003. *The Urban Revolution*. Minneapolis: University of Minnesota Press.
Levenda, A.M., and E. Tretter. 2020. "The Environmentalization of Urban Entrepreneurialism: From Technopolis to Start-up City. *Environment and Planning A: Economy and Space* 52 (3): 490–509. https://doi.org/10.1177/0308518X19889970
Marcuse, P. 2012. "Whose Right(s) to What City?" In *Cities for People, Not for Profit: Critical Urban Theory and the Right to the City*, ed. N. Brenner, P. Marcuse, and M. Mayer, 24–41. New York: Routledge.

McFarlane, C., and O. Söderström. 2017. "On Alternative Smart Cities." *City* 21 (3–4): 312–28. https://doi.org/10.1080/13604813.2017.1327166

McKeithen, W. 2017. "Queer Ecologies of Home: Heteronormativity, Speciesism, and the Strange Intimacies of Crazy Cat Ladies." *Gender, Place & Culture* 24 (1): 122–34. https://doi.org/10.1080/0966369X.2016.1276888

Mitchell, D. 2003. *The Right to the City: Social Justice and the Fight for Public Space*. New York: Guilford Press.

Mitchell, W. 1996. *City of Bits: Space, Place, and the Infobahn*. Cambridge, MA: MIT Press.

Mitchell, D., and N. Heynen. 2009. "The Geography of Survival and the Right to the City: Speculations on Surveillance, Legal Innovation, and the Criminalization of Intervention." *Urban Geography* 30 (6): 611–32. https://doi.org/10.2747/0272-3638.30.6.611

Nelken, D. 2016. "Comparative Legal Research and Legal Culture: Facts, Approaches, and Values." *Annual Review of Law and Social Science* 12: 45–62. https://doi.org/10.1146/annurev-lawsocsci-110615-084950

Pickersgill, M., J. Niewöhner, R. Müller, P. Martin, and S. Cunningham-Burley. 2013. "Mapping the New Molecular Landscape: Social Dimensions of Epigenetics." *New Genetics and Society* 32 (4): 429–47. https://doi.org/10.1080/14636778.2013.861739

Puar, J. 2017. *Terrorist Assemblages: Homonationalism in Queer Times*. Durham, NC: Duke University Press.

Pugliese, J. 2020. *Biopolitics of the More-than-Human: Forensic Ecologies of Violence*. Durham, NC: Duke University Press.

Purcell, M. 2002. "Excavating Lefebvre: The Right to the City and Its Urban Politics of the Inhabitant." *GeoJournal* 58 (2): 99–108. https://doi.org/10.1023/B:GEJO.0000010829.62237.8f

Purcell, M. 2014. "Possible Worlds: Henri Lefebvre and the Right to the City." *Journal of Urban Affairs* 36 (1): 141–54. https://doi.org/10.1111/juaf.12034

Rorty, R. 1979. *Philosophy and the Mirror of Nature*. Princeton, NJ: Princeton University Press.

Rose, G. 1992. "Geography as a Science of Observation: Landscape, the Gaze and Masculinity." In *Nature and Science: Essays in the History of Geographical Knowledge*, Historical Geography Research Series., ed. G. Rose and F. Driver, 8–18. Cheltenham, UK: Institute of British Geographers.

Rose, G. 1993. *Feminism and Geography: The Limits of Geographical Knowledge*. Minneapolis: University of Minnesota Press.

Rose, G. 2017. "Posthuman Agency in the Digitally Mediated City: Exteriorization, Individuation, Reinvention." *Annals of the American Association of Geographers* 107 (4): 779–93. https://doi.org/10.1080/24694452.2016.1270195

Sadowski, J. 2020. *Too Smart: How Digital Capitalism Is Extracting Data, Controlling Our Lives, and Taking Over the World*. Cambridge, MA: MIT Press.

Schmid, C. 2012. "Henri Lefebvre, the Right to the City, and the New Metropolitan Mainstream." In *Cities for People, Not for Profit: Critical Urban Theory and the Right to the City*, ed. N. Brenner, P. Marcuse, and M. Mayer, 42–62. New York: Routledge.

Sedgwick, E.K. 1990. *Epistemology of the Closet*. Berkeley: University of California Press.

Shelton, T., and T. Lodato. 2019. "Actually Existing Smart Citizens: Expertise and (Non) Participation in the Making of the Smart City." *City* 23 (1): 35–52. https://doi.org/10.1080/13604813.2019.1575115

Simon, J. 1999. "Law after Society." *Law & Social Inquiry* 24 (1): 143–94. https://doi.org/10.1111/j.1747-4469.1999.tb00795.x

Travers, A. 2016. "Parallel Subaltern Feminist Counterpublics in Cyberspace." *Sociological Perspectives* 46 (2): 223–37. https://journals.sagepub.com/doi/10.1525/sop.2003.46.2.223

Wainwright, J., and G. Mann. 2018. *Climate Leviathan: A Political Theory of Our Planetary Future*. New York: Verso Books.

Warner, M. 2002. "Publics and Counterpublics." *Public Culture* 14 (1): 49–90. https://doi.org/10.1215/08992363-14-1-49

Whatmore, S. 1999. "Hybrid Geographies: Rethinking the 'Human' in Human Geography." In *Human Geography Today*, ed. D. Massey, J. Allen, and P. Sarre, 22–39. Malden, MA: Polity Press.

Willis, K.S. 2019. "Whose Right to the Smart City?" In *The Right to the Smart City*, ed. P. Cardullo, C. di Feliciantonio, and R. Kitchin, 27–41. Bingley, UK: Emerald Publishing. https://doi.org/10.1108/978-1-78769-139-120191002

Young, I.M. 1990. *Justice and the Politics of Difference*. Princeton, NJ: Princeton University Press.

Zuboff, S. 2019. *The Age of Surveillance Capitalism: The Fight for a Human Future at the New Frontier of Power*. New York: PublicAffairs.

3 Urbanists in the Smart City: Sidewalks, Sidewalk Labs, and the Limits to "Complexity"

MAROŠ KRIVÝ

In September 2020, an opinion piece by Italian urbanist Carlo Ratti (2020) titled "We need more urban innovation projects like the 'Google City'" appeared on the website of the World Economic Forum. Ratti, the founding director of the influential MIT Senseable City Lab (SCL), lamented the demise of Google sibling company Sidewalk Lab's controversial and contested project at Toronto's Quayside, contending that "the failure of this project is sad news." As Ratti, who sat on Sidewalk Toronto's advisory board, further remarked, "the greater cause of urban innovation" was defeated by the "posturing" of #BlockSidewalk activists ignoring that "urban experimentation is needed today more than ever."

In less than three years, between fall 2017 when the Quayside development came to public attention and spring 2020 when Sidewalk Labs abandoned the project, Sidewalk Toronto became something of a barometer of the smart city arena. While boosters celebrated it as the future of urban tech (Florida 2019), critics highlighted issues of data collection and privacy and non-transparent planning process on both sides of the proposed public-private partnership (Flynn and Valverde 2019; Goodman and Powles 2019; Lorinc 2019). The influence of the project extended far beyond Toronto, capturing techno-utopian imagination around the world (Ülemiste City 2019) and functioning as a springboard for numerous startups (Goodman and Powles 2019), while giving rise to international solidarity between urban protest movements (O'Kane 2019). Finally, the case provided a sense of urgency to the rapidly burgeoning critical scholarship around smartness (Cowley and Caprotti 2019; Halpern, Mitchell, and Geoghegan 2018), platform urbanism (Barns 2020; Sadowski 2020), and experimental cities (Evans 2016; Evans, Karvonen, and Raven 2016).

What has gone all but unnoticed in these debates is the role of urbanists such as Ratti, a loose and elusive professional community operating in the interstices between the fields of city planning, urban design, architecture, and many other fields (but see Barns 2020, 67–8; Sauter 2019). As Ratti's statement having been published by a leading neoliberal think tank shows, the influence of urbanists on these elites is not insignificant. Certainly, it cannot be compared to the power of tech corporations such as Google, entrepreneurial policymakers, and finance and real estate capitals. And yet, as I argue through the case of Ratti, urbanists have risen to the status of intermediary elites who sit comfortably within these power structures. This chapter goes beyond the more-just-smart-city bandwagon to ask, how do urbanists understand and intervene in the smart city? What are the slippages of meanings between smartness and adjacent terms such as complexity, self-organization, and experimentation? What kind of city do these imaginaries legitimize and what do they obscure? How do urbanists help perpetuate injustice in the smart city?

In what follows, I first introduce what I mean by urbanism (a profession) and clarify the chapter's contribution to the literature. Next, I review existing scholarship around neoliberal elites, experts, and professions. The third section examines Ratti's work and career, with a focus on Dynamic Street, a modular paving system developed for Sidewalk Toronto. In the conclusion, I elaborate on the limits that urbanists' participation in the smart city pose to the realization of social justice ideals.

Urbanists and the Smart City

Over the past decade and longer, the burgeoning critical scholarship on smart cities has developed from an overt focus on paradigmatic greenfield megaprojects such as Songdo and Masdar to consider – from part-overlapping, part-conflicting, neo-Marxist, feminist, and actor-network theories – how logics, infrastructures, and practices of smartness permeate the urban everyday. Scholars have examined subjectivity (Vanolo 2014), citizenship (Gabrys 2014), and sociality (Rose 2020) as conduits of power in the smart city, analysed the smart city through the lenses of environmental (Gabrys 2014) and algorithmic governmentality (Rouvroy and Berns 2011), and drawn attention to prototyping logics (Halpern, Mitchell, and Geoghegan 2018) and antiplanning motifs (Cowley and Caprotti 2019) underpinning various smart city projects. The messy, contested reality of the "actually existing smart city" (Shelton, Zook, and Wiig 2015) has been linked to tactical potentials and the politics of digital hacking (Chandler 2017), unplugging (Calzada and Cobo 2015), and the glitch (Leszczynski 2020). Last but not least, scholars have situated the smart city in the context of longer and broader mutations within urbanized capitalism (Morozov and Bria 2018), and discussed the excitement around the promise of experimentation (Evans 2016; Evans, Karvonen, and Raven 2016), the strategic roles of the startup city (Levenda and Tretter 2019), the startup state (Moisio and Rossi 2019), and the emergence of platform urbanism (Barns 2020; Sadowski 2020).

This chapter builds on and departs from these debates to discuss the role of urbanism in the smart city. By "urbanism," I refer not to urban life but to a profession and professional expertise, in which sense the term is often used as a synonym for planning. According to the *Oxford English Dictionary*, for example, an urbanist is "a specialist in or advocate of town-planning." Yet, in a narrower sense (and only in English, I believe), "urbanism" refers to the practice of what could be called "antiplanning" (Cowley and Caprotti 2019), meaning a professional expertise premised on abandoning the ambition to shape the city in ways that emphasize comprehensiveness, intentionality, and structural understanding of social problems in favour of celebrating and making room for ostensibly spontaneous urban change. To wit, the list of "The 100 Most Influential Urbanists" compiled by the influential planning website *Planetizen* (2017) includes a number of well-known modernist city planners such as Robert Moses and Le Corbusier, while being topped by Jane Jacobs and Jan Gehl (the Danish advocate for "cities for people"), two vocal critics of the style of "top-down" city planning associated with legacies of Moses and Le Corbusier (Krivý and Ma 2018; Larson 2013). The tension between the descriptive and normative meanings of the term "urbanism" points to a category of urban professionals operating at the interstices of planning and antiplanning, and it is in this group's influence in the smart city arena that this chapter is interested.

The chapter contributes to the critical smart city literature by examining what I call an imaginary of complexity: the belief that the city is a complex, spontaneously self-organizing system. In the field of urbanism, the smart city – in the sense of a city replete with digital sensors – aligns with a broader and older idea that the city is essentially smart, or what various authors refer to as being complex. For example, this is what Jacobs had in mind when she wrote, famously, how "an intricate sidewalk ballet" reveals "a complex order" "under the seeming disorder of the old city" (1961, 65) – the sidewalk reveals complexity (or smartness) not because it has sensors but because its users "are" sensors, meaning that the sidewalk ballet is a spontaneous result of their continually adjusting their actions relative to one another.

The case of Ratti is instructive because he approaches the smart city by adding a digital layer onto this older, liberal narrative so that there is a synthesis between technological and "humanistic" conceptions of complexity. He represents a significant strand within the smart city arena that naturalizes injustice not (only) because it is technocratic (Graham, in this volume), but also because it appeals to a common humanity of individuals as potentially smart, thereby neutralizing difficult and structural questions around class, race, and gender (Burns, in this volume; Listerborn, in this volume). This chapter discusses Ratti's work to reflect critically on the outsized power of elite urbanists to shape smart city imaginaries, and how they translate the elusive "complexity" into everyday, yet unavoidably political, narratives. Before turning to Ratti, however, I should situate the argument within the wider scholarship on professions and neoliberal power.

Neoliberal Elite Power

Scholars of professions and expertise have traced the mutation of elite power on the backs of financialization, digitalization, and the neoliberal diffusion of global authority. According to political geographer Merje Kuus (2021), that power now resides increasingly less in stable hierarchies and more in fragmented, informal networks centred around the production of complex, highly specialized, and high-status knowledge. She focuses on the study of diplomacy to argue that professions are institutionally fluid expert occupations bound to transnational operational spaces.

Kuus draws on the work of anthropologist Janine Wedel (2017), who coined the term "influence elites" to characterize elites defined by flexibility, informality, and their indeterminate and intermediary institutional status. Wedel departs from Charles W. Mills's theory of power elites to explore the rise of new elite "connectors" operating across hierarchies and networks both inside and outside official structures. She warns that influence elites evade visibility and accountability and thus pose challenges to democracy.

A similar departure characterizes political scientist William Davies's (2017) theory of intermediary elites. Davies coined the term "diplomatic intermediaries" to examine the capacity to translate quantitative data and other forms of "asignifying semiotics" (a term he in turn borrows from Maurizio Lazzarato) to political narratives – explanations, justifications, proposals – as a pivot of elite power under advanced neoliberalism. Pointing to the influence of think tanks, advisors, and consultants strategically operating between institutions. Davies throws light on the following paradox: whereas neoliberals such as Friedrich Hayek criticized the elites – planners, experts – in the name of spontaneous market forces, intermediary elites now leverage the capacity to interpret how markets "feel" and "behave" to political and economic agents.

Davies cites an important study by economic historian Philip Mirowski (2009) on the "neoliberal thought collective": a group of thinkers around Hayek bound together by an epistemic commitment to the market's being seen as a complex information processor. According to Mirowski, it is a mistake to identify neoliberalism with a narrowly understood economic theory; rather, it should be seen as a transnational and transdisciplinary discourse community operating in the interstices of academia, public policy, and the private sector, with the goal of reshaping the state around purportedly spontaneous markets.

To sum up, specific complexities of financial, data, and geopolitical structures have allowed a new breed of professional elites to occupy the intermediate and interstitial operational spaces of neoliberalism by translating "complexity" into convincing political narratives, thereby "anchoring" globalizing processes (Larner and Laurie 2010; Rankin 2003). The above observations about consultants, advisors, and other influence elites can be usefully extended to other professions, such as urbanism. Having discussed the entanglements of neoliberalism, complex knowledge, and shifting forms of professional power, in the next section I focus on Ratti to examine the role and power of urbanists in the contemporary smart city arena.

"Jack people into the network and get out of the way"

It is difficult to overstate Carlo Ratti's influence and authority: he has been dubbed "the unconventional smart city philosopher" (Bliss 2018), featured in *Wired*'s (2012) "Smart List: 50 people who will change the world," and trotted the globe as a visionary for humane smart cities. In addition to directing the SCL and advising Sidewalk Toronto, he is MIT Professor of Urban Technologies and Planning, runs an international office (Carlo Ratti Associati, CRA), co-chairs the World Economic Forum's Global Future Council on Cities of Tomorrow, and sits on various advisory boards – for example, advising the European Commission on "the merging of the physical and digital world and smart cities" (*MIT News* 2015).

The premise of Ratti's eclectic and prolific career – spanning master planning, urban design, and product development, along with curating, speaking, and writing[1] – is that digital technology empowers people. The neologism "senseable city" in the laboratory's name evokes a city simultaneously replete with digital sensors and sensitive to people's needs. Ratti's book-length manifesto (with Matthew Claudel), titled *The City of Tomorrow*, abounds with sentences to similar effect, such as "optimization inflected with humanization" and "there can be no smart city without smart citizens" (Ratti and Claudel 2016, 36, 148). The title of the manifesto is a provocative take on Le Corbusier's identically titled 1929 book, and the Swiss architect appears throughout Ratti's work as something of a nemesis poised to destroy cities "without any data" (Ratti et al. 2018).[2] Although such a depiction makes little sense historically, it makes sense if we consider the popular urbanistic representation of Le Corbusier as an emblem of autocratic top-down planning whose legacy destroyed everything good about cities (e.g. Sennett 2018). Put otherwise, the reference works as a foil against which Ratti has positioned his smart (or senseable) city as a form of antiplanning. As Ratti put it pithily, the goal is "to jack people into the network and get out of the way" (Ratti and Townsend 2011, 42) – a statement that registers difficult questions around technology, subjectivity, and the power of urbanists that need to be situated in the context of Ratti's professional career.

After graduating with a dual engineering degree from Turin and Paris, Ratti completed a PhD (Ratti 2001) at the Faculty of Architecture, University of Cambridge, then moved to MIT and founded the SCL in 2004. His career took shape against the backdrop of what historian Mary Louise Lobsinger calls "two Cambridges," a series of exchanges and collaboration between Cambridge, England, and Cambridge, Massachusetts, centred on developing a science-based approach to tackling "complex urban problems" (Lobsinger 2013, 656). Ratti's dissertation explored the potential of digital modelling to contribute to urban design by "decipher[ing] the urban environment" (2001, 251), drawing heavily on the pioneering work of Cambridge's Centre for Land Use and Built Form Study to quantify urban analysis and design (Keller 2006). He was brought to MIT on the invitation of Cambridge-educated William J. Mitchell, Dean of the MIT School of Architecture and Planning and founder of the Smart Cities program in the MIT Media Lab (2003). The SCL was established on the heels of that program, and Ratti followed Mitchell's then-unorthodox belief – communicated in bestsellers such as *City of Bits* (1995) and *e-topia* (1999) – that digital networks will reinforce, rather than diminish, the significance of urban density and place.

The SCL falls into MIT's tradition of experimental interdisciplinary laboratory modelling that combine technoscientific systems thinking with a politically liberal focus on individual user feedback and vaguely stated goals of social betterment (Dutta 2013), while sitting firmly and comfortably within the twenty-first-century academic-industrial complex funded by and accountable to venture capital, private corporations, and entrepreneurial municipalities. The lab was indeed established through seed funding, and within seven years expanded to thirty employees, an externally raised annual budget of US$3 million (Daly 2011), and hundreds of corporate clients, including Audi, Accenture, Cisco, and Uber. The SCL's homepage boasts that "Senseable is as fluent with industry partners as it is with metropolitan governments, individual citizens and disadvantaged communities" (Senseable City Lab 2017), a statement that highlights the lab's multisectoral scope as much as it reveals the order of its priorities. The lab collaborates with influential figures such as Rohit Aggarwala – Sidewalk Labs' Head of Urban Systems and previously a key advisor to Michael Bloomberg's New York mayoralty – as a Visiting Committee member and Richard Florida as an affiliate scholar. Its influence further extends through satellites in Singapore and Stockholm and collaborations with municipalities ranging from Copenhagen to Pristina and Curitiba, and is redoubled by the work of CRA, Ratti's own similarly sized and identically focused "design and innovation office" based in Turin, New York, and London.

The reason for dwelling on these details is to provide a background to Ratti's Dynamic Street project for Sidewalk Labs, to which I now turn. The fact that this was officially a CRA project does not render null my foregoing paragraphs. On the contrary, it illustrates the significance of institutional permeability to elite urbanists, and how their authority derives precisely from being able to operate and transfer reputation across contexts. Ratti's role as the SCL's director gave him the stature to promote CRA so that the two institutions constitute a single operational space.

The Dynamic Street is a reconfigurable paving system prototype consisting of modular, hexagon-shaped pavers with lights and heating – a contribution to the curbless street. The main idea that the allocation of uses on the street would be devised collectively by the users is predicated on an experimental approach to participation. The first iteration, consisting of wooden pavers, was exhibited at Sidewalk Toronto's showroom during summer 2018, and the visitors were invited to "co-create" potential urban scenarios using what Ratti called a "digital reconfigurator" (CRA 2018). This first, rather crude iteration anticipated the future seamless street dynamically responding to city rhythms: "Imagine a city street," the description says, "nestled between buildings with mostly foot and bicycle traffic. During the morning and evening hours, there might be a steady stream of commuters heading to work. In the middle of the day and the evening, families might use the street as a play space. And on the weekend, the street could be cleared for a block party or a basketball game" (CRA 2018). When the Dynamic Street was incorporated into Sidewalk Labs' Street Design Principles, the pavers were equipped with sensors to implement a further degree of flexibility and responsiveness. Referred to as the "dynamic curb," it relies on real-time data to optimize continually the allocation of different functions on the street, motivated by a larger goal: to "recapture space" devoted to parking and to "reallocate [it] to the public realm" (Sidewalk Labs 2019, 5).

The project's premise is that cities should become more responsive and experimental vis-à-vis citizens' ever-changing needs. However, Ratti does not specify who these citizens are, how their needs are constituted, and exactly in what ways they would be emancipated by the experiment beyond reshuffling furniture on the street. It could be argued that, since Sidewalk Labs is no longer pursuing the Quayside development, there is no merit in even discussing the project. Another argument against taking the project too seriously suggests that it is only a part of Sidewalk Labs' strategy to court public opinion or to gain a foothold in the autonomous vehicles and public parking markets (as indeed seems the case; see Harris 2016).

Nevertheless, the paving system highlights the role of the everyday streetscape as an imaginary on which the production of smartness relies – for example, Ratti has already incorporated it in his other projects such as for Paris's Boulevard Périphérique – allowing us to extend the observation that smartness is a process of perpetual experiment or prototype (Evans, Karvonen, and Raven 2016; Halpern, Mitchell, and Geoghegan 2018) beyond the realm of digital infrastructure. For example, the fact that the pavers were equipped with sensors only in the second iteration points to the conception of the smart city as a city replete with digital sensors and as an instance of a broader imaginary that sees cities as fundamentally smart, self-organizing, and complex systems.

What I am trying to argue is that, seen from the vantage point of urbanism as a profession, the smart city belongs to a longer history of seeing the city through the organizing metaphor of "complexity," such that there is a continuity between Jacobs's sidewalk ballet and the ballet (so to speak) on Ratti's Dynamic Street. I have already referred to Jacobs's notion that the sidewalk reveals the city as an essentially complex system, and Ratti has indeed repeatedly expressed an admiration for her work (Ratti and Townsend 2011; Sisson 2016). For example, he refers (in the manifesto's tellingly titled chapter "Wiki City") to Jacobs's insight that "a bottom-up approach begins with the most atomized unit and builds up into increasing complexity" (Ratti and Claudel 2016, 34). The imaginary of complexity shared by Ratti and Jacobs suggests that the smart city is not necessarily technocratic or technophiliac (Graham, in this volume). Rather, as the Dynamic Street project shows, digital technology is a means to realize the inherent "smartness" of the city as a spontaneously evolving, people-centric place, with complexity providing a conceptual bridge between two contrasting schools of thought: technocratic systems thinking and "humanistic" liberalism.

The trouble, crucially, is that it is also an imaginary that relies on the market as a common denominator – driving technological experimentation and conferring democratic legitimacy – so that the human-centric smart city is really a rehashing of the idea that the market is more democratic than democracy (Frank 2000). As Ratti wrote: "[C]ities have a special feature: citizens. By receiving real-time information, appropriately visualised and disseminated, citizens themselves can become distributed intelligent actuators, who pursue their individual interests in co-operation and competition with others, and thus become prime actors on the urban scene. Processing urban information captured in real time and making it publicly accessible can enable people to make better decisions" (Ratti 2009).

What makes the statement striking is that, if we replace "cities" with "markets" and "citizens" with "individuals," it reads as if from the pen of neoliberal thinker Hayek, who theorized the market as a complex adaptive system evolving spontaneously through individual trial and error: "our ancestors," he said, "were really the guinea pigs who experimented and chose the right ways" (Hayek 1983, 7; cf. Slobodian 2018, 218–62). The analogy does not need to be overstretched to highlight an elective affinity between Hayek's neoliberalism and Ratti's (and Jacobs's) urbanism in that, all considered, both place less emphasis on the use of computational technologies in markets/cities than on – a significant nuance – markets/cities themselves being computational technologies or, in Hayek's expression, "information processors" (Mirowski 2009; see also Vanolo, in this volume). In other words, Ratti's smart city is an experimental city that occupies the space between an imaginary of complex self-organizing systems and users internalizing information processing as the default orientation in the city. Urbanism accordingly is understood as a practice of enabling, promoting, and removing barriers to whatever contributes to the realization of complexity, so understood – relative to other practices such as voicing disagreement, expressing discontent, or demanding justice. Yet, as Ratti's taking to the World Economic Forum's website to describe the #Block-Sidewalk initiative as "posturing" makes unmistakably clear, the capacity to translate "complexity" into everyday urban life is itself an expression of neoliberal elite power that needs scrutiny.

Conclusions: Reassuring Complexification?

In the critical smart city literature, the role of urbanism as a profession remains underexplored relative to that of digital platforms or entrepreneurial municipalities. This chapter sheds light on the ideological work that urbanists do when they imagine cities as complex self-organizing systems. Critical scholars usually use "complexity" to refer to messy relations that escape or resist being captured by technologies of power and representation (for example, Olmstead and Spicer, in this volume). In this chapter, however, I have focused on complexity as a frame through which these relations are being ordered – what Isabelle Stengers beautifully refers to as a "reassuring" "fresco of cosmic complexification" (1997, 4; see also Vanolo, in this volume). As I have shown, complexity is an esoteric code and a powerful imaginary that organizes the space in which urbanism intervenes in the smart city. It is especially urgent to politicize this space, as the smart city has morphed from paradigmatic greenfield projects to encompass a series of transformations in the urban everyday around platforms, startups, and experimentation (Barns 2020; Evans 2016; Evans, Karvonen, and Raven 2016; Levenda and Tretter 2019; Moisio and Rossi 2019; Sadowski 2020). The premise of this chapter was not to offer a positive conception of justice, but to highlight how urbanists potentially marginalize disagreement, protest, and social justice when they naturalize the city as complex.

I have also highlighted the rise under neoliberalism of a new type of influence elites operating in transdisciplinary, transnational networks as connectors and intermediaries between the private sector, public policy and academia, whose power derives from being able to translate complex knowledge into convincing political narratives, and whose unaccountability poses formidable challenges to democracy (Davies 2017; Kuus 2021; Mirowski 2009; Wedel 2017). So far, scholars have focused on diplomats, consultants, advisors, and think tanks; in this chapter, I have made a case for considering urbanists – a loose professional group defined by their capacity to explain problems, justify interventions, and propose solutions in an urban context – as members of the neoliberal influence elite.

While few urbanists are card-carrying neoliberals, their providential-like conceptions of urban complexity reflect an antiplanning orientation (Cowley and Caprotti 2019) that scapegoats planners as too top-down and uninterested in everyday people. Yet an emphasis on individuals being ever better informed and more involved is fundamentally misguided, if it is used as a bulwark against the possibilities of comprehensive planning. As geographer and planner Samuel Stein wrote, "while planning is surely a tool of the powerful, it is also essential part of any strategy to challenge them" (2019, 298).

The belief that the purpose of urbanism is to make cities into self-organizing, experimental places aligns with a conspicuous silence about the reality of capitalist institutions and concrete structures of power. The emerging bandwagon of people-centred smart cities (e.g., UN-Habitat 2020) can be seen as an instance of market populism (Frank 2001) that obfuscates key fault lines such as class, race, and gender, along with the imbrication of power and subjectivity. Persuasive narratives such as "no smart city without smart citizens" reinforce rather than challenge the injustice in the technocratic smart city. Certainly, it is not up to individual urbanists to contest the neoliberal smart city. Yet what they can do is stop trying to "get out of the way," and instead rethink what planning could become if it were built on democratic contestation in the urban arena such as #BlockSidewalk.

NOTES

1 As of September 2022, Ratti was listed as a co-author of a staggering 746 SCL papers, presumably following the natural science convention of listing laboratory directors as co-authors of all laboratory output.

2 Thank you to Alan Wiig for drawing my attention to this panel.

REFERENCES

Barns, S. 2020. *Platform Urbanism: Negotiating Platform Ecosystems in Connected Cities*. London: Palgrave.

Bliss, L. 2018. "The sensory city philosopher. Architect, engineer, and inventor Carlo Ratti envisions a future for urban design that's interactive." *Bloomberg*, 13 July. https://www.bloomberg.com/news/articles/2018-07-13/carlo-ratti-the-unconventional-smart-city-philosopher

Calzada, I., and C. Cobo. 2015. "Unplugging: Deconstructing the Smart City." *Journal of Urban Technology* 22 (1): 23–43. https://doi.org/10.1080/10630732.2014.971535

Chandler, D. 2017. "Securing the Anthropocene? International Policy Experiments in Digital Hacktivism: A Case Study of Jakarta." *Security Dialogue* 48 (2): 113–30. https://doi.org/10.1177/0967010616677714

Cowley, R., and F. Caprotti. 2019. "Smart City as Anti-planning in the UK." *Environment and Planning D: Society and Space* 37 (3): 428–48. https://doi.org/10.1177/0263775818787506

CRA (Carlo Ratti Associati). 2018. *The Dynamic Street*. Online at https://carloratti.com/project/the-dynamic-street, accessed 6 January 2021.

Daly, I. 2011. "Data Cycle: Behind MIT's SENSEable Cities Lab." *Wired*, 23 June. Online at https://www.wired.co.uk/article/data-cycle, accessed 23 June 2020.

Davies, W. 2017. "Elite Power under Advanced Neoliberalism." *Theory, Culture & Society* 34 (5–6): 227–50. https://doi.org/10.1177/0263276417715072

Dutta, A., ed. 2013. *A Second Modernism: MIT, Architecture, and the "Techno-Social" Moment*. Cambridge, MA: MIT Press.

Evans, J. 2016. "Trials and Tribulations: Problematizing the City through/as Urban Experimentation." *Geography Compass* 10 (10): 429–43. https://doi.org/10.1111/gec3.12280

Evans, J., A. Karvonen, and R. Raven, eds. 2016. *The Experimental City*. Abingdon, UK: Routledge.

Flynn, A., and M. Valverde. 2019. "Planning on the Waterfront: Setting the Agenda for Toronto's 'Smart City' Project." *Planning Theory & Practice* 20 (5): 769–75. https://doi.org/10.1080/14649357.2019.1676566

Florida, R. 2019. "Sidewalk Labs is the future of urban tech." *Toronto Life*, 4 September. Online at https://torontolife.com/city/sidewalk-labs-is-the-future-of-urban-tech

Frank, T. 2001. *One Market Under God: Extreme Capitalism, Market Populism and the End of Economic Democracy*. New York: Anchor Books.

Gabrys, J. 2014. "Programming Environments: Environmentality and Citizen Sensing in the Smart City." *Environment and Planning D: Society and Space* 32 (1): 30–48. https://doi.org/10.1068/d16812

Goodman, E.P., and J. Powles. 2019. "Urbanism under Google: Lessons from Sidewalk Toronto." *Fordham Law Review* 88 (2): 457–98. https://doi.org/10.2139/ssrn.3390610

Halpern, O., R. Mitchell, and B.D. Geoghegan. 2018. "The Smartness Mandate: Notes toward a Critique." *Grey Room* 68: 106–29. https://doi.org/10.1162/GREY_a_00221

Harris, M. 2016. "Secretive alphabet division funded by Google aims to fix public transit in US." *Guardian*, 27 June. Online at https://www.theguardian.com/technology/2016/jun/27/google-flow-sidewalk-labs-columbus-ohio-parking-transit

Hayek, F. 1983. "Interview with F.A. Hayek." *Cato Institute's Policy Report* 5 (2): 5–9. Online at https://www.cato.org/policy-report/may/june-1984/exclusive-interview-fa-hayek

Jacobs, J. 1961. *The Death and Life of Great American Cities*. New York: Vintage Books.

Keller, S. 2006. "Fenland Tech: Architectural Science in Postwar Cambridge." *Grey Room* 23: 40–65. https://doi.org/10.1162/grey.2006.1.23.40

Krivý, M., and L. Ma. 2018. "The Limits of the Livable City: From Homo Sapiens to Homo Cappuccino." *Avery Review* 30. Online at https://averyreview.com/issues/30/limits-of-the-livable-city, accessed 16 June 2021.

Kuus, M. 2021. "Professions and Their Expertise: Charting the Spaces of 'Elite' Occupations." *Progress in Human Geography* 45 (6): 1339–55. Online at https://journals.sagepub.com/doi/10.1177/0309132520950466

Larner, W., and N. Laurie. 2010. "Travelling Technocrats, Embodied Knowledges: Globalising Privatisation in Telecoms and Water." *Geoforum* 41 (2): 218–26. https://doi.org/10.1016/j.geoforum.2009.11.005

Larson, S. 2013. *Building Like Moses with Jacobs in Mind*. Philadelphia: Temple University Press.

Leszczynski, A. 2020. "Glitchy Vignettes of Platform Urbanism." *Environment and Planning D: Society and Space* 38 (2): 189–208. https://doi.org/10.1177/0263775819878721

Levenda, A., and E. Tretter. 2019. "The Environmentalization of Urban Entrepreneurialism: From Technopolis to Start-up City." *Environment and Planning A: Economy and Space* 52 (3): 490–509. https://doi.org/10.1177/0308518X19889970

Lobsinger, M.L. 2013. "Two Cambridges: Models, Methods, Systems, and Expertise." In *A Second Modernism: MIT, Architecture, and the "Techno-Social" Moment*, ed. A. Dutta, 652–85. Cambridge, MA: MIT Press.

Lorinc, J. 2019. "A Mess on the Sidewalk." *Baffler* 44. https://thebaffler.com/salvos/a-mess-on-the-sidewalk-lorinc, accessed 16 June 2021.

Mirowski, P. 2009. "Postface: Defining Neoliberalism." In *The Road from Mont Pèlerin: The Making of the Neoliberal Thought Collective*, ed. P. Mirowski and D. Plehwe, 417–55. Cambridge, MA: Harvard University Press.

MIT News. 2015. "Carlo Ratti appointed as advisor to European Commission." 6 October. Online at https://news.mit.edu/2015/carlo-ratti-appointed-advisor-european-commission-1006

Moisio, S., and U. Rossi. 2019. "The Start-up state: Governing Urbanised Capitalism." *Environment and Planning A: Economy and Space* 52 (3): 532–52. https://doi.org/10.1177/0308518X19879168

Morozov, E., and F. Bria. 2018. *Rethinking the Smart City: Democratising Smart Technology*. New York: Rosa Luxemburg Foundation.

O'Kane, J. 2019. "Opponents of Sidewalk Labs get advice from German tech protesters." *Globe and Mail*, 24 November. Online at http://theglobeandmail.com/business/article-opponents-of-sidewalk-labs-get-advice-from-german-tech-protesters/

Planetizen. 2017. "The 100 most influential urbanists." *Planetizen*, 9 October. Online at https://www.planetizen.com/features/95189-100-most-influential-urbanists

Rankin, K.N. 2003. "Anthropologies and Geographies of Globalization." *Progress in Human Geography* 27 (6): 708–34. https://doi.org/10.1191/0309132503ph457oa

Ratti, C. 2001. "*Urban Analysis for Environmental Prediction*." PhD diss., University of Cambridge.

Ratti, C. 2009. "Digital cities: 'Sense-able' urban design." *Wired*, 2 October. Online at https://www.wired.co.uk/article/digital-cities-sense-able-urban-design

Ratti, C. 2020. "We need more urban innovation projects like the 'Google City'. This is why." World Economic Forum, 23 September. Online at https://www.weforum.org/agenda/2020/09/google-smart-cities-urban-innovation-technology

Ratti, C., and M. Claudel. 2016. *The City of Tomorrow: Sensors, Networks, hackers, and the Future of Urban Life*. New Haven, CT: Yale University Press.

Ratti, C., C.A. Hidalgo, A. Picon, A. Wiig, and A. Sevtsuk. 2018. "Is Big Data Changing Urban Theory?" Cities and Technology Debate Series, 7 March, Harvard University Graduate School of Design. Online at https://vimeo.com/275111539

Ratti, C., and A. Townsend. 2011. "The Social Nexus." *Scientific American* 305 (3): 42–6. https://doi.org/10.1038/scientificamerican0911-42

Rose, G. 2020. "Actually-Existing Sociality in a Smart City." *City* 24 (3–4): 512–29. https://doi.org/10.1080/13604813.2020.1781412

Rouvroy, A., and T. Berns. 2011. "Algorithmic Governmentality and Prospects of Emancipation: Disparateness as a Precondition for Individuation through Relationships?" *Réseaux* 177 (1): 1–31.

Sadowski, J. 2020. "Cyberspace and Cityscapes: On the Emergence of Platform Urbanism." *Urban Geography* 41 (3): 448–52. https://doi.org/10.1080/02723638.2020.1721055

Sauter, M. 2019. "City planning heaven sent." *e-flux*, 1 February. Online at https://www.e-flux.com/architecture/becoming-digital/248075/city-planning-heaven-sent

Sennett, R. 2018. *Building and Dwelling: Ethics for the City*. New York: Farrar, Straus and Giroux.

Senseable City Lab. 2017. "Senseable City Lab." Online at https://senseable.mit.edu, accessed 6 January 2021.

Shelton, T., M. Zook, and A. Wiig. 2015. "The 'Actually Existing Smart City.'" *Cambridge Journal of Regions, Economy and Society* 8 (1): 13–25. https://doi.org/10.1093/cjres/rsu026

Sidewalk Labs. 2019. "Street Design Principles, Vol. 1." Online at https://sidewalklabs.com/assets/uploads/2019/04/Sidewalk-Labs-Street-Design-Principles-v.1.pdf, accessed 6 January 2021.

Sisson, P. 2016. "The future of smart city technology, from an MIT professor." *Curbed*, 24 May. Online at https://archive.curbed.com/2016/5/24/11761274/smart-city-carlo-ratti-city-of-tomorrow

Slobodian, Q. 2018. *Globalists: The End of Empire and the Birth of Neoliberalism*. Cambridge, MA: Harvard University Press.

Stein, S. 2019. *Capital City: Gentrification and the Real Estate State*. London: Verso.

Stengers, I. 1997. "Complexity: A Fad?" In *Power and Invention: Situating Science*, 3–19. Minneapolis: University of Minnesota Press.

Ülemiste City. 2019. "Robotid ja puitmajad: kuidas Google'i tütarfirma oma esimese targa linna päriselt targaks muudab?" *DigiGeenius*, 28 February. Online at https://digi.geenius.ee/blogi/tuleviku-linn/robotid-ja-puitmajad-kuidas-googlei-tutarfirma-oma-esimese-targa-linna-pariselt-targaks-muudab/

UN-Habitat (United Nations Human Settlements Programme). 2020. "People-Centred Smart Cities." Online at https://unhabitat.org/sites/default/files/2021/01/fp2-people-centered_smart_cities_04052020.pdf, accessed 16 June 2021.

Vanolo, A. 2014. "Smartmentality: The Smart City as Disciplinary Strategy." *Urban Studies* 51 (5): 883–98. https://doi.org/10.1177/0042098013494427

Wedel, J. 2017. "From Power Elites to Influence Elites: Resetting Elite Studies for the 21st Century." *Theory, Culture & Society* 34 (5–6): 153–78. https://doi.org/10.1177/0263276417715311

Wired. 2012. "The smart list 2012: 50 people who will change the world." *Wired*, 24 January. Online at https://www.wired.co.uk/article/the-smart-list

4 The Evolution of Splintering Urbanism in Planetary Information Ecosystems

JOE DANIELS, MICAH HILT, AND ELVIN WYLY

Planetary information ecosystems are evolving. Dramatic yet fragmented advances in information theory and telecommunications underway since the 1940s finally coalesced into an uneven but insistently globalizing infrastructure in the first decade of the twenty-first century. As the world crossed the majority-urban threshold around the year 2006, this infrastructure mediated the flourishing corporate and policy fascination with "smart cities," amid a quickening pace of change in urban movements for social justice energized by "communicative action that involves connections between neural networks from human brains stimulated by signals from a communication environment through communication networks (Castells 2012, 219). Unfortunately, these networks reproduce a complex, confusing, and conspiratorial environment in which the diversity of algorithmically evolutionary identities is easily co-opted, hijacked, and weaponized. In this chapter we examine the paradoxical genealogies of the smart city in the age of fake news, post-truth politics, and infinitely adaptive conspiracy theories.

Contemporary smart city discourses and practices have emerged amid a broad "urban turn," as scholars, practitioners, and corporations explore the implications of the planetary urban age (Batty 2018, 2020; Townsend 2013; West 2017). At the same, urban theory has undergone multiple transformations. In the domain of critical urban studies, where social justice is an explicit priority, "postmillennial spaces of theory" have created a complex intellectual landscape of multiple "directions of diversification" (Leitner, Sheppard, and Peck 2020, 6). New generations of critical urbanists have worked to move beyond what came to be seen as a certain kind of critical orthodoxy that had provided the centre of gravity for the field since the early 1980s: the Marxist urban political economy developed in the previous decade by Manuel Castells and David Harvey. Diversification has produced division. "[D]ebates about urban theory have become unduly heated," Leitner, Sheppard, and Peck (2020, 6) observe, "as new generations of scholars seeking to shift the theoretical conversation have confronted older (established white, often masculine) generations" (see also Catungal 2019; Derickson 2015, 2018; Oswin 2018, 2020; Roy 2016; Roy and Ong 2011). In the rapidly expanding smart cities literature, meanwhile, debates are polarized between advocates and critics (Kitchin et al. 2019). Advocates are led by scientists and technologists whose expertise and inspiration come from the physical and computational sciences, and who are working to create and implement smart cities technologies that are "presented as objective, rational, and apolitical" (Valdez, Cook, and Roby, in this volume). Confronted with critiques on the underlying assumptions of smart cities technologies, advocates "try to side-step the critique," claiming that "they employ a mechanical objectivity in their work, thus ensuring that it is neutral and non-ideological" (Kitchin et al. 2019, 3). Advocates also emphasize that "they are developing what society, the market, and city administrations want or need" (3). Smart cities critics, by contrast, draw inspiration from the postmillennial, poststructural diversification of social theory in the humanities and cultural studies – foregrounding urgent contemporary political and ethical questions of intersectional equity, diversity, inclusion, and justice.

Turn On, Tune In, Hate Cities

Despite a dizzying array of valuable contributions across a diverse panorama of theory and epistemology, crucial insights on the nexus between smart cities and social (in)justice have been misunderstood, marginalized, or forgotten. At the turn of the century, Graham and Marvin (2001) diagnosed how the infrastructures produced through centuries of concentrated human settlement – transport, water, energy, telecommunications – were undergoing a sudden biopolitical "splintering" of segmented, unequal urban experiences. In an eloquent, fascinating dialogue in this volume, Graham analyses how today's smart cities movement is "splintering urbanism on steroids," involving bizarre mutations of the 1960s radical, left-wing, antiwar, and anticapitalist California counterculture into right-wing, libertarian, alt-right movements animated by "profoundly anti-urban, sometimes outright fascist, sometimes white supremacist visions." Graham reminds us of the complex entanglement of advances in communications technologies and cultural politics, vividly embodied in figures such as Timothy Leary and Peter Thiel.

In the 1960s Leary became a charismatic media figure whose "turn on, tune in, drop out" mantra of psychedelic drugs and expanded consciousness was seen as radical and seditious. Richard Nixon called him "the most dangerous man in America." When Leary staged a prison escape and fled after conviction on drug charges, Black Panthers Minister of Information Eldridge Cleaver persuaded Algerian authorities to grant sanctuary to the fugitive by claiming that Leary was a Black university professor fired for his antiwar opinions. By the 1980s Leary was a standard-issue oddity in the Hollywood-Silicon Valley celebrity-industrial complex, mixing with the porn capitalist Larry Flynt, the staunch Republican Arnold Schwarzenegger, and A-list actors like Jack Nicholson and Susan Sarandon. Leary worked with movie producers and software entrepreneurs to develop some of the earliest interactive computer games, such as *The Game of Life*, and *Mind Mirror*, in which players created their own psychological profiles by answering questions that Leary had developed in his master's thesis in the 1940s. Leary also toured the country doing phenomenally successful stage debates with the convicted Watergate felon and right-wing, law-and-order Nixon aide G. Gordon Liddy. "He's Darth Vader to my Luke Skywalker," Leary explained. After initial success on the college lecture circuit, the odd couple performed on Broadway in New York, were interviewed by Andy Warhol, and then, as the *San Jose Mercury News* proclaimed in a headline, "THE DRUG-AND-THUG SHOW HITS THE ROAD" in towns and cities across the country (Greenfield 2006).

"So many discussions about place and digitality are ahistorical," Graham (in this volume) reminds us, and thus we often ignore crucial genealogies. "[P]eople who learn about *smart cities* will never be taught about Nick Land and the *Whole Earth Catalogue*," Graham laments, drawing attention to the way "'hippie' ideologies mutated into neofascist" forms of cyberlibertarian utopianism and anti-urban politics. Graham also notes the influence of Peter Thiel, the first venture capital investor in Facebook and a founder of PayPal. Only a few months after receiving the Friedrich Hayek Lifetime Achievement Award in Vienna, Thiel rallied the crowd for Donald Trump at the 2016 Republican National Convention in Cleveland, Ohio. "This is where I became an American," Thiel told the MAGA delegates, praising the innovation of the industrial metropolis of his childhood in the late 1960s, when "all of America was high-tech." Thiel, a key figure in the transhumanism movement that envisions cloud-computing consciousness as the next stage of human evolution, was appealing to memories of a high-tech Cleveland in the age of Fordist-Keynesian industrial production, but the stage on which he spoke was a portal into the post-industrial consciousness-manipulation infrastructures built by surveillant cognitive capitalism (Moulier-Boutang 2011; Zuboff 2019). Manufacturing consent and false consciousness, a racialized conspiratorial game in US politics for generations (Hackworth 2019; Hofstadter 1964), now requires aggressive strategic intersectionality, even for openly white supremacist coalitions like the Trump cadres that took over the GOP. Thiel spoke the day after former Olympic decathlete, reality television star, and transgender activist Caitlyn Jenner told a Cleveland audience organized by the conservative, pro-LGBTQ American Unity Fund that coming out as trans was easier than coming out as a conservative Republican for Trump. Thiel told the RNC delegates and everyone watching online, "I am proud to be gay. I am proud to be a Republican. But most of all, I am proud to be an American."

The 2016 election involved an industrialized manipulation of identity and cultural politics and the deployment of military-style psy-ops through geographically targeted social media behavioural modelling – exploiting the anti-urban bias of the US Electoral College to counterbalance a 3 million popular vote loss with 77,000 swing votes in three rustbelt counties (Cadwalladr 2018; Wylie 2019). After the election, journalists dubbed Thiel Trump's "PayPal TechPal" and the "Shadow President in Silicon Valley," and subsequent years brought relentless information warfare and cybernetic global geopolitics, along with militant mobilization of anti-urban populism in US domestic politics. Meanwhile, smart cities technologies continued to advance, along with deepening penetration of cybernetic practices in everyday life, creating an uneven yet transnational information infrastructure. It is worth considering a brief case study to illustrate how these planetary information systems mediated the first global pandemic of the planetary urban age.

#FauciFraud and the Gates of Hell

In March, 2015, Bill Gates appeared on a stage in Vancouver, British Columbia, and delivered a TED talk on global epidemiology, "The Next Outbreak? We're Not Ready." "Time is not on our side," Gates explained as he advocated for immediate investment to translate accelerating bioscience innovations into a coordinated worldwide program to develop vaccines. Five years later in Washington, DC, Donald Trump told reporters at a White House Coronavirus Task Force briefing that Mike Pompeo, his Secretary of State, was "extremely busy" and had to get back to his job running the State Department – "or, as they like to call it," Trump grinned, "the Deep State Department." A few seconds later, the cameras caught Dr Anthony Fauci, Director of the National Institute of Allergy and Infectious Diseases (NIAID), standing behind Trump, bow his head in a slow but unmistakable face-palm.

Video segments and meme mashups of the Fauci face-palm went viral through alt-right social media networks, fuelling one of the many virulent conspiracy theories that were co-evolving with COVID-19. Fauci clarified that he had covered his face because a lozenge had momentarily been caught in this throat, but no matter. Twitter and Facebook posts attacking Fauci as an operative of a treasonous "deep state" committed to destroying President Trump reached an audience of more than 1.5 million within days, and New York Times reporters identified more than 70 Twitter accounts pushing the #FauciFraud hashtag. Some of the automated twitterbots spread the conspiracy as many as 795 times per day, quickly becoming one of the most widespread of all pandemic conspiracy theories (Alba and Frenkel 2020; Wakabayashi, Alba, and Tracy 2020). One short video segment appeared on the YouTube channel of Ruptly (a division of RT / *Russia Today*) and soon racked up 581,000 views. Meanwhile, Gates's nine-minute speech from Vancouver in 2015 acquired new life, racking up 25 million new views over a two-week period. Antivaxxers, conspiracy theorists of the "QAnon" movement incubated in the far-right American online extreme discursive frontiers of 4chan and 8chan, and "mainstream" right-wing voices circulated the video as evidence that Gates had created COVID-19 as part of a plot to corner the worldwide market on a vaccine so that elite globalists could control the world's population through governmental surveillance and eugenic culling. Soon NBC's *Meet the Press* included a short clip of a woman at a Pennsylvania protest holding a large placard featuring a caricature of Gates wielding a syringe of poison masquerading as a COVID-19 vaccine: "Bill Gates of Hell."

These events seem to have no coherent relationship. Connecting them is not a "rational" abstraction that "isolates a significant element of the world which has some unity and autonomous force," but is instead a *chaotic conception* that "combines the unrelated or divides the indivisible" (Sayer 1982, 7). And yet combining the unrelated and dividing the indivisible is at the heart of the behavioural surplus extraction that sustains surveillance capital accumulation (Zuboff 2019). Alphabet's market capitalization exceeds $1.5 trillion, and advertising – what Mander (2012) memorably theorized as the "privatization of consciousness" – accounts for four-fifths of the firm's $182 billion annual revenue stream. Identifying and exploiting unexpected, seemingly random connections is pervasive in the non-linear, multidimensional mirror worlds where decentralized, networked, mass self-communication (Castells 2012) interacts with centralized information processing architectures designed by the world's leading minds in artificial intelligence, data mining, evolutionary psychology, and machine learning.

We can glimpse a small part of this nexus by using the open-source tools developed by Bernhard Rieder at the Digital Methods Initiative at the University of Amsterdam (see Rieder, Matamoros-Fernández, and Coromina 2018). Mining the YouTube API allows us to map the connections between videos in the "up next" recommendations, which are provided by the YouTube algorithms that are constantly watching what billions of people are watching. The simplest, first-order network analysis of the Fauci face-palm and Gates's pandemic TED talk reveals a constellation of 195 multicollinear videos, with a total accumulated audience of more than 427 million views, 10.6 million likes, 1.42 million comments, and 481,000 dislikes (Figure 4.1). The correlations are diverse and chaotic across time and space, from "deep state" conspiracies to tech talks on 3D printing to warnings that 5G will make your head explode.

Cognitive Urban Systems

Figure 4.1 is a map of online audience formation, not a map of cities. Yet it reflects and reproduces a fundamental genealogy of urban theory. It is not just that Gates spoke in one city (Vancouver) while Fauci face-palmed in another

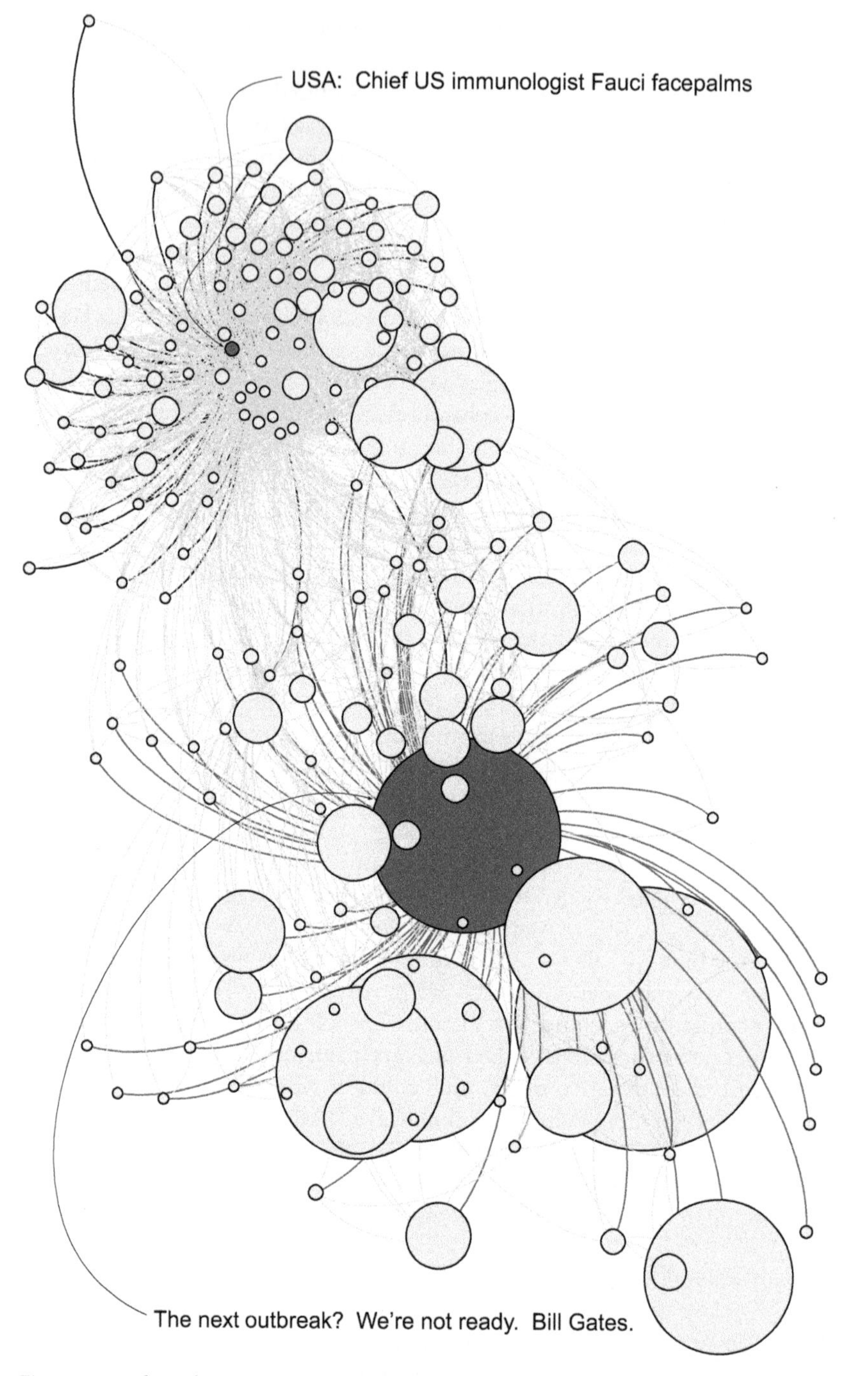

Figure 4.1. Algorithmic Correlations of Conspiracies
Note: "Recommended video" correlations from YouTube videos of Bill Gates's pandemic TED talk and Anthony Fauci's "face-palm." Circle sizes scaled proportional to view count. Detailed colour version available at https://ibis.geog.ubc.ca/~ewyly/gates_fauci_fig_1.pdf
Source: Authors' creation, using tools developed by Bernhard Rieder at the Digital Methods Initiative, University of Amsterdam; data extracted May 2020.

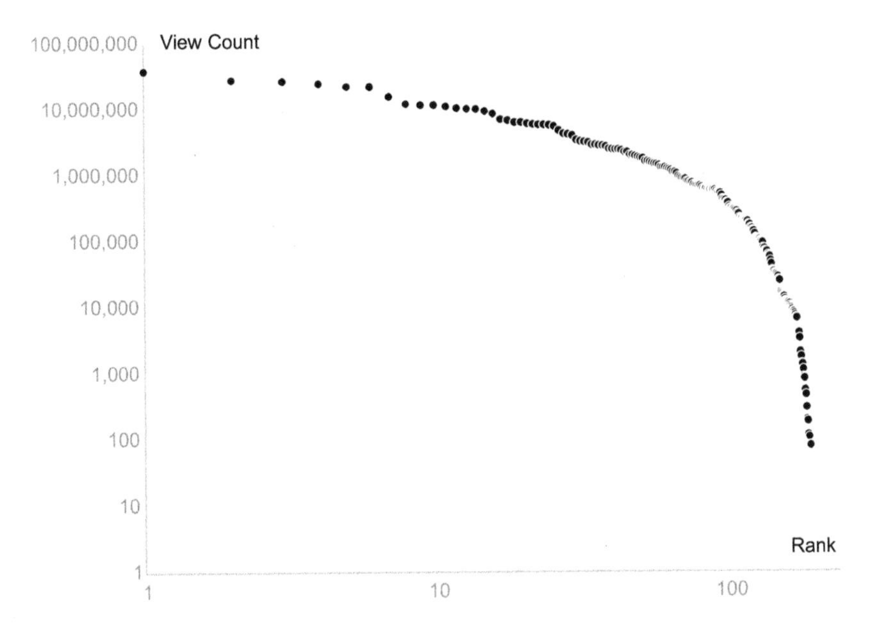

Figure 4.2. Rank-Size Distribution of View Counts for Videos in Figure 4.1

(Washington, DC). YouTube is just the largest of many social media portals that have augmented the traditional role of the metropolis as a giant communications switchboard (Mattern 2017; Webber 1964). Circulating images and narratives of chaotically conceived connections constitute information flows that reflect and reproduce systemic hierarchies of urban communication and cultural change. Note the inset at the bottom right of Figure 4.1, which graphs the log-transformed view counts as a function of the log of the rank. Graphs like these were ubiquitous in urban geography and planning texts in the second half of the twentieth century, as part of a paradigm summarized by "the all-embracing term 'urban system'" (McCann and Simmons 2006, 77; see also Barnes and Minca 2013; Berry 1964; Borchert 1967; Friedmann 1973; Van Meeteren 2016). Berry (1964) used general systems theory (Von Bertalanffy 1950) to interpret the distribution of large cities and small towns as the manifestation of societal cultural evolution, which reproduced a general equilibrium described by Zipf's (1949) rank-size rule:

$$P_r = \frac{P_1}{r^q},$$

where the population P of a city ranked r equals the population of the largest city (P_1) divided by rank r raised to the power of q (which usually approximates 1.0).

In our case, the view count hierarchy of the Gates-Fauci conspiracy correlations is non-linear in the tail, but well described by

$$\log(\text{Views}) = 7.458 - 0.0213\ (\text{Rank}) + \text{m}$$
with R^2=0.93.

These hierarchies prevail regardless of content. YouTube discloses few details about its recommendation algorithm, which is periodically revised in response to legal and legislative pressure over online radicalization, incitement to violence, and other matters. The details do not alter the epistemology: when humans search the Internet, the Internet searches humans (Dyson 2012), and the resulting hybrid human-algorithmic noumena (Smith 1989) performs more and more of what was once the monopoly of the physical concentrations of human settlement we call cities: sifting the chaotic infinity of information into coherent, embodied human meaning and understanding (Mattern 2017; Mirowsky and Nik-Khah 2017). One lesson of urban systems theory is that individual cities rise and fall in the rankings – Cleveland has lost half its population since Thiel arrived there as an infant with his immigrant parents in 1968 – but the hierarchy itself endures.

The smart city age is a mirror world of parallel urban systems, of material and mental hierarchies of communication and cultural evolution. While few in Silicon Valley know of Brian Berry, there is widespread fascination with Zipf (Aiden and Michel 2013; West 2017), and all technology theorists and venture capitalists understand the antecedent to the persuasive, interdisciplinary general systems theory that helped Berry become the world's most widely cited geographer for the first quarter-century of the Social Sciences Citation Index, after Friedrich Hayek's (1941, 1945, [1968] 2002) theorization of the free market price system as a collective human mind created by millions of years of evolution. Hayek, who famously fought with Keynes and built the ideology of neoliberalism – "the idea that swallowed the world" (Metcalf 2017) – saw himself first and foremost as a biologist (Hayek 1982). He translated the Darwinian revolution into a comprehensive theory of information (Mirowsky and Nik-Khah 2017). For Hayek (1982, 289), the individual human mind is an embodied "mental universe" in which neural pattern-recognition algorithms are engaged in "a process of continuous and simultaneous classification and constant reclassification" as "the organism constantly changes from one set of dispositions to interpret and respond to what is acting upon it and in it, to another such set of dispositions." When individual human minds are networked in relations of exchange, the system of price signals becomes a collective, transcendent consciousness beyond the design or understanding of any individual or group of experts. The market is the most powerful information processor ever produced by evolution, "the most complex structure in the universe" (Hayek 1988,

127). US Republicans carefully avoid this central feature of Hayek's philosophy for fear of alienating creationist Christian conservatives, while poststructural, postmillennial critical social theorists now regard the *critique* of neoliberalism as conservative and obsolete. What is often ignored is how well Silicon Valley cognitive capitalism has implemented Hayek's (1982, 291) biological vision of the human brain as a matrix of "neural impulses, largely reflecting the structure of the world in which the central nervous system lives," which he compared to "a stock of capital being nourished by inputs and a continuous stream of outputs" (see also Mirowsky 2011; Mirowsky and Nik-Khah 2017). "Fortunately," Hayek explained, "the stock of this capital cannot be used up" (1982, 291), because there is a combinatoric infinity of networked human neural networks adapting to price signals that are always changing.

Splintering Cognitive-Capitalist Urbanism

What does this all mean for the relations between smart cities and social justice? Three implications are most important.

First, today's smart cities visions are not entirely new. Genealogies matter. While Peter Thiel was in diapers in Cleveland, Brian Berry (1970) was explaining to British geographers how virtual reality technologies would transform the hierarchical structure of the US urban system. George Kingsley Zipf's (1949) rank-size analyses of city sizes – and of word frequencies across multiple languages, from English to Hebrew, Plains Cree, and Norwegian – provided the inspiration for Silicon Valley dreams that attract millions of TED talk views and billions of web-page clicks (Aiden and Michel 2013; West 2017). Jane Jacobs's urban theory was deeply influenced by an address to the Rockefeller Foundation by Warren Weaver, today best remembered for the Shannon-Weaver information theory in cybernetics (Batty 2018, 6; Mirowsky and Nik-Khah 2017).

Second, the discourse of "smart cities" is misleading. This is not an attack on smart city advocates in the dualism outlined by Kitchin et al. (2019). It is just a reminder of what Shannon Mattern's (2017) eloquent archaeology of media demonstrates: cities have been *smart* for thousands of years. What *is* new are the adaptive, non-Cartesian scales of space-time produced by the dynamic geographies of billions of interactive sensors, bots, auto-recommend algorithms, and friendly surveillance tracking devices (smartphones) with humans attached. While smart cities technologies are portrayed as digital, the most significant transformations do not involve the unambiguous, 0/1 binaries of digital computing. Like postmillennial social justice commitments to diverse, performative sexuality and gender identity, today's smart cities are non-binary. Non-binary communication and computation are also known as analogue. The logics of previous generations of cognitive machines – telegraph, radar, AM/FM radio, vacuum tube computers, landline telephones, VHF and

UHF television – have returned with a vengeance (Dyson 2020). In contrast to the sharp, discrete codes of the digital world, analogue signals are continuous, probabilistic, metaphorical, and often noisy or interrupted – just like human communication.

An analogue perspective illuminates the puzzling co-evolution of post-truth conspiracy politics with technologically advanced smart cities. This is the third implication. Algorithmic binary digital infinities are channelled through the limited and uniquely embodied neural pattern-recognition algorithms of individual human brains, creating an analogue interface between computers communicating with other computers in digital code. YouTube's digital algorithm suggests videos for you to watch, but you can ignore them, watch them, misunderstand them, love them, or hate them. Hayek would love the Silicon Valley corporate ecosystem where Thiel became Trump's Shadow President, and he would love it even more when young progressives on social media called him out for being a cisheteropatriarchal Eurocentric settler-colonial white supremacist who does not understand the ironies of TERFs allying with religious conservatives in the *Harris Funeral Homes v. EEOC* US Supreme Court case amid the backlash to Caitlyn Jenner's headline-grabbing transition. Language, culture, politics, memes, and acronyms like those for trans-exclusionary radical feminists accelerate the "psycho-social" (Hayek 1988, 25) evolution of capitalism, and you too have the opportunity to win the Hayek Lifetime Achievement Award just like Thiel did, if your channel on YouTube, Instagram, or TikTok appeals to the right combination of market signals in the ever-expanding networked neural networks in the latest evolutionary stage of the most powerful information processor in the universe. The only arbiter of the diversity of linguistic representations of neural impulses is The Market, and thus planetary informational ecosystems are entirely compatible with the dialectical co-evolution of radical intersectionality and the cyberlibertarian fascism that runs in a straight line from Hayek's admiration for Augusto Pinochet to Thiel's cheerleading for Trump. Those involved in social movements must, unfortunately, devote an increasing share of strategic and cognitive resources to vigilance against the way the "factory of fragmentation" of cognitive capitalism (Harvey 1992) co-opts and hijacks movements for equality into diverse, confusing, and divisive struggles in which difference is monetized, weaponized, and reproduced.

Thus it is little surprise that, while Thiel-backed startups pursue IPOs on the NASDAQ so investors can speculate on the same magic mushrooms that got Leary in trouble with his Harvard Psilocybin Project, Arnold Schwarzenegger gives interviews on his assessment of Caitlyn Jenner's chances to win the governor's office in California, as YouTube suggests you watch the old Leary-Liddy debates or Liddy fanning the flames of the Obama birther lie that Trump rode to the White House, while the Gates-Fauci conspiracies continue to draw evolutionary attentional hierarchies (Figure 4.3) summarized by

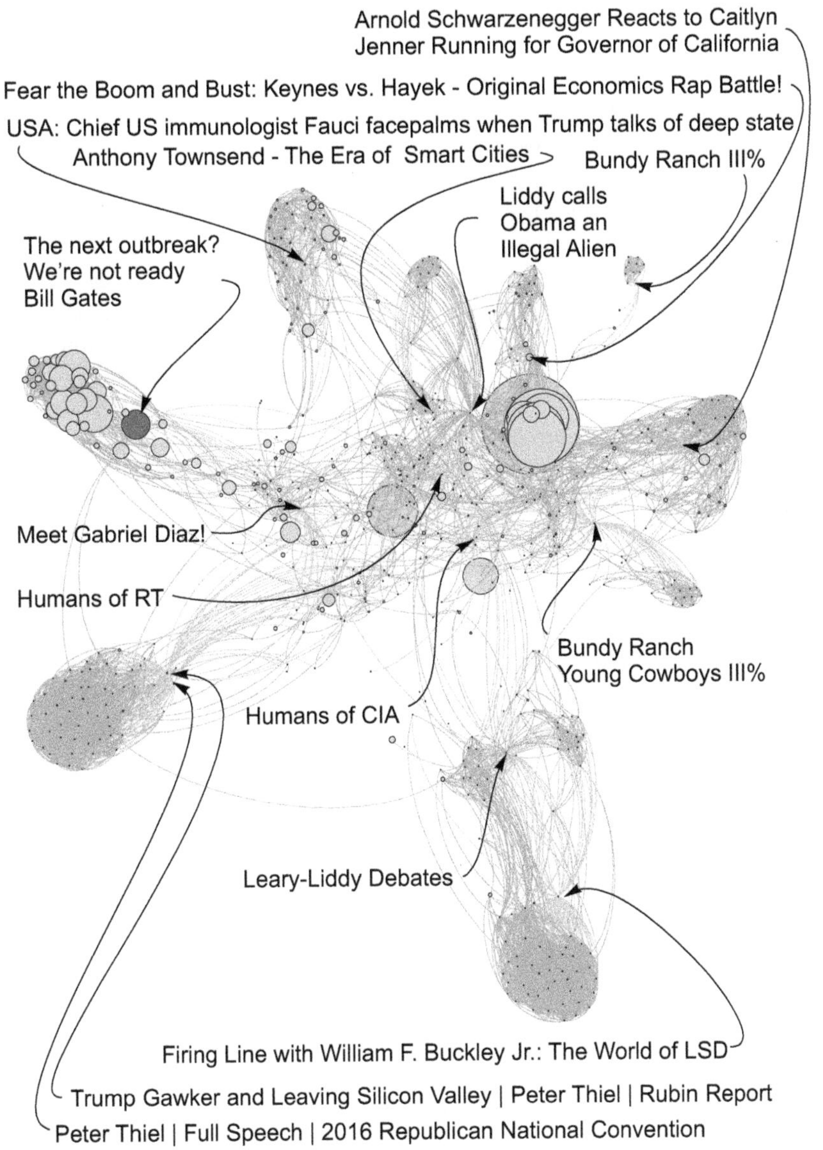

Arnold Schwarzenegger Reacts to Caitlyn Jenner Running for Governor of California

Fear the Boom and Bust: Keynes vs. Hayek - Original Economics Rap Battle!

USA: Chief US immunologist Fauci facepalms when Trump talks of deep state

Anthony Townsend - The Era of Smart Cities

Bundy Ranch III%

Liddy calls Obama an Illegal Alien

The next outbreak? We're not ready Bill Gates

Meet Gabriel Diaz!

Humans of RT

Bundy Ranch Young Cowboys III%

Humans of CIA

Leary-Liddy Debates

Firing Line with William F. Buckley Jr.: The World of LSD

Trump Gawker and Leaving Silicon Valley | Peter Thiel | Rubin Report

Peter Thiel | Full Speech | 2016 Republican National Convention

Figure 4.3. The Gates-Fauci Correlations Evolve

Note: "Recommended video" correlations from YouTube videos of Bill Gates's pandemic TED talk and Anthony Fauci's "face-palm," *and old and new connections in US anti-urban politics.* Circle sizes scaled proportional to view count. Detailed colour version available at https://ibis.geog.ubc.ca/~ewyly/gates_fauci_fig_3.pdf

Source: Authors' creation, using tools developed by Bernhard Rieder at the Digital Methods Initiative, University of Amsterdam; data extracted May 2021.

$$\log(\text{Views}) = 7.244 - 0.00652\,(\text{Rank}) + \text{m},$$

with an R^2 explaining 95 per cent of the rank-size variation among 642 videos with a cumulative audience of 1.74 billion. The CIA pursues SJW recruits with their "Humans of CIA" series, featuring an overachieving Latinx quoting Zora Neale Hurston and declaring, "I am a woman of colour. I am a mom. I am a cisgender millennial who's been diagnosed with generalized anxiety disorder. I am intersectional, but my existence is not a box-checking exercise." The "woke CIA" ad is immediately lampooned – attracting more attention – while *Russia Today* produces a spoof so viewers can enjoy comedy segments between clips of Victory Day Parade in Moscow, or Anthony Townsend's *RT* interview when he was promoting his *Smart Cities* book. Expect the unexpected. Not far from the Gates-Fauci coronavirus conspiracies is activists' revelation of "Gabriel Diaz," a Dominican Black man who appeared on social media as a Latino New York cab-driver wearing a swastika armband on the job, before he posed as a leftist Muslim to infiltrate and betray an antifa group. Elsewhere in what journalists call the "multiracial far right," a diasporically intersectional BIPOC former Marine whose ancestry is Chinese, Guamanian, Inuit, Mexican, Greek, and Scandinavian, has participated in five armed standoffs with the US federal government, including the violent Bundy Ranch episode (Allam and Nakhlawi 2021). All who are committed to social justice must remember that, in the cognitive hierarchies of the surveillance capitalist smart city bequeathed by Hayek and Thiel, intersectionality is complex, multidimensional, non-linear, and evolutionary.

REFERENCES

Allam, H., and R. Nakhlawi. 2021. "Black, brown, and extremist: Across the far-right spectrum, people of color play a more visible role." *Washington Post*, 16 May.

Aiden, E., and J.-B. Michele. 2013. *Uncharted: Big Data as a Lens on Human Culture.* New York: Riverhead Books.

Alba, D., and S. Frenkel. 2020. "Medical expert who corrects Trump is now a target of the far right." *New York Times*, 28 March. Online at https://www.nytimes.com/2020/03/28/technology/coronavirus-fauci-trump-conspiracy-target.html, accessed 11 January 2021.

Barnes, T., and C. Minca. 2013. "Nazi Spatial Theory: The Dark Geographies of Carl Schmitt and Walter Christaller." *Annals of the Association of American Geographers* 103 (3): 669–87. https://doi.org/10.1080/00045608.2011.653732

Batty, M. 2018. *Inventing Future Cities.* Cambridge, MA: MIT Press.

Batty, M. 2020. "The Smart City." In *The City Reader*, 7th ed., ed. R. LeGates and F. Stout, 503–15. New York: Routledge.

Berry, B. 1964. "Cities as Systems within Systems of Cities." *Papers in Regional Science* 13 (1): 147–63. https://doi.org/10.1111/j.1435-5597.1964.tb01283.x

Berry, B. 1970. "The Geography of the United States in the Year 2000." *Transactions of the Institute of British Geographers* 51: 21–53. https://doi.org/10.2307/621761

Borchert, J.R. 1967. "American Metropolitan Evolution." *Geographical Review* 57 (3): 301–32. https://doi.org/10.2307/212637

Cadwalladr, C. 2018. "I made Steve Bannon's psychological warfare tool: Meet the data war whistleblower." *Guardian*, 17 March. Online at https://www.theguardian.com/news/2018/mar/17/data-war-whistleblower-christopher-wylie-faceook-nix-bannon-trump, accessed 11 January 2021.

Castells, M. 2012. *Networks of Outrage and Hope: Social Movements in the Internet Age*. Cambridge, UK: Polity Press.

Catungal, J.P. 2019. "Classroom." In *Keywords in Radical Geography: Antipode at 50*, ed. Antipode Editorial Collective, 45–9. Hoboken, NJ: John Wiley & Sons.

Derickson, K. 2015. "Urban Geography I: Locating Urban Theory in the 'urban age.'" *Progress in Human Geography* 39 (5): 647–57. https://doi.org/10.1177/0309132514560961

Derickson, K. 2018. "Masters of the Universe." *Environment and Planning D: Society and Space* 36 (3): 556–62. https://doi.org/10.1177/0263775817715724

Dyson, G. 2012. *Turing's Cathedral: The Origins of the Digital Universe*. New York: Vintage.

Dyson, G. 2020. *Analogia: The Emergence of Technology Beyond Programmable Control*. New York: Farrar, Straus, and Giroux.

Graham, S., and S. Marvin. 2001. *Splintering Urbanism: Networked Infrastructures, Technological Mobilities, and the Urban Condition*. London: Routledge.

Greenfield, R. 2006. *Timothy Leary: A Biography*. New York: Harcourt.

Friedmann, J. 1973. "The Urban Field as Human Habitat." In *The Place of Planning.*, ed. S.P. Snow. Auburn, AL: Auburn University; rev. and repr. in *Systems of Cities: Readings on Structure, Growth, and Policy*, ed. L.S. Bourne and J.W. Simmons, 42–60. New York: Oxford University Press, 1978.

Hackworth, J. 2019. *Manufacturing Decline: How Racism and the Conservative Movement Crush the American Rust Belt*. Cornell, NY: Cornell University Press.

Harvey, D. 1992. "Capitalism: The Factory of Fragmentation." *New Perspectives Quarterly* 9 (2): 42–5.

Hayek, F. 1941. *The Pure Theory of Capital*. London: Macmillan.

Hayek, F. 1945. "The Use of Knowledge in Society." *American Economic Review* 35 (4): 519–30.

Hayek, F. [1968] 2002. "Competition as a Discovery Procedure," trans. and repr. in *Quarterly Journal of Austrian Economics* 5 (3): 9–23. https://doi.org/10.1007/s12113-002-1029-0

Hayek, F. 1982. "The Sensory Order after 25 Years." In *Cognition and the Symbolic Processes*, vol. 2, ed. W.B. Weimer and D.S. Palermo, 287–93. Hillsdale, NJ: Lawrence Erlbaum Associates.

Hayek, F. 1988. *The Fatal Conceit: The Errors of Socialism*. Chicago: University of Chicago Press.

Hofstadter, R. 1964. "The Paranoid Style in American Politics. *Harper's*, November. Online at https://harpers.org/archive/1964/11/the-paranoid-style-in-american-politics/, accessed 11 January 2021.

Kitchin, R., C. Coletta, L. Evans, and L. Heaphy. "Creating Smart Cities." In *Creating Smart Cities*, ed. C. Coletta, L. Evans, L. Heaphy, and R. Kitchin, 1–18. New York: Routledge.

Leitner, H., E. Sheppard, and J. Peck. 2020. "Urban Studies Unbound: Postmillennial Spaces of Theory." In *Urban Studies Inside/Out: Theory, Method, Practice*, ed. H. Leitner, E. Sheppard, and J. Peck, 3–20. Los Angeles: SAGE.

Mander, J. 2012. "Privatization of Consciousness." *Monthly Review* 64 (5): 18–41. https://doi.org/10.14452/MR-064-05-2012-09_2

Mattern, S. 2017. *Code + Clay, Data + Dirt: Five Thousand Years of Urban Media*. Minneapolis: University of Minnesota Press.

McCann, L.D., and J. Simmons. 2000. "The Core-Periphery Structure of Canada's Urban System." In *Canadian Cities in Transition*, 2nd ed., ed. T. Bunting and P. Filion, 76–96. Don Mills, ON: Oxford University Press Canada.

Metcalf, S. 2017. "Neoliberalism: The idea that swallowed the world." *Guardian*, 18 August. Online at https://www.theguardian.com/news/2017/aug/18/neoliberalism-the-idea-that-changed-the-world, accessed 11 January 2021.

Mirowski, P. 2011. *Never Let a Serious Crisis Go to Waste: How Neoliberalism Survived the Financial Meltdown*. London: Verso.

Mirowski, P., and E. Nik-Khah. 2017. *The Knowledge We Have Lost in Information: The History of Information in Modern Economics*. Oxford: Oxford University Press.

Moulier-Boutang, Y. 2011. *Cognitive Capitalism*. Cambridge, UK: Polity Press.

Oswin, N. 2018. "Planetary Urbanization: A View from Outside." *Environment and Planning D: Society and Space* 36 (3): 540–6. https://doi.org/10.1177/0263775816675963

Oswin, N. 2020. "An Other Geography." *Dialogues in Human Geography* 10 (1): 9–18. https://doi.org/10.1177/2043820619890433

Rieder, B., A. Matamoros-Fernández, and Ò. Coromina. 2018. "From Ranking Algorithms to 'Ranking Cultures': Investigating the Modulation of Visibility in YouTube Search Results." *Convergence* 24 (1): 50–68. https://doi.org/10.1177/1354856517736982

Roy, A. 2016. "What Is Urban about Critical Urban Theory?" *Urban Geography* 37 (6): 810–23. https://doi.org/10.1080/02723638.2015.1105485

Roy, A., and A. Ong, eds. 2011. *Worlding Cities: Asian Experiments and the Art of Being Global*. Chichester, UK: Wiley-Blackwell.

Sayer, A. 1982. "Explanation in Economic Geography: Abstraction versus Generalization." *Progress in Human Geography* 6 (1): 68–88. https://doi.org/10.1177/030913258200600103

Smith, N. 1989. "Geography as Museum: Private History and Conservative Idealism in *The Nature of Geography*." In *Reflections on Richard Hartshorne's The Nature of*

Geography, ed. J. Entrikin and S. Brunn, 91–120. Washington, DC: Association of American Geographers.

Townsend, A. 2013. *Smart Cities: Big Data, Civic Hackers, and the Quest for a New Utopia*. New York: W.W. Norton.

Van Meeteren, M. 2016. *From Polycentricity to a Renovated Urban Systems Theory: Explaining Belgian Settlement Geographies*. Ghent: Department of Geography, Ghent University

Von Bertalanffy, L. 1950. "An Outline of General Systems Theory." *British Journal for the Philosophy of Science* 1 (2): 134–65. https://doi.org/10.1093/bjps/I.2.134

Wakabayashi, D., D. Alba, and M. Tracy. 2020. "Bill Gates, at odds with Trump on virus, becomes a right-wing target." *New York Times*, 17 April. Online at https://www.nytimes.com/2020/04/17/technology/bill-gates-virus-conspiracy-theories.html, accessed 11 January 2021.

Webber, M. 1964. "The Urban Place and the Non-place Urban Realm." In *Explorations into Urban Structure*, ed. M. Webber, J. Dyckman, and D. Foley, 79–153. Philadelphia: University of Pennsylvania Press.

West, G. 2017. *Scale: The Universal Laws of Innovation, Sustainability, and the Pace of Life in Organisms, Cities, Economies, and Companies*. New York: Penguin Press.

Wylie, C. 2019. *Mindf*ck: Cambridge Analytica and the Plot to Break America*. New York: Random House.

Zipf, G.K. 1949. *Human Behavior and the Principle of Least Effort: An Introduction to Human Ecology*. Cambridge, MA: Addison-Wesley Press.

Zuboff, S. 2019. *The Age of Surveillance Capitalism: The Fight for a Human Future at the New Frontier of Power*. New York: Public Affairs.

5 Cybernetic Urbanism: Tracing the Development of the Responsibilized Subject and Self-Organizing Communities in Smart Cities

GÜNEŞ TAVMEN

In 2013 a UK based think tank, Future Everything, released a publication with contributions from well-known smart city critics, including Dan Hill, Anthony Townsend, Martijn de Waal, and Adam Greenfield. Titled *Smart Citizens*, this collection of essays aimed at shifting the focus from the large technology companies and governments that are fostering innovation and efficiency in cities towards the aspirations and abilities of individuals and social businesses in smart cities. With this, the overall aim of the publication was to discuss the ways in which a more participatory and innovative society could be engendered. Most of the essays in the collection encapsulated the issue of citizenship and participation in the smart city as a matter of top-to-bottom versus bottom-up decision-making, and in pursuit of the latter, the discussion is geared towards creating self-organizing communities. For instance, in his essay, Greenfield (2013) says that, while the everyday struggles of city dwellers are largely unrecognized in the smart city literature, if we build technological frameworks that support the processes of self-organization through adaptive and dynamic, real-time local intelligence, then there could be an urban order that is bottom-up. Another common thread in these essays is that a smart citizen is an active citizen who, for example, makes use of open data to participate in the management of their city. Maltby (2013) writes that open-data initiatives provide "a raw material for citizen-led innovation in communities" and, therefore, to benefit from this raw material, citizens should be provided with opportunities to develop further skills in making use of it. Similarly, de Waal (2013) points out a few critical issues regarding open data such as the fact that most citizens might not have the necessary technical skills to make use of data. Subsequently, he argues that open-data platforms need to be paired with online tools and intuitive interfaces to enable effective participation by citizens. Consequently, all these writers agree that citizens should be the active constituents of smart cities by claiming data-driven innovation instead of leaving it solely to a few selective companies. Thus, they celebrate a vision of decentralized and adaptive technologies to foster responsible and self-organizing community groups.

As I will show in this chapter, smart citizenship in such formulations is envisioned as a subjectivity at the intersection of cybernetic urbanism and the responsibilized individual in neoliberal forms of governance. This means that, although the writers of these essays are critical of corporate visions of smart cities, the views they establish about the smart citizen are not a critique of the way smart cities work. As Vanolo (2014) observes, for instance, smart city discourses describe the urban as a site of responsibilization where the smart citizen is in production together with the smart city. As such, despite the attention to how smart cities perpetuate a top-to-bottom form of governance, smart cities are indeed developed through the very participation of their citizens in various ways. The "2.0 version of urban citizenship" is embedded in a participatory, open-source, and DIY digital urbanism whereby citizens are mainly environmentally distributed unwitting data agents (Gabrys 2014). Furthermore, hackathons, which are based on the voluntary work of participants with technical, design, and business skills, have also been essential to exploiting open data with an entrepreneurial drive that is essential to smart cities (Perng, Kitchin, and Mac Donncha 2018). Subsequently, the articulation of the ideal smart citizen as tech-savvy, responsible, and active meant a lack of critical understanding of the body-subjects that are involved in the (re)production of the smart city through participation and that operate within existing power structures relying on unpaid or low-paid care work (Burns and Andrucki 2021). Therefore, although citizens have indeed been active (and passive) participants in the making of smart cities, participation does not necessarily facilitate a more just city. What these forms of participation mean for (in)justice in the smart city needs more detailed attention to who is empowered to what ends, instead of a binary discussion of top-to-bottom and bottom-to-top. This chapter aims to do this by providing a historical background to cybernetic and neoliberal forms of governance in the urban environment.

I am not alone in suggesting that the construction of the smart citizen embeds a cybernetic vision. For example, Zandbergen and Uitermark (2020) observe that, in Amsterdam, citizen participation in air-pollution sensing is articulated within a *republican* and *cybernetic* framework. By *republican*, they mean informed citizens who use data to leverage their participation in decision-making, whereas *cybernetic* citizenship refers to project participants' contribution and immersion in the sensing activities. Similarly, Gillian Rose (2020) makes a case for understanding forms of sociality proposed in smart city planning in three groupings: *sociological, neoliberal,* and *cybernetic*. She observes that *neoliberal* accounts refer to the characterization of social groups as "autonomous," "self-directing," and "self-improving," while a *cybernetic* form of sociality emphasizes "informational feedback loops" created by enhanced data flows at the service of citizens. In that vision, whether through data or not, the main pursuit is to create unimpeded information flows to and from citizens. In her conclusion, Rose argues that the cybernetic form is distinct from the neoliberal form due to its lack of attention to "self-improving autonomy."

Departing from the suggestion that cybernetic sociality or citizenship is distinct from neoliberal or republican views, in this chapter I argue that a cybernetic form of citizenship has always been articulated in entanglement with neoliberal aspirations in which citizens are imagined as self-organizing and autonomous. By tracing a historical lineage to the implementation of cybernetics on urbanism – especially following later developments in cybernetics (i.e., second-order cybernetics and the emergence of autopoiesis) – I will show that city managers have long sought to build self-organizing, decentralized, autonomous, and responsibilized communities that overlap purviews of both cybernetics and neoliberalism. To do this, I provide an example from the late 1960s to elaborate on an example of cybernetics in action in the design of New York City's governance, together with Jay Forrester's pioneering work that applied cybernetics on urban planning. Through these examples, I intend to show the ramifications of cybernetic urbanism when it is put into action in pursuit of neoliberal governance, and discuss what this means for (in)justice in the smart city.

Cybernetic Urbanism and Self-Governance

IBM's smart city technologies and marketing plans have attracted critical scholarly attention; the cybernetic disposition of these technologies and plans has also been part of these critiques. For example, Söderström, Paasche, and Klauser (2014) note that IBM resurrected the urban cybernetics of the 1970s by "travelling back to the heroic times of post-war cybernetics." Similarly, Goodspeed (2015) starts his analysis of IBM's Urban Operation Centre in Rio de Janeiro by drawing a brief historical lineage between corporate smart city initiatives and urban cybernetics, and argues that corporate definitions of the smart city are the equivalent of the *failed* urban cybernetics of the 1960s and 1970s. As an antidote to this, he proposes a strategy of collaborative and participatory planning by using information technologies, which, according to Goodspeed, is not a cybernetic approach. This is because, for him, cybernetic thinking has been found to be incompatible with urban planning since the 1960s. As this chapter shows, however, most of these suggestions that are given as counter to smart cities indeed overlap with ideas rooted within cybernetics such as decentralization, horizontal decision-making (e.g., community networks), adaptivity, and self-organizing systems.

Although these scholars have rushed to dismiss the urban application of cybernetics, the field has had long and intricate relationships with urban planning, architecture, space, media technologies, and design (see, for example, Halpern

2015; Krivý 2016; Martin 2003). Furthermore, cybernetics has involved a great variety of disciplines, making it a highly complex area of thinking. Although it was established as a universal field, the ideas in the early days of cybernetics (i.e. first-order cybernetics) were later challenged in the 1960s, when many fundamental theories within the field were contested (i.e. second-order cybernetics). Initially developed by a mathematician, Norbert Wiener, in the late 1940s, cybernetics aimed at presenting a new paradigm in information and communications with an emphasis on informational and spatial decentralization. This first wave studied how to maintain stability through feedback loops in machines in the same way as living organisms that regulate themselves so as to maintain a steady state (homeostasis) (Hayles 1999). With the involvement of many social scientists, there have been efforts to apply these theories to societal relations and processes, although Wiener was not blind to the drawbacks of translating biological systems to social systems. According to Medina (2011, 37), for example, for Wiener, cybernetics was "ill-suited for the study of social systems because they could not generate the long-term datasets under the constant conditions that his statistical prediction techniques required."

Emerging during the 1960s, second-order cybernetics redefined systems by locating the observer within the system, rather than outside, in contrast to first-order cybernetics. This meant an epistemological shift in the field that brought the observer's "objectivity" into question. Humberto Maturana, a biologist and a pioneering thinker in second-order cybernetics, defined the observer's inference as the individual's own interpretation – in other words, as an "embodied action" (Wolfe 1995) – and suggested a different understanding of how systems are organized. According to this idea, "a living system is not a goal-directed system" (Maturana, cited in Hayles 1999, 139); therefore, it would always function insofar as its organizational capacity allows, but always in relation and in reference to structurally coupled outsiders (i.e., observers), meaning that systems respond to their environment in ways determined by their internal self-organization. As Hayles (1999) argues, substituting homeostasis with autopoietic systems put an emphasis on the *process*, which makes it readily adaptable to the analysis of social systems. This meant that the application of these ideas to governance occurred around the same period as the emergence of second-order cybernetics, although the conceptual relationships among architecture, urban planning, and cybernetics was present from the very beginning. This is partly due to the convenience of autopoietic theory's adaptability to social systems, since the idea of self-organizing systems was not particular to cybernetics but had been a feature of urban studies for a while. As Krivý (in this volume) argues, urbanists have enthusiastically embraced the idea that cities are complex and self-organizing systems, and have offered urban governance models that address these characteristics, often in a way that is auxiliary to the way (smart) urbanism is imagined by capitalistic objectives.

Urban Cybernetics and Self-Organizing Community Visions

In 1968, in a paper delivered at the Annual Symposium of the American Society for Cybernetics, the First Deputy City Administrator at the Office of the Mayor, City of New York, Steve Savas, expounded on the nexus of cybernetics and urban government. Savas, manager of urban systems at IBM before moving to the New York City mayor's office, was one of the "analytically trained administrators" who was hired to carry out governmental reform in order to tackle the high levels of social unrest at the time (Green and Kolesar 2004). In the revised version of his paper entitled "Cybernetics in City Hall" (1970), Savas answers the question as to what one could tell if one were to apply the principles of cybernetics to cities. Although Goodspeed (2015) suggests that Savas's paper is a "pessimistic prognosis for the political feasibility of urban cybernetics," Savas was indeed embracing cybernetic principles as an antidote to political and systemic shortcomings together with a short account of the possible challenges. For instance, in Savas's view, local government employees are "mediocre" and "inadequate" due to the current political system; however, their incompetency could be overcome by using a "systems analysis" approach in the urban management model. In fulfilment of his duty to reform government, Savas sought to introduce the principles of "management science" to city government, an idea arguably shaped during his time at IBM. To make up for the failings of political institutions, Savas suggested implementing cybernetics-informed systems to address New York's structural problems in a technocratic manner.

Savas was critical of first-order cybernetics due to its linear assumptions, which tend to simplify the complexities of urban life. His assertion, however, did not dismiss the use of cybernetic principles in governance; rather, it was about praising the novelties anticipated by second-order cybernetics. For instance, expanding on the time constraints that an elected administration faces, Savas proposed a cybernetics-inspired participatory democracy model based on a decentralized feedback control function realized by using minor-loop controls. By means of such controls, "[g]etting decision-making down into the community offers hope of getting more rapid response and more effective performance of the system" (Savas 1970, 1067). He also drew attention to the complications of complete decentralization by referring to a current "school turmoil" in New York City that stemmed from failing to handle the high-level regulative diversity of decentralized decision-making. He therefore suggested a "cascade control" type of administration: a terminology and technique borrowed from systems engineering in order to coordinate independent minor-loop controllers. According to this model, the high-level controller sets the goals, yet the means to achieve these goals are determined by local action. Subsequently, the means to the end may be diverse and designed hyperlocally – or "crowdsourced," to use a current term – whereas the end is determined by the central authority. Thus, participatory practices through decentralized cybernetic systems are envisioned as those where participation becomes an instrument to fix problems that are identified in a top-to-bottom fashion. In other words, participation is not envisioned as an act for citizens to contest the governance of their urban environments; rather, it is imagined as turning citizens into agents who maintain the city's operation *efficiently*. Thus, his vision is about enabling a pragmatic urban management system that would make up for the shortcomings of public administration, instead of finding ways to build a more just city by inviting citizens to voice their priorities and everyday struggles.

Furthermore, Savas pointed out a problem regarding the information systems that are available to a city mayor. Using a cybernetic lexicon, he identified possible information channels: members of the mayor's party, bureaucrats at city hall, civil disorder, and elections. He then noted that these channels are of "high-impedance." For this reason, he suggested, "[t]he cyberneticist can immediately identify ways to improve the quality, quantity, and flow of usable information to the mayor: increase the sampling rate, open more feedback channels, increase the bandwidth, enhance weak signals, match impedances, suppress noise, and correct biased signals" (Savas 1970, 1068). He said this was the sort of innovation New York City was undertaking by building up "Neighbourhood City Halls" to open up more information channels and increase sample sizes until the day when the computer became the "electronic equivalent" of the mobile neighbourhood hall. And when this happens, he added, it would be much more convenient to decentralize data acquisition and service delivery, while also maintaining centralized coordination and control – a vision that is perpetuated by today's smart cities.

Predictably, Savas's paper contains several references to Jay Forrester. A computer engineer who later turned into a "systems scientist" at MIT, Forrester applied his dynamic systems theories to cities in his highly influential book *Urban Dynamics* (1969). One of Forrester's influential projects was to build systems by mathematical modelling that focused on avoiding time delays by feedback loops – one of the urban management problems Savas mentions. For this project, Forrester first analysed a General Electric plant and subsequently wrote *Industrial Dynamics* in 1961. Later, in search of another complex system on which to apply his "cybernetic tool kit" that he called "system dynamics," his path coincidentally crossed with that of a former Boston mayor at MIT (Townsend 2013). Inspired by this collaboration, Forrester looked into housing and labour markets, and consequently created computer simulations to abstract a generic system that explained how cities worked. Highly criticized for his lack of reference to the research carried out in urban studies at the time, his response was that it would be a time-consuming and a separate set of work to go through relevant studies on urban behaviour and dynamics. According to Townsend (2013), instead of studying relevant research on urban studies,

Forrester merely relied on his computer simulation to propose the demolition of slums and subsidized social housing, describing these systems as "poverty traps." He also concluded that job training and job creation resulted in higher levels of unemployment because urban systems demonstrated "counterintuitive behaviour" due to their self-organizing nature, just like an autopoietic organization (Savas 1970). Thus, Forrester demonstrates that problems such as unemployment and house shortages are the yields of the system's internal forces and cannot be dealt with by acting upon the external symptoms (Birch 1970). In other words, because they are self-organizing (i.e., autopoietic), urban systems are structurally open to policy intervention, but are organizationally closed to changes resulting from them.

In addition to his lack of reference to current research on urban studies, Forrester was also attacked for simplifying urban systems and jumping to quick conclusions despite his claim that urban systems were complex organizations that had an intrinsic logic. Birch, who was working on economic development in cities and suburbs at Harvard University, writes in his book review that "Forrester's model is dangerous," since "a man with a good will" but lacking in technical knowledge could take the book seriously by ignoring the fact that Forrester constructed a model "which will make *any* set of policy recommendations winners" (Birch 1970, 69). Reminding readers that such a model had never been tested in an actual city, but rather consisted of the speculative ideas of a few technicians sitting in a room, Birch believed the validity of the model's assumptions was highly questionable. In another review of the book, James Hester (1970), a researcher at the Joint Center for Urban Studies at MIT and Harvard University, says that *Urban Dynamics*, whose primary authors are "Jay Forrester and an IBM 360/67 computer," is essentially about a computer simulation model and not the urban reality itself. Although Forrester (1970) believed his computer modelling was inclusive of humans – unlike engineers' approaches to the city – his critics argued that his computerized methodology to understand city dynamics was based on observations detached from the city itself.

Paradoxically, in effect, Forrester (1970) argued that many urban policies were designed to attack symptoms, but not the causes, of problems due to the technocratic and linear tendencies of urban regulations designed to give results in the short term, although causalities require long-term solutions. He further stated that engineers tend to favour economic and technological improvements and to ignore intangible factors such as social values and quality of life; they therefore create stagnation and decline in urban areas. This is especially remarkable given that Forrester was (and has since been) criticized for simplifying cities with his mathematical modelling, although, in theory, he believed that cities and their problems were much too complex to solve with short-term technocratic solutions. Moreover, according to him, since engineers had long been interested in their systems and technologies without enough consideration of people, only cybernetics could solve the problem of human interaction. Strikingly similar to why smart cities are criticized, Forrester also blamed technocratic urban management systems that overlook causalities and focus on symptoms.

On the other hand, what is profound in Forrester's work is his application of self-organizing systems approach to urban dynamics, together with his dismissal of contingency and the embodiment of observations that were fundamental to second-order cybernetics. By employing the view of self-organization in this manner, he was able to suggest a set of essential attributions to slums and social housing as "poverty traps," as if these were closed systems that exist outside any social or political context. Although he acknowledged that short-term technological solutions would not solve structural problems, and that the cause of these problems should be addressed instead of the symptoms, he also suggested cutting off policy interventions designed to eradicate these problems. This was because, in his formulation, political interventions emerge outside the system in which slums and social housing emerge. The ideas of decentralization and self-organization do not necessarily indicate a progressive governing system – quite the opposite: instead of tackling the major problem of social inequality, they entrench it by implying that it is a natural phenomenon due to the autopoietic nature of urban systems. This is mainly why the "systems approach" fails to address "wicked problems" of cities, which are malignant, tricky, and malicious, as opposed to the benign and tame problems that engineers choose to deal with, as Rittel and Webber (1973) argue.

As a result of developing housing projects inspired by Forrester and his computer-simulated housing modelling, Savas was later appointed assistant secretary of the US Department of Housing and Urban Development by President Ronald Reagan. According to his profile page on the website of Baruch College, City University of New York, where he is currently a professor, Savas "is an internationally known pioneer in, and authority on, privatization."[1] Indeed, he is the writer of many books, including *Privatization: The Key to Better Government* (1987) and *Teaching Children to Use Computers* (1985). His book on privatization has been translated into twelve languages, from Spanish to Korean, Turkish to Polish. A former IBM manager of urban systems, throughout his time in various managerial roles in the public sector, Savas has played an important role in transforming New York City's and, indeed, America's housing, welfare, and public service systems. As Martin (2003) puts it, "[c]orporations like IBM linked modulated flexibility with organicist notions of open-ended yet controlled growth, correlated with pseudo-freedoms of self-realization within a flexible framework" (159). Pseudo freedoms and self-realization were consequences of the convergence of the cybernetic and neoliberal agendas of the era: defining urban realms as self-organizing, thereby suggesting inequality and other structural problems were inevitable by keeping politics out of the equation. Savas's appointment to the New York City mayor's office was already within the scope of a "governmental reform" that was essentially prompted by the neoliberal agenda of the era. Embracing and adapting a selective set of cybernetic principles to urban management, he then became an internationally renowned expert in the privatization of public services.

Smart Citizenship in Open-Data-Driven Smart Cities

Cyberneticians were not oblivious to the potential of adverse social consequences of the systems they designed and materialized. Those especially who were involved in establishing second-order cybernetics thought of their main principles outside the paradigm of efficiency and optimization. With the emergence of unintentional adverse effects over time, however, this did not result in an overall critical understanding of cybernetic thinking. Instead, these were addressed as "flaws" to be overcome through *technical progress*. This meant a lack of critical overview of the application of cybernetics to social systems, including urban governance. In a recursive system of its own creation, built on validation from multiple disciplines, cyberneticians and administrators who adhered to their principles therefore ended up entrenching these flaws deeper within the mechanisms they designed through cybernetic principles (see, for example, Turner 2006, 24–6).

Similarly, in smart citizen visions, policymakers and even some critical writers continue to frame the lack of equitable participation in (smart or otherwise) urban planning as a technical issue, whereby the decentralization of data collection, access, and use is the main goal. This is reflected particularly in open-data-driven smart city discourses, since the main idea here is that citizens inexorably would be empowered once they had free access to data about their city. Although it has been argued that open data in smart city panning are mobilized towards an entrepreneurial agenda (see Barns 2016), there is a lack of critique that it is also geared towards a cybernetic urban vision coupled with neoliberal tendencies. In addition to the pressing question of who will be able to use the open data, we should also think critically about the responsibilization of citizens and the narrative of self-organizing groups in an environment characterized by stark inequality. Otherwise, our understanding and formulation of the empowerment and participation of citizens will be highly limited and eventually will perpetuate social and digital injustice in open-data-driven smart cities. Decentralization, horizontalization, and self-organization do not inexorably pave the way to an equitable society in a world of inequalities. Cyberneticians have failed to acknowledge this fundamental issue by taking politics out of the equation in their application of cybernetics to social systems, as we have seen with the example of Forrester's urban dynamics. Despite his effort to go beyond technocratic management models and his neglect of conclusions from urban studies and reliance instead on mathematical models, Forrester's work has been instrumental in infusing efficiency and productivity into the equation, instead of empowering citizens towards progressive ends, such as eradicating inequality. I argue that, since open-data-driven smart cities tend to propagate a similar approach, we need more critical outlook that goes beyond the entrepreneurial dimension to take the perils of self-organization into account.

NOTE

1 See the website at http://www.baruch.cuny.edu/mspia/faculty-and-staff/full-time-faculty/es-savas.html

REFERENCES

Barns, S. 2016. "Mine Your Data: Open Data, Digital Strategies and Entrepreneurial Governance by Code." *Urban Geography* 37 (4): 55–71. https://doi.org/10.1080/02723638.2016.1139876

Birch, D. 1970. "Urban Dynamics-Forrester, JW." *Sloan Management Review*: 67–9.

Burns, R., and M. Andrucki. 2021. "Smart Cities: Who Cares?" *Environment and Planning A: Economy and Space* 53 (1): 12–30. https://doi.org/10.1177%2F0308518X20941516

de Waal, M. 2013. "Open Data: From 'Platform' to 'Program.'" In *Smart Citizens*, ed. D. Hemment and A. Townsend, 79–82. Manchester: Future Everything.

Forrester, J. 1969. *Urban Dynamics*. Cambridge, MA: MIT Press.

Forrester, J. 1970. "Systems Analysis as a Tool for Urban Planning." *IEEE Transactions on Systems Science and Cybernetics* 6 (4): 258–65. https://doi.org/10.1109/TSSC.1970.300299

Gabrys, J. 2014. "Programming Environments: Environmentality and Citizen Sensing in the Smart City." *Environment and Planning D: Society and Space* 32 (1): 30–48. https://doi.org/10.1068/d16812

Goodspeed, R. 2015. Smart Cities: Moving beyond Urban Cybernetics to Tackle Wicked Problems." *Cambridge Journal of Regions, Economy and Society* 8 (1): 79–92. https://doi.org/10.1093/cjres/rsu013

Green, L., and P. Kolesar. 2004. "Improving Emergency Responsiveness with Management Science." *Management Science* 50 (8): 1001–114. https://doi.org/10.1287/mnsc.1040.0253

Greenfield, A. 2013. "Recuperating the Smart City." In *Smart Citizens*, ed. D. Hemment and A. Townsend, 9–12. Manchester: Future Everything.

Halpern, O. 2015. *Beautiful Data: A History of Vision and Reason since 1945*. Durham, NC: Duke University Press.

Hayles, K. 1999. *How We Became Posthuman: Virtual Bodies in Cybernetics, Literature, and Informatics*. Chicago: University of Chicago Press.

Hemmett, D., and A. Townsend, eds. 2013. *Smart Citizens*. Manchester: Future Everything.

Hester, J. 1970. "Systems Analysis for Social Policies." *Science* 168 (3932): 653–4. https://doi.org/10.1126/science.168.3932.693

Krivý, M. 2016. "Towards a Critique of Cybernetic Urbanism: The Smart City and the Society of Control." *Planning Theory* 17 (1): 8–30. https://doi.org/10.1177/1473095216645631

Maltby, P. 2013. "Open Data and Beyond: How Government Can Support a Smarter Society." In *Smart Citizens*, ed. D. Hemment and A. Townsend, 59–62. Manchester: Future Everything.

Martin, R. 2003. *The Organizational Complex: Architecture, Media, and Corporate Space*. Cambridge, MA: MIT Press.

Medina, E. 2011. *Cybernetic Revolutionaries: Technology and Politics in Allende's Chile*. Cambridge, MA: MIT Press.

Perng, S., R. Kitchin, and D. Mac Donncha. 2018. "Hackathons, Entrepreneurial Life and the Making of Smart Cities." *Geoforum* 97 (December): 189–97. https://doi.org/10.1016/j.geoforum.2018.08.024

Rittel, H.W.J., and M. Webber. 1973. "Dilemmas in a General Theory of Planning." *Policy Sciences* 4: 155–69. https://doi.org/10.1007/BF01405730

Rose, G. 2020. "Actually-Existing Sociality in a Smart City." *City* 24 (3–4): 512–29. https://doi.org/10.1080/13604813.2020.1781412

Savas, E.S. 1970. "Cybernetics in City Hall." *Science* 168 (3935): 1066–71. https://doi.org/10.1126/science.168.3935.1066

Savas, E.S. 1985. *Teaching Children to Use Computers: A Friendly Guide*. New York: Teachers' College Press.

Savas, E.S. 1987. *Privatization: The Key to Better Government*. Chatham, UK: Chatham House.

Söderström, O., T. Paasche, and F. Klauser. 2014. "Smart Cities as Corporate Storytelling." *City* 18 (3): 307–20. https://doi.org/10.1080/13604813.2014.906716

Townsend, A. 2013. *Smart Cities: Big Data, Civic Hackers and the Quest for a New Utopia*. New York: Norton.

Turner, F. 2006. *From Counterculture to Cyberculture: Stewart Brand, the Whole Earth Network, and the Rise of Digital Utopianism*. Chicago: University of Chicago Press.

Vanolo, A. 2014. "Smartmentality: The Smart City as Disciplinary Strategy." *Urban Studies* 51 (5): 883–98. https://doi.org/10.1177/0042098013494427

Wolfe, C. 1995. "In Search of Post-Humanist Theory: The Second-Order Cybernetics of Maturana and Varela." *Cultural Critique* 30 (Spring): 33–70. https://doi.org/10.2307/1354432

Zandbergen, D., and J. Uitermark. 2020. "In Search of the Smart Citizen: Republican and Cybernetic Citizenship in the Smart City." *Urban Studies* 57 (8): 1733–48. https://doi.org/10.1177/0042098019847410

PART TWO

Data Decisioning and Data Justice

A Dialogue with Rob Kitchin

Part Two of this volume examines the logics and rationalities of urban data, a now integral and interwoven element of urban life. Authors in this part challenge the discourse of "data as fuel," by explicating processes of smartification and the profit motives *fuelling* data capture and extraction. Building on critical data studies and data justice, the authors identify (in)justice within these data-based sociotechnical arrangements. Specifically, interrogating disjunctures between dominant conceptions of rights and data-based assemblages, rationalities, and relationalities of smart urbanism.

To discuss these entanglements, here we dialogue with Dr Rob Kitchin, Professor and European Research Council Advanced Investigator at Maynooth University. Kitchin has held numerous prestigious positions and has published nearly two hundred articles and book chapters and thirty authored or edited books. His research focuses on the politics and impacts of data, software, and related digital technologies.

Editors: We are really intrigued by these emerging conversations around social justice in the smart city and, in particular, you, Taylor Shelton, and a few other people have raised this idea of the right to the smart city. Your latest collaboration (Cardullo, Di Feliciantonio, and Kitchin 2019) *inspired us to think more precisely about social justice and what frameworks we can draw on in that conversation. The right to the smart city, as you know, is one of dozens of frameworks that one could draw in for social justice conversations, and so in this book contributors grapple with the benefits and drawbacks that come with different kinds of frameworks for thinking about social justice in the smart city. For instance,*

Miguel Valdez, Matthew Cook, and Helen Roby forward theories of epistemic justice to navigate these uneven and unequal power relations.

What do different frameworks omit? What do they include or assume, and what sort of work can they do for us? So, we are taking a step back to ask critical questions about the right to the smart city, about other frameworks and so on.

Over the past several years of the critical smart cities research agenda, we have been intrigued by the emergence of a few competing notions of social justice. As noted by contributors, there are calls for smart cities from the bottom up; there are calls for reappropriating smartness language; there are calls for attention to difference such as race, gender, coloniality, and so on. One of your recent interventions has been to emphasize the potentialities of the right to the smart city as theorized by Henri Lefebvre. The title of your chapter in your recent book is "Towards a Genuinely Humanizing Smart Urbanism" (Kitchin 2019), *which we understand is riffing off David Harvey. But it divulges one of our key concerns with the framework. We wonder what are the limitations of rights language? As Ryan Burns contends in the previous part, are we able to reappropriate rights languages for more-than-human frameworks? To think about non-human actors, non-human animals, ecologies, or even machines, possibly – computational devices.*

RK: The first thing you must do is unpack social justice. At the highest philosophical level, where you come from will shape where you're going with it. If you are an egalitarian, you will think about this differently than if you're a utilitarian, or if you're a libertarian, and so on. One is everybody is equal; one is greatest number for the greatest good; one is the market is naturally just, and if you get screwed, that's your problem. So, the top level makes a difference, and you can argue that there are kinds of rights within each of those frameworks. I have a problem with discussions around social justice that uncritically take the implicit notion that they all mean people will be treated fairly, equitably, and equally, without actually stating how they will get there. Depending on your worldview, your notion of fairness, equality, equity, and so on is different.

There could be instrumental notions of justice, which is about outcomes. There could be procedural justice that pays attention to how mechanisms work. There could be distributional forms of justice around the fair distribution of resources. There could be recognition – that is to say, equal respect, the same treatment across subjects. Or there could be representational justice that seeks equal voice and ability to challenge data, power, and so on. We might say that representational concerns are more political in nature, recognition more cultural, distributions more spatial, procedural more process-related, instrumental more about outcomes, and so on. One of the things we have to consider is whether technology is part of any sort of solution, or even if it should be figured into the framing of the problem. Is the solution *technological* justice or is the solution just *justice*, full stop? I've been known to make the case that we shouldn't really be thinking about the right to the *smart* city. It should just be the right to the *city*. I said a little bit about that in my "Genuinely Humanizing" chapter (Kitchin 2019).

The right to the city is really rooted in a kind of political economy that goes back to a Marxist critique of urbanism, with an eye towards notions of citizenship and governmentality. This is an important shift in how people are governed, managed, treated, regulated, controlled, etc. Within these new forms of citizenship, there is a rights and entitlement side, but there is also a more neoliberal notion of citizenship based on acting responsibly, consumerism and choice based on ability to pay. Other people will come at this through a feminist, or a postcolonial, approach; Catherine D'Ignazio's work is a good example of the first, and Ayona Datta's of the second. Within data justice there are people like Evelyn Ruppert and a whole bunch of Indigenous scholars such as Tahu Kukutai, Stephanie Russo, and Maggie Walter. This work is about reconfiguring power as opposed to rights in a pure sense.

When we were doing the smart city stuff, we were linking it into a kind of political economy reading of the smart city, which would emphasize capital, property, uneven development, and so on. That was partly where rights language was informing our social justice framework. And this political economy reading would be in opposition to popular entrepreneurial urban thought. In a lot of ways, the smart city is the third wave of entrepreneurial urbanism: we had entrepreneurial planning in the '90s, we then had the creative city in the 2000s, and now we have the smart city. It has gone from economic planning through cultural economy, towards tech. Smart cities are a kind of tech version of that kind of entrepreneurial urbanism.

Editors: Some of these frameworks are far more amenable to thinking about the recognition of non-human actors. In this part, Torin Monahan highlights how ride-hailing platforms have inserted themselves as obligatory passage points in urban transportation systems. Through privatization, urban mobility platform companies, such as Uber, "capture and capitalize data," limiting accessibility often for the most marginalized in our communities. With cases like these, it is often easier to consider property or the distribution of resources; however, as noted by Lorena Melgaço and Lígia Milagres and others, extending rights to animals or to the environment becomes more challenging. To this point, Nathan Olmstead and Zachary Spicer further politicize the data assemblage, highlighting its co-constitution and invoke ideas of around "biodegradability" to capture the fleeting nature of data. Can you speak more to navigating these frameworks moving towards justice?

RK: I think within ethics of care it is probably easier to get there than with a rights-based framework. Catherine D'Ignazio and Lauren Klein, in their book *Data Feminism*, have this nice distinction between data ethics and data justice. They say that the problem with ethics, which is centred on rights, compliance, and regulation, is that it accepts the system as it is; it is about whether you match the system, and tweaks to the system, not about radically changing it. In other words, it locates the problem in individuals and the technology, rather than locating problems in structural conditions. This can be a bit of a problem with rights-based justice: rights operate within that political economy. It does not challenge the political economy; it just formulates rights within it. D'Ignazio and Klein talk about data ethics versus data justice, but it could just as easily be reformulated as smart city ethics versus smart city justice. There is an important distinction there. I think we say that a little bit in the first chapter of *The Right to the Smart City*.

I would say the second side – the justice side – is more likely to be useful for more-than-human stuff. I am seeing more of that in the data literature than I am seeing in the smart city literature. In the data literature you find people like Deborah Lupton and Ash Watson (2021), for example, writing on the more-than-human with new materialist ideas around data ethics and data justice. Data justice research in some ways overlaps with the smart city research, given that all city operations are data driven, but they really are separate conversations.

Editors: In approaching this volume, we have been interested in varied understandings of "smart." You have called for the reframing, reimagining, and remaking of the smart city, and you have conceptualized the effort as "an emancipatory and empowering project; one that works for the benefit of all citizens and not just selected populations" (Kitchin 2019, 4). *And indeed, this is one of our preoccupations: first, whether smartness is capable of being reappropriated for emancipatory and just ends, and related, what would the conception of emancipatory justice entail? More specifically, is smartness an inherently worthwhile pursuit, or should it be abandoned altogether? What is to be gained from saving the smart city, or smartness as a concept?*

RK: It fundamentally comes down to whether you think that technology is useful at all. Can you imagine driving around Calgary with no traffic control system? You would probably be gridlocked the entire time. So, technology is useful. That is our premise in *Slow Computing: Why We Need Balanced Digital Lives* (Fraser and Kitchin 2020). We start by saying, "Look, there are lots of problems with technology, but it's also quite productive, you can get a lot of joy out of using it, and you can do lots of useful things with it." Getting the balance between its beneficial use and its pernicious, exploitive uses is the biggest challenge. There are tons of technology in the city actually being used for social good and for our benefit. Whether that's streetlighting, or whether it is traffic control, or something else, there's loads of stuff that is actually useful. The problem is when it is used as a mechanism of exploitation or for profit, or to create a certain power differential that can be exploited in a particular way.

So, I think this notion of being against the smart city is wrong-headed in a sense because technology is so embedded into the fabric of our cities now. If we were to take it out, we would just have non-functioning cities. I think the goal should instead be to configure it in a way that it works for the benefit of citizens. Several people, certain city councils, and even some European projects have done quite a bit of work around this. Barcelona is the classic example that nearly everybody talks about, where they moved from a right-wing neoliberal government in 2015 to a left-leaning government. They adopted the notion of technological sovereignty, which is a particular form of rights in relation to the city: technology has to serve citizens, not corporations and states. They had a whole series of efforts to transform how they were using technology: they began disinvesting from big multinational companies, they started to shift over to open-source platforms and tools, invested in open data, and so on. They are using a collective decision-making platform to foster public debate, and this is not a small-town hall meeting with fifty people. This is tens of thousands of people giving their opinion on what they think should happen in a district or on a street in relation to municipal renewal, or water delivery, for example. And this is feeding into other debates about things like Airbnb, Uber, and platform economies – how to regulate them, and so on. In Medellín, Colombia, for example, the city mandates that the smart district can't gentrify and displace the existing community, and that it has to serve the existing community.

It is about saying that there might be some useful stuff with this technology, but on what basis do we think it's useful? If you set things up in a different way, then you will get a different conversation. If you say the aim of the smart city is fairness, equity, fair distribution, justice, and so on, you will get a different kind of city than if you say the drivers should be efficiency, optimization, profit, and so on. You will get a different city. The technology is neither good nor bad. It is productive, and the consideration should be in how it's used.

Editors: Many of the technologies and practices of datafication, like those discussed by Valdez, Cook, and Roby, have been around for decades; however, calling them "smart" is relatively new. It sounds like you might not think that there is anything necessarily transformative about designating them with the term "smart." As we discuss in the Introduction, calling them smart does some discursive work, at least, but we wonder if, materially, you might not see it as particularly transformative.

RK: It can be transformative. Depending on what it is aimed at and how it is implemented. We can potentially make a difference to an area or people's quality of life and so on – in terms of how it tackles an issue. The smartness thing is a marketing label to a large degree. And it is trying to play off this notion that gaining optimization or efficiency can be done through machine learning, artificial intelligence, or automated and algorithmic forms of governance. That is where the smart bits are coming from. But smartness is a tricky term, you are right.

Now a lot of it is actually not that smart, right? Being able to tap in, tap out to get on to the Underground and whatever else – there's not a lot going on there. Some of it is more sophisticated. Traffic control rooms are pretty sophisticated, handling data in real time and using them to phase the traffic lights and react in real time to what is going on. Martin Dodge always uses this example of the control room in *The Italian Job*. I do not know if you saw the original film, where they drive Minis [Mini Coopers] through the city; Michael Caine's in the film. It is made in 1969, but one of the ways they get the cars out of the city really fast is they hack the traffic light system. So, even back in 1969 they were hacking the system. Of course, they had to hack it a different way: you had to break in, take the magnetic tape out, and put your magnetic tape in. But it was still a hack, if you think about the cybernetic stuff and the late '60s, early 1970s.

Using digital technology to manage what's going on in the city, then, has a longer history, and we have cycled through many different terms. Back in the 1980s, I think it was *wired cities*, and then we went into *cyber cities* and *network cities* and *knowledge cities* and *innovation cities*. Now we are at smart, right? And we have forgotten that some of these conversations were happening. Bill Mitchell, Mike Batty, Steve Graham, Simon Marvin, Matt Zook – many people were already talking about some of this stuff in the 1990s.

Editors: So, on the one hand, you could say smartness is here to stay until the next term takes over, so as scholars it would work in our favour to recognize that it is here, and try to achieve just smart cities. On the other hand, it also would make quite a bit of sense to say that smartness is as ephemeral as digital cities, intelligent cities, creative cities, and so on, and therefore we need to think outside of those terms to work towards just cities regardless of the technological assemblage.

RK: The problem with using smart as the link to justice in relation to the city is that it tends to suggest that the solution will also be technological. This is why you might want to decentre that. It is difficult even with great ideas like technological sovereignty: technology is still in the term. It is still the solution.

Editors: Smartness is a marketing term, and as Alberto Vanolo, Vincent Mosco, and others in this volume note, it did not originate from good intentions or a positive social impact. Alison Powell (2021,4) *suggests that "[w]hat began as an idea about improving citizens' access to knowledge by expanding access to the internet built up into a set of systems oriented toward extracting, modelling, and optimizing systems based on data." David Murakami Wood and Torin Monahan* (2019, 1) *take this further, suggesting that "digital platforms fundamentally transform social practices and relations, recasting them as surveillant exchanges whose coordination must be technologically mediated and therefore made exploitable as data." With these and other critiques in mind, we have been sceptical of the long-term viability of saying we need to work within the smart framework and make it just, for precisely the reasons you just said.*

RK: Sure, but it is a little bit of a game and it is a bit more nuanced than that. You need to be in this space challenging what the actors are saying, as opposed to talking from outside. This is always the dilemma: do you get in the room and try to influence from the inside, or do you shout over the barricades from the outside? Which is going to get you further along in terms of transforming how cities work? I'm one of these people who likes to be in the room. I'll sit on a government committee, go to industry events, and try to influence the conversation within those spaces, as opposed to writing from the outside and hope that they read it and it might influence them. In fact, we developed the city dashboards to work from the inside. Now, I know that that's a tactical choice, but that's the way that we've tried to do it. It's how we've tried to shift some of the thinking around citizen inclusion and citizen-centric ways of operating cities. Whereas, if we were not there, and we were just generally advocating in a broad sense, I'm not sure how much we would be able to influence what was going on.

This is what we have been writing around ethics washing, particularly in smart cities, where they will adopt principles or guidelines, or they will plug it into their corporate social responsibility. All the time, it is really more of an exercise in marketing and visioning than it is about fundamentally shifting their ethos and practice. Compliance as an idea is a bit like that. Catherine D'Ignazio and Laura Klein critiqued data ethics because its compliance bits can help some pernicious groups say, "Well, we're complying. Even if the system itself is structurally unjust and creates oppression, if we're complying with the law, then everything is fine."

Editors: Authors in this volume offer a range of critical approaches to studying the intersections of technology and society. As detailed in the chapters on data and infrastructure, there is no shortage of people who can use data and develop technology. However, as both you and Alison Powell in Dialogue 5 note, more needs to be done in these technological spaces.

RK: The technologists are the people who need to do the heavy work of learning to think through critical theory. But it is our job to get this into the technology space in a way that makes sense to data scientists. That is who I mostly teach. I only teach first-year and a master's course in geography; my other teaching is in maths and computer science, introducing them to critical thinking around data and technology. It has raised many issues with how computer science and data science are taught. Students do not get an ethical grounding. It is just not really part of an orthodox curriculum, and if it is included, it is a very deontological form of ethics that mostly orients around compliance. It is similar in the smart city space, and when you go and talk to chief information officers or chief technology officers, they are coming out of that data science background and they're largely not grounded in social science or ethical thinking.

Editors: Nor do they have social science or humanities scholars on their research teams. Obviously, it is hard to specialize in everything; computer scientists are often trained to be programmers, and not much else. But when a team of computer scientists develop "smart" technology to address complex social issues – such as using facial recognition to detect intoxication for use in shelters, without anyone on the team who specializes in addiction, homelessness, or social intervention – we get very concerned. We argue that tech-focused and "smart" research teams need humanities and social science scholars.

RK: You see it when you go to hackathons as well, some of the weird and wonderful projects they come up with without deeply thinking about it.

Of course, it does differ in different contexts. For example, Ireland is a sort of postcolonial context, and surveillance is frowned upon. One implication is that the closed-circuit television cameras in Dublin don't record. The video they capture is just there and gone. We interviewed one of the camera operators, who said, "Why the hell would I want to record it? I live here, right? I don't want a surveillance grid!" It is linked to this colonial history of the British state surveilling the Irish population. We don't have the same cultural sensibility around it, and we're much more resistant to it. Also, Barcelona, being in Catalonia, which is in opposition to the Spanish state, does have a very different set of – you could call it *smart-mentalities* or *data-mentalities* or some way of thinking about this kind of infrastructure. People in Hong Kong have a very different notion of this, because of what China can introduce into Hong Kong's surveillance grid and its implications for democracy. So, depending on where you are, you'll find that the local authorities, or even the tech specialists, will have a different view of it because of their history and their culture and their systems of governance and so on.

Editors: In this part on data decisioning, contributors interrogate the development and utility of urban data. Jonathan Gray and Noortje Marres stress that "[u]rban data technologies present a critical site of experimentation in rendering cities legible, inhabitable, and liveable on the level of the collective: as the arrangements for articulating urban collectives with data technologies continue to be captured, appropriated, and repurposed by a variety of actors, we must analyse critically not only by whom and to what end urban experiments are organized, but also the methodologies *for the articulation of urban collectives with data they implement." Elsewhere in the book, Orland Woods explores how data can problematize the terms of "smart." Much of your work has centred around data: Big Data, open data, the data revolution. What role do data play for your conception of a socially just smart city?*

RK: One of the most important considerations is in collecting the data that you need as opposed to collecting everything you can get, as well as using them in fair and sensible ways. Anything that is indexical or identifiable to a person or an object – for instance, a transaction or a person's location – is sensitive data and can create a system of dataveillance or surveillance or geoveillance, however we want to phrase it. We must be very careful about how we use that data and what sorts of restrictions we put around them.

Many of the debates going on now are around privacy, and I think that is important, but the bigger issue is governmentality and how the data are used. Now, privacy feeds into that because it's the conditions under which you can get the data, but what's more important is how the data are shared, processed, and so on. We need to mandate that only data necessary to a system's working should ever be collected. We write about this in *DIS Magazine*: we reproduce what's collected off an Uber app, and it's pretty much everything on your phone. It pulls data from all over the phone, even when there's no relationship to your getting a taxi. It measures about five different things related to the battery on your phone, but there's no need for it to know your battery temperature, your battery signature, or your battery type. Interestingly, this is just an alternative form of indexical data: you can identify phones from their battery signature alone. So, even if you turn your other pieces of identity off, they can identify your phone based on how your battery decays or how the energy is pulled off of it.

There are really two reasons why this is happening. The first one, with particular significance for smart cities (but also lots of other domains), is data capitalism. Some people call it surveillance capitalism, but I think data capitalism is a bigger set of processes and practices. One part of data capitalism is basically asking how to make money off of data. Most smart city technologies create a lot of data: you're tapping in and tapping out, it's stored on an identifiable card, you're getting your licence plate scanned as you drive around the city, your phone is getting pinged as you walk around shopping malls. This is all in the name of optimization or efficiency, but it raises the question of whether we want those kinds of systems, and if we do, how are they to be regulated in a sensible fashion?

We're seeing lots of debates falsely framed as trade-offs. Is it *privacy or convenience* or *privacy or security*? We've also seen in the debates around COVID-19, a false framing of the trade-off between privacy and public health. It frames privacy as harmful for addressing the pandemic: contact tracing apps are all about location, movement, and data, but really they're intended to control and discipline movement and location.

The adoption of smart cities, by the way, is all very fractured and messy. Companies are pushing technologies, and governments are often adopting them, but there's also a lot of resistance inside local authorities or municipalities. It is really an adoption gap. Companies like IBM and Cisco to a certain degree have kind of turned away from *smart city* or they have rebranded themselves on the Internet of Things, because they were finding that a lot of cities weren't adopting or only partly adopting.

And this varies with geographic context, as well. One of the problems with popular smart city discourses is that they create a universalism: a universal discourse about how the smart city is going to save every city on the planet. On the contrary, they're actually going to look different within Europe, North American, African, and Asian contexts. There is a wealth of different political economies and histories and everything else going on.

This tension means that smart cities are not a done deal at all. This is playing out in quite messy ways, and that's how these discussions around justice and citizenship and so on can actually be productively used. We can push things onto a different trajectory. We can change the future vision.

FURTHER READING

Cardullo, P., C. Di Feliciantonio, and R. Kitchin, eds. 2019. *The Right to the Smart City*. Bingley, UK: Emerald Publishing.

Datta, A. 2015. "New Urban Utopias of Postcolonial India: 'Entrepreneurial Urbanization' in Dholera Smart City, Gujarat." *Dialogues in Human Geography* 5 (1): 3–22. https://doi.org/10.1177/2043820614565748

D'Ignazio, C., and L.F. Klein. 2020. *Data Feminism*. Cambridge, MA: MIT Press.

Isin, E., and E. Ruppert. 2020. *Being Digital Citizens*. Lanham, MD: Rowman & Littlefield.

Kitchin, R. 2019. "Towards a Genuinely Humanizing Smart Urbanism." In *The Right to the Smart City*. Bingley, UK: Emerald Publishing.

Kitchin, R., and A. Fraser. 2020. *Slow Computing: Why We Need Balanced Digital Lives*. Bristol, UK: Bristol University Press.

Lupton, D., and A.A Watson. 2021. "Towards More-than-Human Digital Data Studies: Developing Research-Creation Methods." *Qualitative Research* 21 (4): 463–80. < https://doi.org/10.1177%2F1468794120939235/other>

Murakami Wood, D., and T. Monahan. 2019. "Platform Surveillance." *Surveillance & Society* 17 (1/2), 1–6. https://doi.org/10.24908/ss.v17i1/2.13237

Powell, A.B. 2021. *Undoing Optimization: Civic Action in Smart Cities*. New Haven, CT: Yale University Press.

Ruppert, E., E. Isin, and D. Bigo. 2017. "Data Politics." *Big Data & Society* 4 (2). https://doi.org/10.1177/2053951717717749

6 Articulating Urban Collectives with Data

JONATHAN GRAY AND NOORTJE MARRES

What is a city? Who lives there, and how? The life of a city may be gleaned from a wide range of accounts and artefacts of urban life: maps, conversations, music, histories, novels, newspaper clippings, photographs, architectural drawings, policy documents, statistical records, statues, road signage, and street furniture (Figure 6.1).

Throughout history, such items have been used to make sense of cities by artists, scholars, policymakers, and residents, with the effect of highlighting certain aspects and leaving others out, addressing some audiences and contexts rather than others. The schematic abstraction of the tube map displays sequences of stops for transport users (Vertesi 2005). Monuments organize public space by invoking historical events or establishing relations with famous figures (Rev 2005). Photographs portray scenes in accordance with touristic, political, or aesthetic conceptions of the city. The significance of such "conventionalized representations" (Becker 2007) depends on how they are used and who uses them. Data and data technologies can be involved in various ways in the production of such accounts, organizing relations and shaping the dynamics of cities, and they may also be deployed to intervene in these processes (Figure 6.2).

Records of city life have been kept for centuries, enabling different regimes for quantifying, monitoring, operating, managing, and ruling urban activity: counting city dwellers, recording who owns what, documenting births, deaths, marriages, and crimes.[1] Digital technologies have the potential to multiply the sites where data are created, but at the same time, as governments, organizations, and citizens increasingly generate their data digitally, the multiplicity of forms, genres, logics, and practices of data-making comes under threat. As digital data technologies are grafted onto existing urban infrastructures, architectures, and devices, ways of monitoring, analysing, intervening, and doing urban life are both altered and invented.

One distinctive feature of today's computational, networked technologies is that they enable the generation of data through the logging of *in situ* activity

Figure 6.1. Diorama of London and Panorama of London
Sources: Diorama, Sohei Nishino, 2010, © Sohei Nishino; Panorama, Claes Visscher, 1616

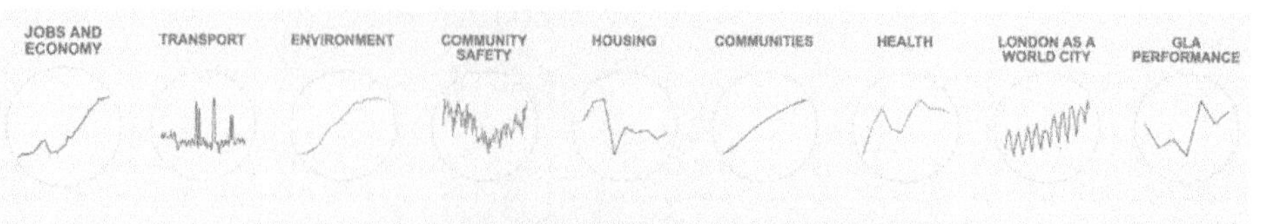

Figure 6.2. City Dashboard, London Datastore, 2018
Source: London Datastore, online at http://data.london.gov.uk

and interactions, by means of transport smartcards, mobile phones, motion sensors, credit card readers, smart energy grids, or the growing repertoire of network-enabled devices associated with the Internet of Things. These new kinds of digital uploads, pings, scans, pushes, swipes, and triggers make for a different kind of data than the records that are traditionally associated with bureaucracies: while data collection by bureaucracies generally relies on the identification of named individuals, the digital devices that are used and installed across cities today enable data capture pertaining to activities that are not necessarily tied to named individuals.

Urban data, then, can render visible shared patterns in activity *without* at the same time rendering transparent who is involved in these activities. As we will go on to note below, this affordance of digital data devices opens up distinctive opportunities for the organization of urban collectives through data. However, it does *not* necessarily follow from this that urban data generation is inherently collectivizing, or generative socially, culturally, and politically speaking. Many of the forms of *in situ* urban data capture that are practised today in digital cities produce individuating – and disciplining – effects that are associated with the bureaucratic gaze. Studies (for example, Brayne 2017) of digitally enabled neighbourhood policing have shown that the interpellation of named individuals *in the street* is one of the key affordances of data-driven policing noted by police officers such as those of the Los Angeles Police Department. These individuating, disciplining effects of the embedding of data technologies in the city do not only arise from institutional practices such as policing; they also affect wider social relations in the city. As one of us overheard one morning through an open window, while street sweepers paused in the road below: "I need to take a picture of the street every day now. I'm not sure I want to come to work anymore ... I'm not here to check on you, I am just here to hear how you're doing."

While the contexts of creating and using urban data might seem primarily practical and mundane, the visions and aesthetics associated with them can be sweeping, panoramic, perhaps even sublime (Davies 2015; Gray 2020a), as Figure 6.3 shows. Just as panoramas were said to give rise to new perspectives on

Figure 6.3. Selfiecity London
Source: Selfiecity London, online at http://selfiecity.net/london/

the city (Ford 2016), so today's data technologies provide apparently comprehensive insights into the intimate particulars of urban living: a sudden increase in "footfall" in a given street, a neighbourhood where the lights stay on at night longer than elsewhere. Data technologies suggest ways of viewing urban life at a distance as millions of transactions are depicted as constellations on sprawling maps or pulses in minutes of animated video.[2] The fruits of data capture end up in boardrooms and on dashboards, in exhibitions (e.g., the Museum of London) and coffee table books (e.g., *The Information Capital*), local government training workshops, and real estate developers' brochures. Many of these visualizations of urban data reduce the enactment of data collectives with urban data to a performative spectacle, a form of impression management (Currie 2020) in front of (imagined) audiences of decisionmakers, investors, and stakeholders. In this short text, we review what opportunities remain available, in this context of the widespread instrumentalization of data technologies for urban government, for critical and creative use of digital data to articulate urban collectives and issues.

Machines for Living Otherwise: Articulating Urban Collectives with Data

Contemporary data-driven and algorithmic governance can reinforce and reactivate past uses and abuses of data to exclude, exploit, and divide urban communities. One of the author's maternal grandparents arrived in San Francisco from Southeast Asia against a background of "redlining" – discriminatory lending practices that refused loans to "lower grade" or "undesirable populations," shaping who could live where.[3] These lending practices relied on various urban maps (Figure 6.4) that identified the neighbourhoods where lending was and was not available, which was determined based on the backgrounds of those who lived there. In London, it was recently found that the colonial-era term "Oriental" is apparently still being used in information systems of the Metropolitan Police.[4] As Wendy Chun (2021, xi) puts it, "eugenic and segregationist thinking have become integrated into our machines." The practice of mapping urban collectives is thus entangled with a history of the uptake of techniques of exclusion, repression, and segregation used against incoming and minority communities.

The idea of the city as machine can help us to make sense of recent histories of its digitalization. Cities in and of themselves have long been considered "machines for living otherwise," characterized by the movement and mixing of people, making possible new and non-traditional kinds of associations, cultures, and forms of coexistence. In line with this vision, artists, activists, and researchers have taken up data technologies to render the city and those who live with it more "aware," "responsive," or even "intelligent": to enable modes of dwelling in the city that are *more attuned* in terms of an openness to encounters with strangers and other inhabitants, *more expansive* in terms of the range of entities acknowledged as critical to the enactment of urban life, and, possibly, *more "creative"* as they enable feedback and active readjustment of relations between people, environments, and institutions (Halpern 2015).

In contrast to what sociologist Ulrich Beck called "methodological nationalism" – a way of framing society that is enabled by research methods such as the sample survey that assumes the national population as the dominant unit of analysis – the uptake of data-based computational methods across government, industry, science, activism, arts, and culture has inaugurated a turn towards "methodological urbanism." Computational arrangements for monitoring and analysing the city enable the inscription of new types of populations and collectives into data: through the logging of activity, the users of a particular service or participants in a public activity may assemble into data collectives (Marres 2017). For example, in the "Street-level City Analytics" project undertaken by the Citizen Data Lab and DensityDesign Lab, web and social media data pertaining to a particular street were used to elicit conversation and engagement around the future of the area with local residents (Niederer et al. 2015).

However, such projects also highlight how cities are implicated in wider circuits and regimes of data making that are coordinated from outside the city (e.g., through national agencies, transnational regulatory bodies, and multinational technology companies), rendering city dwellers as urban data subjects

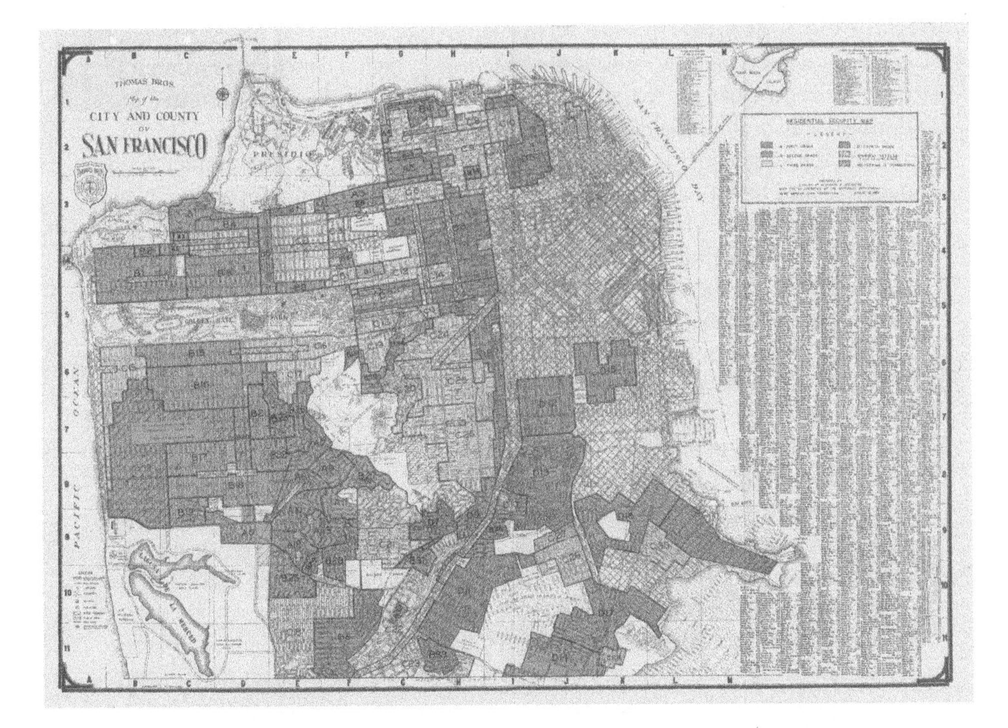

RESIDENTIAL SECURITY MAP

– LEGEND –

A - FIRST GRADE
B - SECOND GRADE
C - THIRD GRADE
D - FOURTH GRADE
SPARSELY SETTLED (Color Indicates Grade)
INDUSTRIAL & COMMERCIAL

PREPARED BY
DIVISION OF RESEARCH & STATISTICS
WITH THE CO-OPERATION OF THE APPRAISAL DEPARTMENT
HOME OWNERS' LOAN CORPORATION APRIL 15, 1937

Figure 6.4. Redlining in San Francisco, 1937, and Detail of the Legend
Source: Stanford University Center for Spatial and Textual Analysis, online at http://purl.stanford.edu/pc204zy5923

with active parts to play in realizing the "smart city." One of the dominant "governance engineering" narratives about the role of digital data in the city tempts us to envision cities as "living laboratories," a term that has been adopted enthusiastically around the world (Engels, Wentland, and Pfotenhauer 2019; Evans and Karvonen 2010).[5] This use of the lab metaphor invites us to understand our lives and living environments as open to experimentation, participation, and change through knowledge and awareness (Lezaun, Marres, and Tironi 2016). At the same time, the data arrangements that are implemented and legitimized under this banner tend to operate at scales beyond the city, in extended regional, transnational, and global infrastructures. In the face of this metaphorical appropriation of urban experimentation as part of transnational projects of infrastructure development, it is important to emphasize that the role of the city dweller in the city is very different from that of the laboratory subject in a lab: the urban laboratory, as it has been envisioned since the 1920s by sociologists from the Chicago School onwards, deviates from a scientific laboratory in crucial ways, not least in that the encounter with the unfamiliar in the city, as a condition for learning and conviviality, depends critically on the absence of central oversight, on immersion in a complex environment, and a *lack of control* over this environment (Gross 2009; Marres and Stark 2020).

Furthermore, a key feature of the progressive vision of the city as a laboratory for living is the reflexive capacities of participants, which derive from their experience of living the city *from the inside*: city dwellers make sense of their own lives by actively participating in collectives that exceed the patterns of behaviour that analysts identify in the populations they observe *from the outside*. While the appropriation of the living laboratory metaphor in infrastructure development tends to flatten this difference between a behavioural population and a reflexive collective, this does not in any way invalidate the uptake of the metaphor of the "city as machine" by experimental collectives for reflexive purposes: urban data offer opportunities for city dwellers to make sense of the city, and they can be appropriated in creative ways to make visible unobvious urban collectives in which we – inhabitants, passengers, visitors, and carers – participate.

Urban data do not just represent city life; they render it intelligible, experienceable, and actionable. Data are generated in the course of different forms of urban activity – travelling, visiting, managing, engineering, and selling things – that inform which entities, relations, and environments can be considered salient. Data enable transactions and boundaries, identifying "targets," aspirations, trends, and concerns. To succeed in the tasks assigned to them, data technologies must gather people as users, analysts, witnesses, and allies. For this reason, it has been suggested that infrastructures should be understood in terms of "relations," rather than as "things" (Star and Ruhleder 1996). In line with this, to appreciate how the work of assembly performed through urban data technologies can serve publics, not just governing populations, we should move beyond the metaphor of the "gaze." Having access to urban data is not the same as "seeing" the city. Today's urban data technologies enable the articulation of collectives by collectives, specifying the entities, relations between them, and modes of attunement and mutual intervention that make up the city. Data infrastructures can be understood not just optically, by thinking of lenses, telescopes, or microscopes, but also as coordination devices, which, through being used, shape and reassemble social life in specific ways.

Public Experiments with Urban Data

What, then, might urban digital data do for city dwellers? What counts and who decides? Who, how, and what do data infrastructures assemble, in what capacity and to what end? Could data and data technology enable different forms of participation in the city, of envisioning, coordinating, and transforming city life? As well as feeding narratives and projects of smartening and optimizing cities, data are also being used as the basis for other kinds of public experimentation.

Many experiments using Big Data have their parameters and questions fixed in advance. Data, however, can also be used to broaden and diversify the kinds of concerns, experiences, expertise, and collectives that are brought into research and enquiry. By involving urban constituents in the assembly and reassembly of data, they might become not just subjects of data capture, but also co-inquirers in experimentation: questioning which questions are asked and problematizing which problems are considered important. As well as enabling citizens to take positions on the issues of the day through the ballot box, data technologies can also play a role in eliciting, articulating, and evaluating new concerns.

In evaluating urban data experiments, however, it is key *not* to consider them only from the perspective of participation. Urban subjects figure in multiple roles in data experiments, and we should differentiate among users, subjects, beneficiaries, and those affected by "knowing data." That is, it is a mistake to equate involvement in data experiments with active participation in them, as this risks reproducing the exploitative model in which public engagement with urban experiments is designed to create value through data generation for external agencies. In today's context, participation is increasingly used as an opportunity to generate data, and it thus is instrumentalized, mobilized as a *resource* for industry and government (Kelty 2019; Powell 2021). As a consequence, digital participation might become one of the *least* productive modes of engagement in urban experimentation.

A primary aim of collective experimentation with urban data today is the articulation of alternative modes of engagement – the trying out of models of digital experimentation in the urban environment that do not reproduce the instrumentalism of smart city business models or Big Data analytics consultancies. Instead, the goal is to curate occasions for *in situ* enquiry and attunement, for recomposing collectives and respecifying conditions of relevance for the (re)production of city life. Urban experimentation with digital devices could serve as a way to assemble, attend to, and imagine responses from within the city to urban issues across space, time, and social settings. Could urban experiments with data be a way to diversify how issues, things, and actors are made relevant, salient, relatable *in situ*?

While data technologies, techniques, and infrastructures today serve the ends of government, management, and commerce (e.g., surveys, advertising, nudge), there are many alternative ways in which data can be put to work to civic ends. Some projects inventively repurpose official data as an invitation to care for urban spaces, infrastructure, and non-human city dwellers. For example, Code for America's Adopt-a-Hydrant project enables "community members to volunteer to take care of local infrastructure like fire hydrants in severe weather."[6] The project is open source, and the code has been used in other cities, giving rise to other projects looking after trees, paths, drains, and emergency sirens. Such projects signal a deeply worrying reluctance on the part of some governments to take responsibility for running the city. "Adopt-a" models can be traced back to cash-strapped cities in the United States, where such models were critiqued for treating residents as "public servants of last resort" (Cheyfitz 1980). But given how difficult it can be to force the state to care, such projects also demonstrate the commitment of citizens to make the environment liveable otherwise, in ways that are doable under conditions of neoliberal governance, austerity, inequality, disaster, and rolling crises. Using a similar mechanism, the 596 Acres initiative repurposes official data to encourage city dwellers to (re-)appropriate land, lots, and other spaces, bringing them into "resident stewardship," by highlighting potential sites and facilitating community organizing efforts to gain access (Figure 6.5).[7]

Official data can also be repurposed to address different questions and concerns than those for which they were intended – from gathering memories on living with trees in Bogotá to seeing how segregation is operative in the United States (Bounegru and Gray 2021). In New York, data originally gathered to monitor crime have been repurposed by journalists and activists to highlight problematic and discriminatory policing practices (Didier 2018). In the United Kingdom, a network of local data journalists has gathered data from PDF documents to identify and report on cuts to council spending.[8] In Manchester, the TaxHack initiative has combined data on public procurement with company ownership information to identify which city contractors use tax havens.[9] Activists have used scraped data from Airbnb's website to show how official datasets had been "photoshopped" to remove unwanted listings.[10]

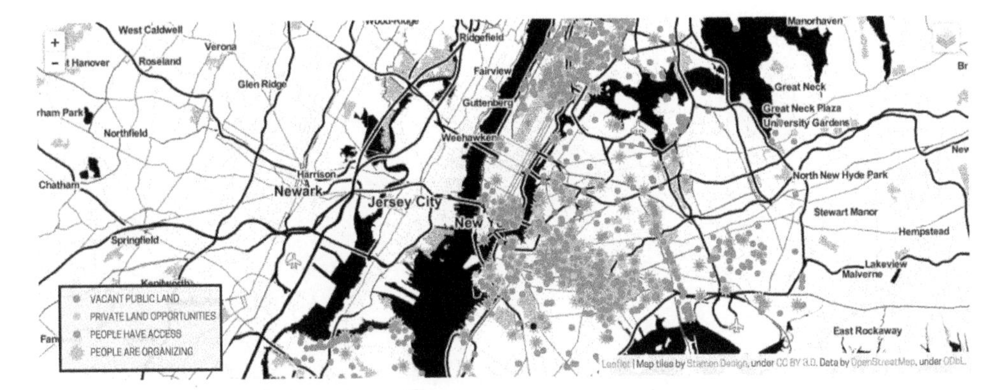

Figure 6.5. Living Lots NYC Project by 596 Acres
Source: Living Lots NYC by 596 Acres, online at https://livinglotsnyc.org/

Other projects make their own data. Perhaps one of the best-known examples in the UK context is the FixMyStreet project, which maps and reports "street problems to the councils responsible for fixing them."[11] Another project, called Doodoowatch, made headlines with its "poo-shaming map," mobilizing residents to highlight and respond to dog fouling.[12] In Chicago, a similar map-based reporting system is used to highlight the lack of residential recycling facilities.[13] These projects demonstrate how data reuse works as a form of issue articulation, surfacing, activating, and acting on collective concerns through the assembly of databases, interfaces, and infrastructures (Gray, Gerlitz, and Bounegru 2018).

As well as articulating local issues, distributed data-gathering operations can also be used to create what anthropologist Helen Verran describes as "enumerated entities": matters of concern that arise through diverse practices of numbering, counting, and datifying (Verran 2015). Participatory counting initiatives enable the enumeration of aspects of city life that might be overlooked or underrepresented in official data, including urban ecology (such as trees,[14] hedgehogs,[15] butterflies,[16] and bees[17]), cyclists and pedestrians,[18] wheelchair-accessible features,[19] public space usage,[20] and rough sleepers.[21] Such data can subsequently be mobilized to demand institutional recognition and response to issues and matters of concern (Gray, Lämmerhirt, and Bounegru 2016). These projects are not exhausted by the numbers, lists, or rows that are produced. For example, the attention to details required for biodiversity-counting exercises or for noting which parts of the city are not wheelchair accessible might cultivate sensibilities for noticing these things *in situ*, beyond data-gathering exercises.

Finally, data can enable different ways of experiencing and envisioning cities. Experiments in "citizen sensing" explore public involvement in the use of digital sensors to attend to environmental phenomena such as pollution, flora and

Figure 6.6. Yellow Dust Project, Seoul
Source: Yellow Dust, online at http://yellowdust.intheair.es/; photo courtesy of Nerea Calvillo.

fauna, and damp in homes (Calvillo 2018; Gabrys 2016; Powell 2021).[22] Other projects aim to materialize data beyond the screen, such as the Yellow Dust project (Figure 6.6),[23] which translated air-quality measurements into vapor clouds of yellow mist in Seoul, South Korea. Practices of "data walking" (van Es and de Lange 2020) encourage collective enquiry into urban datafication, including observing the role of data infrastructures in the shaping of city life (e.g., credit-card-based bike lockers), as well as reflecting on what data are and how they might enable doing things differently (e.g., simplifying access to public facilities).[24] In "listening walks," field recordings and data mapping are combined to enable different modes of listening to the dynamics of urban life – such as how a street sounded before and after the pandemic or through the sonification of electromagnetic signals of networked domestic devices such as babyphones that are normally outside the range of human hearing.[25]

Urban Experiments as Critical Sites: Which Data? What Collectives? Whose City?

To what ways of relating – what modes of engagement – do the urban data experiments foregrounded here give rise? What kinds of collectivity do they support and enact? Even as urban experiments involving data devices are being used to manage and enact administrative, economic, and technical schemes, they enlist a widening range of actors: citizens as sensors, auditors, entrepreneurs, witnesses, hackers, users, custodians, carers, listeners, and investigators (Gray 2020b). These emergent collectives suggest different models for the ownership and governance of data, from decentralized data platforms to open-data commons. These forms of collectivization are at risk of appropriation, and the social, moral, and political ends they serve might shift in the process of their implementation. This is one reason it is vital to describe the different forms of experimental knowledge and politics that are enabled through the articulation of urban collectives with data.

Urban data technologies present a critical site of experimentation in rendering cities legible, inhabitable, and liveable on the level of the collective: as the arrangements for articulating urban collectives with data technologies continue to be captured, appropriated, and repurposed by a variety of actors, we must analyse critically not only by whom and to what end urban experiments are organized, but also the *methodologies* for the articulation of urban collectives with data they implement. Amid growing concerns about the societal, political, and environmental costs of privatization, surveillance, discrimination, extractivism, and exploitation of cities enabled by platforms and algorithms (Chun 2021; Couldry and Mejias 2019a, 2019b; Helmond 2015; Zuboff 2019), the disability activist slogan "nothing about us without us"[26] takes on an additional, different significance: with us, but in what capacity? Digital data technologies require critical and creative engagement if they are to be implemented responsibly and generatively. More than that, they open up the very methods of articulating with data who and what dwells in the city, and whom and what it is for.

NOTES

1 Sociologists have studied how these types of urban data arise from bureaucratic, scientific, and social practices of classifying and "sorting things out" (Bowker and Star 2000; Bruno, Jany-Catrice, and Touchelay 2016; Espeland and Stevens 2008; Rottenburg et al. 2015).

2 Visualizations of London's transport systems are a good example: J. Gordon, "London in Motion," online at https://www.youtube.com/watch?v=4FrnF2HlBGg; and O. O'Brien, "Oyster Card Touch Ins & Touch Outs," map animation, online at https://www.youtube.com/watch?v=QQV3UHsZ_u4

3 Hoodline, "A History of Redlining in San Francisco Neighborhoods, online at https://hoodline.com/2014/06/a-history-of-redlining-in-san-francisco-neighborhoods/; R. Marciano, D. Goldberg, and C. Hou, "T-RACES: a Testbed for the Redlining Archives of California's Exclusionary Spaces," online at https://web.archive.org/web/20150902044004/http://salt.umd.edu/T-RACES/mosaic.html

4 End Violence and Racism against East and Southeast Asian Communities, "Data on Hate Crimes against ESEA People in the UK," online at https://evresea.com/data
5 See, for example, the European Network of Living Labs, online at https://enoll.org/
6 See Adopt-a-Hydrant, online at http://www.adoptahydrant.org/
7 See 596 Acres, "Championing Resident Stewardship of Land to Build Just and Equitable Cities," online at 596acres.org
8 See Bureau of Investigative Journalism, "Is Your Council about to Cut a Vital Local Service? Help Us Find Out," online at https://www.thebureauinvestigates.com/blog/2017-12-19/is-your-council-about-to-cut-a-vital-local-service
9 TaxHack, "Open Data Day Manchester 2014 – TaxHack Review and Findings." #ODD14 #taxjustice #opendata, online at https://taxhack.wordpress.com/
10 See M. Cox and T. Slee, "How Airbnb's data hid the facts in New York City," online at http://insideairbnb.com/reports/how-airbnbs-data-hid-the-facts-in-new-york-city.pdf
11 See FixMyStreet, "Report, View, or Discuss Local Problems," online at https://www.fixmystreet.com/
12 See Hunt (2018); and BBC News, "'Poo-shaming' map hopes to tackle dog fouling," 10 March 2018, online at https://www.bbc.co.uk/news/uk-england-cambridgeshire-43334730
13 See C. Micklin, B. Wilhelm, and A. Kahn, "Open City, My Building Doesn't Recycle," online at http://mybuildingdoesntrecycle.com/
14 Opentreemap, "Inventory Your Trees," online at https://www.opentreemap.org/
15 See Warwick (2017).
16 Butterfly Conservation, "Big Butterfly Count," online at https://bigbutterflycount.butterfly-conservation.org/
17 Friends of the Earth, "Great British Bee Count," online at https://friendsoftheearth.uk/bee-count/great-british-bee-count-2017-map
18 LACBC, LA Bike + Ped Count, https://web.archive.org/web/20180311081524/http://la-bike.org/our-work/bike-ped-count/
19 Open Street Map Blog, "Disability Mapping with OpenStreetMap,: online at https://blog.openstreetmap.org/2013/12/03/disability-mapping-openstreetmap/
20 Gehl, "Do You Live in a Bubble?" online at https://gehlpeople.com/tools/public-life-data-protocol-beta/
21 See Gentleman and Butler (2017).
22 See, for example, Citizen Sense, https://citizensense.net/, http://making-sense.eu/; Knowle West Media Centre, "The Bristol Approach," online at http://kwmc.org.uk/projects/bristolapproach/
23 C+arquitectos/In The Air, "Yellow Dust," online at http://yellowdust.intheair.es/
24 See Data Walking UK, "Data Walking," online at https://www.datawalking.uk/
25 See, for example, MIT Senseable City Lab, "Sonic Cities," online at http://senseable.mit.edu/sonic-cities/; Warwick University, Centre for Interdisciplinary Methodologies, "Sampling Sounds of the Future," online at https://warwick.ac.uk/fac/cross

_fac/cim/research/sampling-sounds-of-the-future/projects/. For more listening projects, see Data Sonification Archive, online at https://sonification.design/

26 See Wikipedia, "Nothing About Us Without Us," online at https://en.wikipedia.org/wiki/Nothing_About_Us_Without_Us

REFERENCES

Becker, H.S. 2007. *Telling about Society*. Chicago: University of Chicago Press.

Bounegru, L., and J. Gray. 2021. *The Data Journalism Handbook: Towards a Critical Data Practice*. Amsterdam: Amsterdam University Press.

Bowker, G.C., and S.L. Star. 2000. *Sorting Things Out: Classification and Its Consequences*. Cambridge, MA: MIT Press.

Brayne, S. 2017. "Big Data Surveillance: The Case of Policing." *American Sociological Review* 82 (5): 977–1008. https://doi.org/10.1177/0003122417725865

Bruno, I., F. Jany-Catrice, and B. Touchelay, eds. 2016. *The Social Sciences of Quantification: From Politics of Large Numbers to Target-Driven Policies*. New York: Springer. http://www.springer.com/gp/book/9783319439990

Calvillo, N. 2018. "Political Airs: From Monitoring to Attuned Sensing Air Pollution." *Social Studies of Science* 48 (3): 372–88. https://doi.org/10.1177/0306312718784656

Cheyfitz, K. 1980. "Self-service: The city that governs least governs best." *New Republic*, 15 November.

Chun, W. 2021. *Discriminating Data: Correlation, Neighborhoods, and the New Politics of Recognition*. Cambridge, MA: MIT Press.

Couldry, N., and U.A. Mejias. 2019a. *The Costs of Connection*. Stanford, CA: Stanford University Press.

Couldry, N., and U.A. Mejias. 2019b. "Data Colonialism: Rethinking Big Data's Relation to the Contemporary Subject." *Television & New Media* 20 (4): 336–49. https://doi.org/10.1177/1527476418796632

Currie, M. 2020. "Data as Performance – Showcasing Cities through Open Data Maps." *Big Data & Society* 7 (1). https://doi.org/10.1177%2F2053951720907953

Davies, W. 2015. "The Data Sublime." *New Inquiry*, 12 January. Online at http://thenewinquiry.com/essays/the-data-sublime/

Didier, E. 2018. "Globalization of Quantitative Policing: Between Management and Statactivism." *Annual Review of Sociology* 44: 515–34. https://doi.org/10.1146/annurev-soc-060116-053308

Engels, F., A. Wentland, and S.M. Pfotenhauer. 2019. "Testing Future Societies? Developing a Framework for Test Beds and Living Labs as Instruments of Innovation Governance." *Research Policy* 48 (9): 103826. https://doi.org/10.1016/j.respol.2019.103826

Espeland, W.N., and M.L. Stevens. 2008. "A Sociology of Quantification." *European Journal of Sociology / Archives Européennes de Sociologie* 49 (3): 401–36. https://doi.org/10.1017/S0003975609000150

Evans, J., and A. Karvonen. 2010. "Living Laboratories for Sustainability: Exploring the Politics and Epistemology of Urban Transition." In *Cities and Low Carbon Transitions*, ed. H. Bulkeley, V.C. Broto, M. Hodson, and S. Marvin, 142–57. London: Routledge.

Ford, L. 2016. "'Unlimiting the bounds': The Panorama and the Balloon View." *Public Domain Review*, 3 August. Online at https://publicdomainreview.org/essay/unlimiting-the-bounds-the-panorama-and-the-balloon-view, accessed 7 June 2021.

Gabrys, J. 2016. *Program Earth: Environmental Sensing Technology and the Making of a Computational Planet*. Minneapolis: University of Minnesota Press.

Gentleman, A., and P. Butler. 2017. "'This is their bedroom': Counting rough sleepers in England's homeless hotspots." *Guardian*, 22 December. Online at https://www.theguardian.com/society/2017/dec/22/this-is-their-bedroom-counting-rough-sleepers-england-homeless-hotspots

Gray, J. 2020a. "The Data Epic: Visualisation Practices for Narrating Life and Death at a Distance." In *Data Visualization in Society*, ed. H. Kennedy and M. Engebretsen, 313–28. Amsterdam: Amsterdam University Press.

Gray, J. 2020b. "The Datafication of Forests? From the Wood Wide Web to the Internet of Trees." In *Critical Zones: The Science and Politics of Landing on Earth*, ed. B. Latour and P. Weibel, 362–69. Cambridge, MA: MIT Press.

Gray, J., C. Gerlitz, and L. Bounegru. 2018. "Data Infrastructure Literacy." *Big Data & Society* 5 (2): 1–13. https://doi.org/10.1177/2053951718786316

Gray, J., D. Lämmerhirt, and L. Bounegru. 2016. *Changing What Counts: How Can Citizen-Generated and Civil Society Data Be Used as an Advocacy Tool to Change Official Data Collection?* CIVICUS and Open Knowledge International. https://doi.org/10.2139/ssrn.2742871.

Gross, M. 2009. "Collaborative Experiments: Jane Addams, Hull House and Experimental Social Work." *Social Science Information* 48 (1): 81–95. https://doi.org/10.1177/0539018408099638

Halpern, O. 2015. *Beautiful Data: A History of Vision and Reason since 1945*. Durham, NC: Duke University Press Books.

Helmond, A. 2015. "The Platformization of the Web: Making Web Data Platform Ready." *Social Media + Society* 1 (2): 1–11. https://doi.org/10.1177/2056305115603080

Hunt, E. 2018. "Doodoo watch: A crowdsourced solution to our cities' dog mess minefield?" *Guardian*, 4 May. Online at https://www.theguardian.com/cities/2018/may/04/poo-patrol-how-doodoowatch-could-solve-our-cities-dog-mess-problems

Kelty, C. 2019. *The Participant: A Century of Participation in Four Stories*. Chicago: University of Chicago Press.

Lezaun, J., N. Marres, and M. Tironi. 2016. "Experiments in Participation." In *Handbook of Science and Technology Studies*, 4th ed., ed. U. Felt, R. Fouche, C.A. Miller, and L. Smitt-Doer, 195–222. Cambridge, MA: MIT Press.

Marres, N. 2017. *Digital Sociology: The Reinvention of Social Research*. London: Polity Press.

Marres, N., and D. Stark. 2020. "Put to the Test: For a New Sociology of Testing." *British Journal of Sociology* 71 (3): 423–43. https://doi.org/10.1111/1468-4446.12746

Niederer, S., G. Colombo, M. Mauri, and M. Azzi. 2015. "Street-level City Analytics: Mapping the Amsterdam Knowledge Mile." In *Hybrid City 2015: Data to the People*, ed. I. Theona and D. Charitos, 215–20. Athens: URIAC.

Powell, A. 2021. *Undoing Optimization: Civic Action in Smart Cities*. New Haven, CT: Yale University Press.

Rév, I. 2005. *Retroactive Justice: A Prehistory of Post-Communism*. Stanford, CA: Stanford University Press.

Rottenburg, R., S.E. Merry, S-J. Park, and J. Mugler, eds.2015. *The World of Indicators: The Making of Governmental Knowledge through Quantification*. Cambridge: Cambridge University Press.

Star, S.L., and K. Ruhleder. 1996. "Steps toward an Ecology of Infrastructure: Design and Access for Large Information Spaces." *Information Systems Research* 7 (1): 111–34. https://doi.org/10.1287/isre.7.1.111

van Es, K., and M. de Lange. 2020. "Data with Its Boots on the Ground: Datawalking as Research Method." *European Journal of Communication* 35 (3): 278–89. https://doi.org/10.1177/0267323120922087

Verran, H. 2015. "Enumerated Entities in Public Policy and Governance." In *Mathematics, Substance and Surmise*, ed. E. Davis and P. Davis, 365–79. New York: Springer. https://doi.org/10.1007/978-3-319-21473-3_18

Vertesi, J. 2005. "Mind the Gap: The Tube Map as London's User Interface." Conference proceedings, Human Factors in Computing Systems (CHI 05) workshop on "Engaging the City." Online at https://www.semanticscholar.org/paper/Mind-The-Gap-%3A-The-%E2%80%98-Tube-Map-%E2%80%99-as-London-%E2%80%99-s-User-Vertesi/75f8acafd96663ca368442d1cd5f55dc125a1c6e, accessed 7 June 2021.

Warwick, H. 2017. "There's a way to save hedgehogs – and all of us can help." *Guardian*, 15 August. Online at https://www.theguardian.com/commentisfree/2017/aug/15/citizen-scientists-map-wildlife-hedgehog-housing-census-vital-difference

Zuboff, S. 2019. *The Age of Surveillance Capitalism: The Fight for a Human Future at the New Frontier of Power*. London: Profile Books.

7 Coding Out Justice: Digital Platforms' Enclosure of Public Transit in Cities

TORIN MONAHAN

In May 2019, Uber enfolded the City of Denver's transit services into its smartphone app. Heralded as a synergistic partnership that would improve customer convenience and possibly increase ridership on public transit, the collaboration generated enormous amounts of publicity in news outlets throughout the United States and beyond (see, for example, Conger 2019; Saltman 2019). A dramatic black commuter train, which was used as a prop for press events, marked the transition with a proclamation in bold white font, "This train, now available on Uber" (Bosselman 2019a) (see Figure 7.1). The moment signalled the next step in Uber's goal – in the words of the company's CEO – of becoming the "Amazon for Transportation" (Recode 2018).[1] Uber's growth strategy thus hinges not on ride hailing or autonomous vehicles alone, but instead on becoming the dominant platform facilitating all transportation exchanges in urban spaces. With similar transit partnerships afoot in Sydney, Cairo, New Delhi, and other cities around the world (Sisson 2019), as well as with the company's foray into micromobility services such as bike and scooter rentals (Giambrone 2019; Wilhelm 2019), it is clear that these totalizing moves are central to Uber's plans for long-term success. The implications of these trends for cities, however, and especially for social justice in cities, are deeply troubling.

This chapter explores the question of what happens when city functions are captured by private digital platforms and aligned with their capitalist logics. As smart urbanism discourses direct attention to the technological refashioning of cities, even critiques of those *smart* visions can neglect adjacent and more insidious transformations taking place in the provision of services for people (Goodman and Powles 2019). When platform companies such as Uber partner with cities to enfold public transit options into their apps, they establish themselves as obligatory passage points for individuals seeking city information or accessing city services. In the process, private platforms are rapidly becoming the interfaces through which people experience the city and themselves in it: citizens are transformed into "users" who can be steered towards

Figure 7.1. Uber Ad on a Denver Regional Transportation District Train, 2019
Source: Photo courtesy of Lamar Advertising Company.

private transportation options while generating massive amounts of profitable data to assist companies with their further monopolization efforts. Drawing upon a two-year study of large-scale digital platforms in US cities,[2] this chapter approaches platforms as components of smart urbanism developments and probes their effects upon city services and functions. I argue that potentials for justice are eroded through the platformization of cities, particularly as rights are recast as consumer choices and collective resources are depleted.

The Conceptual Frame

A great deal of scholarship has focused on the ways digital platforms operate as infrastructures that impose protocols on interaction and extract profit from the exchanges they enable (Plantin and Punathambekar 2019; Srnicek 2017; van Dijck, Poell, and de Waal 2018). A complementary awareness has been growing, though, about the capacity of private platforms to extend into and poach upon existing *public* infrastructures, converting them, in the process, into resources for capitalist expansion (Murakami Wood and Monahan 2019; Plantin et al. 2018; Sadowski 2020a). In some respects, this can be seen as a continuation of long-standing neoliberal patterns in "splintering urbanism," where collective services and goods are unbundled as public assets and rebundled as scarce commodities sold to those who can afford them (Graham and Marvin 2001; see also Marvin and Luque-Ayala 2017). Although such (re)distributions produce profit, they also exacerbate social inequalities by stripping poorer communities of resources necessary for survival (e.g., electricity, clean water) and fuelling a racialized policing apparatus against those communities (Benjamin 2019; Monahan 2017), all the while offering augmented amenities and services to the relatively affluent and white (Cardullo, Di Feliciantonio, and Kitchin 2019; Heaphy and Wiig 2020; Monahan 2018).

In the present moment, smart urbanism developments are characterized by similar platform logics with an emphasis on data extraction and mobilization to achieve control over city systems, function, and populations (Barns 2020; Kitchin 2014; Sadowski and Pasquale 2015; Shapiro 2020).[3] Platform operators – be they Uber, Amazon, Airbnb, or others – are engaged in agonistic struggles over data in order to maintain dominance or sustainability. For instance, Uber and Lyft are notoriously uncooperative in sharing data with city transportation departments (Monahan 2020), but they are perfectly willing to capture city data for their own gain, as the example of the Denver partnership suggests. Viewed through the prism of "the platform," data are necessary to achieve organizational missions, so primacy is given to producing, acquiring, and capitalizing upon those data. This includes efforts to mould *data subjects* (through gamification, convenience, threats of exclusion) who gravitate to platform interfaces and filter their activities through those interfaces (Gabrys 2014; Lupton 2015; Smith 2018). As Sarah Barns writes, "by instituting relational dynamics between code, commerce, and corporeality, platforms remediate the 'technological everyday' in powerful ways" (Barns 2019, 1). Such subject entrainment occurs, in other words, not through conspiracy, but instead by design, as a product of systems structured to encourage, or script, those responses (Akrich 1992; Schüll 2012; Winner 1980). The net result might be the wide-scale, undemocratic refashioning of cities and their populations in the data images that harmonize best with the objectives of platform companies. Jathan Sadowski (2020b, 1) posits that such a development could portend "fundamental shifts in urban sovereignty as technology companies move beyond treating the city merely as a place to extract value from and start thinking of it as also a space to exercise dominion over." In light of how patterns of splintering urbanism historically have harmed economically precarious and racialized groups disproportionately, such shifts in urban sovereignty, even if only partially successful, would impede social justice.

In the realm of urban mobility in particular, platformization processes introduce additional challenges for city planners, engineers, and others struggling to meet the needs of people. In the US context, for instance, the encroachment of platform-based ride-hailing services has catalyzed a *downward spiral in transit*: "decreased public transit use leads to decreased funding, which in turn leads to fewer transit options and makes platform alternatives more attractive" (Monahan 2020, 2). The current COVID-19 pandemic, while also negatively affecting ride-hailing companies, is further decimating transit systems that are facing billions of dollars in lost revenue from passengers, while simultaneously needing to operate as if they were close to full capacity in order to allow for adequate social distancing (Gelinas 2020). Whereas cities throughout the world have been gradually confronting the negative effects of ride hailing (e.g., congestion, pollution, labour exploitation) and attempting to push back through regulatory means (Collier, Dubal, and Carter 2018; Griswold 2019), the data hegemony of platform companies hinders these efforts. As John Stehlin and colleagues explain: "If the data needed to govern urban mobility is increasingly accumulated, processed, and packaged by private entities, particularly evident with route-planning apps like CityMapper and Uber's "Uber Movement" dashboard, then democratic governance of infrastructure planning is likely to be eroded" (Stehlin, Hodson, and McMeekin 2020, 13). This trend is advanced through the emerging platform modality of "Mobility as a Service" (MaaS), wherein platforms like those operated by Uber seek to manage and interlink all existing transportation modes (e.g., bus, bike, ride hailing, train) to organize users' entire trips from origin to destination (Cottrill 2020). The example of the Denver partnership is one such articulation of MaaS, especially because the presumption is that, even if riders use public transit, they will also take Uber ride-hailing vehicles or micromobility rentals to connect the "first-mile, last-mile" segments of their trips. As cities enter into such partnerships, however, they contribute to a "concerning dynamic where city administration becomes 'locked in' to specific corporate products and interests, and thereby 'locked out' from alternatives" (Lee et al. 2020, 116). Given that transit services are vital for correcting mobility inequities in cities and providing a mobility safety net for the poor, people of colour, the elderly, and those with disabilities (Clark 2017; Lubin and Deka 2012), recasting public transit as simply one MaaS option within a larger, privately owned platform ecosystem might threaten the public-good orientation of transit, especially as transit systems begin to adapt to these private platforms.

Platform Enclosures of Transit

The developments underway with platforms' colonization of public spaces and services can be understood as aspects of more general processes of *digital enclosure*. As Mark Andrejevic (2007, 2) explains, digital enclosure implies the capture of public resources within a private "interactive realm wherein every action and transaction generates information about itself." Such enclosures enforce divisions between private owners who set the terms for access and use and individuals who "submit to particular forms of monitoring in order to gain access to goods, services, and conveniences" (3). Digital enclosures are not stable achievements, but rather ongoing struggles over the definition, management, and commodification of "the public." While public institutions might challenge and resist these efforts, they also often accept the terms of competition for "customers" and adopt enclosure efforts themselves – or outsource to (or partner with) private entities. As Gregory Donovan (2020, 6) relates, the dominant approach is one of constructing *proprietary ecologies* that "help corporations and governments to occupy the interface between people and places so as to commodify both sociospatial interactions and the data that can be extracted from them."[4] Efforts to impose a proprietary frame over sociospatial actions and interactions are at the root of Uber and other ride-hailing companies' various partnerships with public transit systems.

In keeping with the monopolistic tendencies of large-scale digital platforms, partnerships like Uber's experiment with Denver's Regional Transportation District (RTD),[5] should be viewed as efforts to enclose and appropriate a public service (i.e., transit) that the company has explicitly admitted is its competition (Bosselman 2019a). To achieve this goal, such companies build all-encompassing systems – MaaS – that become what Michel Callon (1984) would call *obligatory passage points* for customers, even if those customers are primarily interested in public transit options. In the words of an Uber spokesperson celebrating the Denver partnership: "With this step, we are moving closer to making Uber's platform a one-stop shop for transportation access, from shared rides to buses and bikes" (Kbidy 2019). One city planner in my study compared this to the model previously adopted by cable companies: "Uber and Lyft are trying to become the cable companies of transportation. It's why they are buying bike-share companies; it's why they are buying scooter-share companies; it's why they are investing in these municipal and agency contracts [as in Denver], because *they are trying to create a privatized world where all of your transportation goes through them*" (emphasis added).

To the extent they succeed in these efforts, platform companies can benefit parasitically from the labour and infrastructural investments of the public sector without incurring any associated costs. Moreover, public transit can become wrapped in relations of dependency with platform companies such that their sustainability could be tied to the market dominance of the platform companies to which they are subordinated.

Once users are routed through a central private platform for all transportation choices, they become a captive audience whose behaviour can be shaped by algorithmic protocols. One city planner explained:

> Uber and Lyft are starting to provide information about transit ... which I think is in part to sort of continue *to own the customer and to own their experience*. So that even if you are going to take a transit service, you're doing so because Uber told you to, not because you made the decision ... The other thing too, just to keep in mind, is that when Uber tells you to take transit, it is first offering you a bunch of other options, and then you decided to take transit. So, they were able to advertise to you as best as they could before you chose a competitor, so to speak. (emphasis added)

As has been well documented with other digital platforms, the algorithms that shape users' experience are far from neutral, even if they mask their underlying prejudices and politics (Benjamin 2019; Monahan 2016; Noble 2018). In this instance, my informant is highlighting how the app interface prioritizes the most profitable transportation modes for the company, thereby steering users to value and select those modes. Even if transit options *are* inferior to other offerings, if some riders who previously would have selected transit instead choose a different mode, this only accelerates transit's downward spiral.

The Denver-Uber partnership reveals some of the frictions of platform enclosure, especially with respect to the different missions of private and public sector organizations. As one story critical of the partnership provocatively asked: "Why would RTD partner with a company that will show you, in real time, how much faster you would get to work by clicking the 'send to Uber' button versus pushing the 'tell me about my lame bus options' button" (Bosselman 2019b). On one hand, the answer appears to be an attempt to distinguish the city as especially innovative (Kbidy 2019), which is a pattern seen with other smart city publicity stunts (Monahan 2018; Sadowski and Bendor 2019); on the other hand, the city might have been attempting to slow transit's downward spiral by curtailing the loss of riders to ride hailing (Conger 2019). The early ridership data did not bear this out: "Uber sold 1,217 RTD tickets through its app [in two months], according to RTD. The number is small compared to the agency's 97.6 million annual boardings ... RTD could find itself outfoxed and losing riders to the Silicon Valley Goliath [Uber] whose drive for profit aligns poorly with the goals of the region RTD serves" (Bosselman 2019a). My informants in city transportation departments described this struggle as one over the public good: "I think the bottom- line vision of having integrated transportation services that are not siloed is a good one, but I think the question is, do we really want that to be in the hands of private companies? ... Are we giving them [platform companies] access to our public right of way in a way that doesn't benefit the public good?" Although the partnership did generate impressive publicity and has improved upon previous ticket-purchasing interfaces for Denver's transit riders (Bosselman 2019b), if viewed as a longer-term struggle over the public good and the right to the city, then platform enclosure undoubtedly cedes territory to private companies and their agendas.

To return to the data-extraction element of digital enclosures, by routing possible transit riders through their app, Uber is able to capture and capitalize upon those data. With such data, they can construct highly granular user profiles; analyse transport choices by demographics, time of day, day of week, weather conditions, and competing services nearby; adjust pricing in real time to guide users to options that are more profitable for the company; and concoct longer-term strategies for ensuring market (and mode) dominance. Thus, there is an anticipatory or future-oriented dimension to data analysis. As a city planner in my study reflected, this anticipatory modality can be operational as a form of passive surveillance, even in the absence of users' selecting a transportation option: "The data that these companies are collecting is fascinating. And one of the things that they're able to collect is when you open the application – it pings their servers. So, they not only know how much you are riding on the Uber app generally, they also know when you've opened and closed the application. So, they basically understand latent demand in addition to actual, observed behaviour." In sum, irrespective of whether partnerships such as the one in Denver increase or decrease transit riders, they are forms of *platform surveillance* that generate data that can be used to solidify platform companies' proprietary ecologies and secure platform dominance.[6] They allow for the further directed influence of data subjects to align their preferences and behaviours to platform protocols and values. In the process, however, cities become less equipped to meet the mobility requirements of their neediest residents and less able to ensure – or work towards – social justice in the city.

Conclusion

With the majority of the world's population now residing in urban settings (Ritchie and Roser 2019), it is vital that cities provide mobility services that meet the basic needs of all residents and visitors. In the absence of adequate transportation options, the most marginalized members of society suffer, be they the elderly, the young, the poor, ethnic minorities, or people with disabilities (Hine 2012). Social justice in cities depends upon maintaining a transportation safety net, at the very least, but the provision of a robust and ecologically friendly transit system that offers equal service for all riders would be even better. With the encroachment of ride-hailing services and their emphasis, to date, on automobility, many cities have been struggling to combat the negative externalities of these platforms (congestion, pollution, labour exploitation, etc.), while also contending with lost revenue and associated transit reductions. In the present crisis sparked by the COVID-19 pandemic, threats to basic transit services are amplified in the United States and elsewhere, as transit departments face massive budget shortfalls at a time when many "essential workers" – a large percentage of whom are racial minorities who experience higher mortality rates (Godoy and Wood 2020; Rho, Brown, and Fremstad 2020) – rely on public transit to get to work (George 2020). The basic transit safety net, in other words, is fraying, and the most vulnerable members of society are suffering because of it.

In this context, the platform-enclosure moves of companies such as Uber further threaten the long-term viability of urban transit systems and their capacity to provide equitable service. As transit is enfolded into private MaaS apps, transit riders are captured and their experiences are modulated by the platforms. Riders are converted into profit- and data-generating customers, and transit is recast as a subordinated option in a larger portfolio of transportation modes. It has been theorized that earlier crises in capital accumulation were postponed through various spatial fixes of colonizing other territories or bodies (Harvey 2001). Similarly, platform enclosure of transit is an articulation of data capitalism that affords a "modal fix" that colonizes publicly managed transportation modes to forestall companies' reckoning with their own fiscally insecure and unprofitable operations. Ultimately, urban transit systems could become a casualty of these arrangements or be transformed into systems that cater to elite riders (e.g., those taking commuter trains to work) while they reduce lines and routes for the poor (e.g., on buses). When cities such as Denver commit to industry partnerships of the sort described here, they strike a Faustian bargain that could compromise the public good in the pursuit of novelty.

ACKNOWLEDGMENTS

This material is based upon work supported by the National Science Foundation under grant number SES-1826545. Any opinions, findings, conclusions, or recommendations expressed in this material are those of the author and do not necessarily reflect the views of the National Science Foundation. Special thanks to Caroline Lamb and Wayne Powell for assisting with the data collection for the project. I am also grateful to the participants of Data & Society's "Against Platform Determinism" 2021 workshop for providing generous feedback on an earlier draft of this chapter.

NOTES

1 Although the YouTube clip includes the phrase "Amazon of Transportation" in its title, Uber CEO Dara Khosrowshahi actually says, "Amazon for transportation" in the video.
2 This study focused on the ways that US cities are responding to and mediating large-scale digital platforms (e.g., Uber, Amazon, Airbnb). Methods included document analysis, GIS mapping, observation, and semi-structured interviews with expert key informants in government agencies, community advocacy groups, industry, unions, and university research centres. We concentrated on three case-study cities (Boston, Austin, and San Francisco), but our document analysis extended to other cities throughout the United States, and our informants frequently spoke of developments in other cities, such as the Denver-Uber partnership analysed in this chapter. For more information about the study and its methods, see Monahan (2020).
3 This is an articulation of what Rob Kitchin calls "data capitalism" in the Dialogue introduction to this section. Data capitalism is operationalized by platform companies working to engender and exploit power differentials to achieve influence, market domination, and profit.
4 Related patterns can be seen across disparate platform types, where various forms of enclosure and capture have become staple mechanisms by which platform

companies achieve and maintain monopoly status. For example, in the realm of livestreaming on Amazon's Twitch platform, as William Partin (2020, 1) has shown, the capture of accessory innovations by third-party developers generates "a cycle in which the technical architecture of platforms evolves through the exploitation of power asymmetries between platform owners and dependents."

5 Technically, this partnership also includes the platform company Masabi, which manages ticket purchases for Denver's RTD.

6 The term *platform surveillance* references "the manifold and often insidious ways that digital platforms fundamentally transform social practices and relations, recasting them as surveillant exchanges whose coordination must be technologically mediated and therefore made exploitable as data" (Murakami Wood and Monahan 2019, 1).

REFERENCES

Akrich, M. 1992. "The De-scription of Technological Objects." In *Shaping Technology / Building Society: Studies in Sociotechnical Change*, ed. W.E. Bijker and J. Law, 205–24. Cambridge, MA: MIT Press.

Andrejevic, M. 2007. *iSpy: Surveillance and Power in the Interactive Era*. Lawrence: University Press of Kansas.

Barns, S. 2019. "Negotiating the Platform Pivot: From Participatory Digital Ecosystems to Infrastructures of Everyday Life." *Geography Compass* 13 (9): 1–13. https://doi.org/10.1111/gec3.12464

Barns, S. 2020. *Platform Urbanism: Negotiating Platform Ecosystems in Connected Cities*. Singapore: Springer.

Benjamin, R. 2019. *Race after Technology: Abolitionist Tools for the New Jim Code*. Medford, MA: Polity Press.

Bosselman, A. 2019a. "Irony! The best way to buy Rtd tickets is from Uber." *StreetsBlog-Denver*, 9 July. Online at https://denver.streetsblog.org/2019/07/09/the-best-way-to-buy-an-rtd-ticket-is-from-uber-but-the-company-wants-to-wipe-out-public-transit/, accessed 3 August 2020.

Bosselman, A. 2019b. "Uber partners with Denver Transit – What could go wrong?" *StreetsBlog-Denver*, 31 January. Online at https://denver.streetsblog.org/2019/01/31/use-uber-for-denver-transit-company-is-our-overlord-and-savior-as-it-crushes-the-epic-failure-that-is-rtds-mobile-app/, accessed 5 August 2020.

Callon, M. 1984. "Some Elements of a Sociology of Translation: Domestication of the Scallops and the Fishermen of St Brieuc Bay." *Sociological Review* 32 (1): 196–233. https://doi.org/10.1111/j.1467-954X.1984.tb00113.x

Cardullo, P., C. Di Feliciantonio, and R. Kitchin, eds. 2019. *The Right to the Smart City*. Bingley, UK: Emerald Publishing.

Clark, H. 2017. *Who Rides Public Transportation*. Washington, DC: American Public Transportation Association. Online at https://www.apta.com/wp-content/uploads/Resources/resources/reportsandpublications/Documents/APTA-Who-Rides-Public-Transportation-2017.pdf, accessed 15 July 2020.

Collier, R.B., V.B. Dubal, and C.L. Carter. 2018. "Disrupting Regulation, Regulating Disruption: The Politics of Uber in the United States." *Perspectives on Politics* 16 (4): 919–37. https://doi.org/10.1017/S1537592718001093

Conger, K. 2019. "Uber wants to sell you train tickets. And be your bus service, too." *New York Times*, 7 August. Online at https://www.nytimes.com/2019/08/07/technology/uber-train-bus-public-transit.html, accessed 4 August 2020.

Cottrill, C.D. 2020. "Maas Surveillance: Privacy Considerations in Mobility as a Service." *Transportation Research Part A: Policy and Practice* 131: 50–7. https://doi.org/10.1016/j.tra.2019.09.026

Donovan, G. 2020. *Canaries in the Data Mine: Understanding the Proprietary Design of Youth Environments*. London: Palgrave.

Gabrys, J. 2014. "Programming Environments: Environmentality and Citizen Sensing in the Smart City." *Environment and Planning D: Society and Space* 32 (1): 30–48. https://doi.org/10.1068/d16812

Gelinas, N. 2020. "Mass transit, and cities, could grind to a halt without federal aid." *New York Times*, 3 July. Online at https://www.nytimes.com/2020/07/03/opinion/coronavirus-mass-transit.html?referringSource=articleShare, accessed 14 July 2020.

George, J. 2020. "For many 'essential workers,' public transit is a fearful ride they must take." *Washington Post*, 11 April. Online at https://www.washingtonpost.com/local/trafficandcommuting/for-many-essential-workers-public-transit-is-a-fearful-ride-they-must-take/2020/04/11/8dec874a-79ad-11ea-a130-df573469f094_story.html, accessed 5 August 2020.

Giambrone, A. 2019. "Uber's electric jump scooters come to D.C." *Curbed.*, 9 April. Online at https://dc.curbed.com/2019/4/9/18302036/dc-uber-jump-electric-scooters-dockless-mobility, accessed 4 August 2020.

Godoy, M., and D. Wood. 2020. "What Do Coronavirus Racial Disparities Look Like State by State?" *NPR.org*, 30 May. Online at https://www.npr.org/sections/health-shots/2020/05/30/865413079/what-do-coronavirus-racial-disparities-look-like-state-by-state, accessed 14 July 2020.

Goodman, E.P., and J. Powles. 2019. "Urbanism under Google: Lessons from Sidewalk Toronto." *Fordham Law Review* 88: 457–98 https://doi.org/10.2139/ssrn.3390610

Graham, S., and S. Marvin. 2001. *Splintering Urbanism: Networked Infrastructures, Technological Mobilities and the Urban Condition*. New York: Routledge.

Griswold, A. 2019. "Uber's free-wheeling era of growth is coming to an end." *Quartz*, 3 December. Online at https://qz.com/1759492/uber-faces-tighter-regulations-in-its-most-important-cities/, accessed 30 January 2020.

Harvey, D. 2001. "Globalization and the 'Spatial Fix.'" *Geographische Revue* 2: 23–30.

Heaphy, L., and A. Wiig. 2020. "The 21st-Century Corporate Town: The Politics of Planning Innovation Districts." *Telematics and Informatics* 54: 1–10. https://doi.org/10.1016/j.tele.2020.101459

Hine, J. 2012. "Mobility and Transport Disadvantage." In *Mobilities: New Perspectives on Transport and Society*, ed. M. Grieco and J. Urry, 21–40. Farnham, UK: Ashgate.

Kbidy, L. 2019. "RTD, Uber and Masabi launch first-ever Uber Transit ticketing for riders in Denver." *Masabi*, 2 May. Online at https://www.masabi.com/2019/05/02/rtd-uber-and-masabi-launch-first-ever-uber-transit-ticketing-for-riders-in-denver/, accessed 12 January 2020.

Kitchin, R. 2014. "The Real-Time City? Big Data and Smart Urbanism." *GeoJournal* 79 (1): 1–14. https://doi.org/10.1007/s10708-013-9516-8

Lee, A., A. Mackenzie, G.J.D. Smith, and P. Box. 2020. "Mapping Platform Urbanism: Charting the Nuance of the Platform Pivot." *Urban Planning* 5 (1): 116–28. https://doi.org/10.17645/up.v5i1.2545

Lubin, A., and D. Deka. 2012. "Role of Public Transportation as Job Access Mode: Lessons from Survey of People with Disabilities in New Jersey." *Transportation Research Record* 2277 (1): 90–7. https://doi.org/10.3141/2277-11

Lupton, D. 2015. *Digital Sociology*. New York: Routledge.

Marvin, S., and A. Luque-Ayala. 2017. "Urban Operating Systems: Diagramming the City." *International Journal of Urban and Regional Research* 41 (1): 84–103. https://doi.org/10.1111/1468-2427.12479

Monahan, T. 2016. "Built to Lie: Investigating Technologies of Deception, Surveillance, and Control." *Information Society* 32 (4): 229–40. https://doi.org/10.1080/01972243.2016.1177765

Monahan, T. 2017. "Regulating Belonging: Surveillance, Inequality, and the Cultural Production of Abjection." *Journal of Cultural Economy* 10 (2): 191–206. https://doi.org/10.1080/17530350.2016.1273843

Monahan, T. 2018. "The Image of the Smart City: Surveillance Protocols and Social Inequality." In *Handbook of Cultural Security*, ed. Y. Watanabe, 210–26. Cheltenham, UK: Edward Elgar.

Monahan, T. 2020. "Monopolizing Mobilities: The Data Politics of Ride-Hailing Platforms in US Cities." *Telematics and Informatics* 55 (December): 101436. https://doi.org/10.1016/j.tele.2020.101436

Murakami Wood, D., and T. Monahan. 2019. "Editorial: Platform Surveillance." *Surveillance & Society* 17 (1/2): 1–6. https://doi.org/10.24908/ss.v17i1/2.13237

Noble, S.U. 2018. *Algorithms of Oppression: How Search Engines Reinforce Racism*. New York: New York University Press.

Partin, W.C. 2020. "Bit by (Twitch) Bit: 'Platform Capture' and the Evolution of Digital Platforms." *Social Media + Society* 6 (3): 1–12. https://doi.org/10.1177/2056305120933981

Plantin, J.-C., C. Lagoze, P.N. Edwards, and C. Sandvig. 2018. "Infrastructure Studies Meet Platform Studies in the Age of Google and Facebook." *New Media & Society* 20 (1): 293–310. https://doi.org/10.1177/1461444816661553

Plantin, J.-C., and A. Punathambekar. 2019. "Digital Media Infrastructures: Pipes, Platforms, and Politics." *Media, Culture & Society* 41 (2): 163–74. https://doi.org/10.1177/0163443718818376

Recode. 2018. "Uber CEO Dara Khosrowshahi says he wants to be the 'Amazon of transportation.'" YouTube. Online at https://www.youtube.com/watch?v=5Yo32U4mtE0, accessed 3 August 2020.

Rho, H.J., H. Brown, and S. Fremstad. 2020. "A Basic Demographic Profile of Workers in Frontline Industries." Washington, DC: Center for Economic and Policy Research. https://cepr.net/a-basic-demographic-profile-of-workers-in-frontline-industries/, accessed 14 July 2020.

Ritchie, H., and M. Roser. 2019. "Urbanization." *Our World in Data*. Online at https://ourworldindata.org/urbanization, accessed 5 August 2020.

Sadowski, J. 2020a. "The Internet of Landlords: Digital Platforms and New Mechanisms of Rentier Capitalism." *Antipode* 52 (2): 562–80. https://doi.org/10.1111/anti.12595

Sadowski, J. 2020b. "Who Owns the Future City? Phases of Technological Urbanism and Shifts in Sovereignty." *Urban Studies*, online. https://doi.org/10.1177/0042098020913427

Sadowski, J., and R. Bendor. 2019. "Selling Smartness: Corporate Narratives and the Smart City as a Sociotechnical Imaginary." *Science, Technology, & Human Values* 44 (3): 540–63. https://doi.org/10.1177/0162243918806061

Sadowski, J., and F. Pasquale. 2015. "The Spectrum of Control: A Social Theory of the Smart City." *First Monday* 20 (7). Online at http://firstmonday.org/ojs/index.php/fm/article/view/5903/4660, accessed 17 December 2016.

Saltman, J. 2019. "Can Translink adapt to ride hailing? Uber, Lyft ... and Translink say yes." *Vancouver Sun* 11 August. Online at https://vancouversun.com/news/local-news/ride-hailings-effects-on-transit, accessed 4 August 2020.

Schüll, N.D. 2012. *Addiction by Design: Machine Gambling in Las Vegas*. Princeton, NJ: Princeton University Press.

Shapiro, A. 2020. *Design, Control, Predict: Logistical Governance in the Smart City*. Minneapolis: University of Minnesota Press.

Sisson, P. 2019. "Uber for trains? New feature allows Denver riders to book transit with app." *Curbed*, 31 January. Online at https://www.curbed.com/2019/1/31/18205143/app-denver-uber-transit, accessed 3 August 2020.

Smith, G.J.D. 2018. "Data Doxa: The Affective Consequences of Data Practices." *Big Data & Society*, online. https://doi.org/10.1177/2053951717751551

Srnicek, N. 2017. *Platform Capitalism*. Malden, MA: Polity.

Stehlin, J., M. Hodson, and A. McMeekin. 2020. "Platform Mobilities and the Production of Urban Space: Toward a Typology of Platformization Trajectories." *Environment and Planning A: Economy and Space* 52 (7): 1250–68. https://doi.org/10.1177/0308518X19896801

van Dijck, J., T. Poell, and M. de Waal. 2018. *The Platform Society: Public Values in a Connective World*. Oxford: Oxford University Press.

Wilhelm, A. 2019. "Uber doubles down on micromobility." *TechCrunch*, 17 December. Online at https://techcrunch.com/2019/12/17/uber-doubles-down-on-micromobility/, accessed 4 August 2020.

Winner, L. 1980. "Do Artifacts Have Politics?" *Daedalus* 109 (1): 121–36. http://dx.doi.org/10.4324/9781315259697-21

8 Epistemic (In)Justice in a Smart City: Proto-Smart and Post-Smart Infrastructures for Urban Data

MIGUEL VALDEZ, MATTHEW COOK, AND HELEN ROBY

In this chapter we use the notion of epistemic (in)justice to investigate critically the mechanisms through which smart cities and the people in them are made visible and are represented and acted upon (or not) as a result of specific modes of data production and utilization (D'Ignazio and Klein 2020; Origgi and Ciranna 2017; Taylor 2017). Smart cities represent a resurgence of data-driven technoscientific planning, promising technological solutions for social, environmental, and economic problems (Shelton, Zook, and Wiig 2015), with algorithmically collected and processed data often regarded as the new standard of objective and impartial knowledge in urban planning and governance. Data infrastructures that sort, categorize, and affect urban phenomena can render urban issues visible and actionable, but there is also a growing risk that structural inequalities embedded in data structures will render injustices and inequalities invisible or make them appear inevitable. Here we use the notion of epistemic injustice (Fricker 2007) to reveal that such city practices and institutions are profoundly political, and might be deployed and structured in ways that introduce epistemic disfunction (Pohlhaus Jr 2017). Epistemic institutions might be unjust when they suppress the testimony of particular groups or individuals (testimonial injustice) or when gaps in collective interpretive resources put groups or individuals at a disadvantage when it comes to making sense of their own experiences (hermeneutic injustice). As the editors reflect in the Introduction to this book, there are various frameworks through which justice in a digital world may be understood. Different justice frameworks use different informational bases to evaluate whether a decision, society, or distribution is fair (Sen 1999), but in every case knowledge and scrutiny precede the making of just decisions. Epistemic injustice therefore can invisibly permeate all the mechanisms through which justice claims are made, through which justice is pursued, and through which institutions are evaluated and held accountable (or not).

In the dialogue that introduces Part Two of this book, Rob Kitchin discusses how smart city infrastructures set up to pursue for data-driven efficiency, optimization, and profit might be unsuitable for addressing inequity, unfairness, and injustice. Pressing urban problems become invisible because addressing them would not serve the powerful interests that shape the *smart* epistemic institutions through which cities are understood and acted upon. Following D'Ignazio and Klein (2020), Kitchin also contrasts the compliant approach of data ethics, which accept the system as it is, and the challenging approach of data justice, whose aim is to change the system by reflexively acknowledging structural power differentials and working towards dismantling them. In that spirit, the purpose of this chapter is not to react and reject the use of data for urban planning and governance, but to interrogate critically the power differentials embedded in the data practices of smart cities and to reflect on how to achieve outcomes that are more just and inclusive. We use a case study as an entry point to investigating a smart city project and to engage with the most pressing questions in the field of epistemic justice, such as "who has voice and who does not?"; "are voices interacting with equal agency and power?"; "in whose terms are they communicating?"; "who is being understood and who is not (and at what cost)?"; "who is being believed?"; and "who is even being acknowledged and engaged with?" (Kidd, Medina, and Pohlhaus Jr 2017, 1). The case study documents the transition between two data infrastructures in the English town of Milton Keynes. The first one is the Milton Keynes Intelligence Observatory (MKiO), launched in 2004 as part of a wider movement calling for the provision of place-specific information, context, and perspective to improve health outcomes and reduce inequality and social exclusion across the United Kingdom. The second infrastructure, the Milton Keynes Data Hub, was deployed in 2014 in the context of the smart city program MK:Smart, and eventually came to be seen as a replacement for the observatory. However, the observatories of the early 2000s and the smart cities of the late 2010s reflected very different logics and political rationalities. Although the datasets from the observatory were readily transferred to the hub, data were inflected by the infrastructures through which they were stored and analysed as well as by the constituencies that coalesced around them, ultimately developing very different relationships with the places and bodies constituting the city.

Our Theoretical Approach

Throughout this chapter we apply the notion of epistemic (in)justice to explore how long-standing traditions of data-driven urban planning might have been transformed as they were co-opted by and became indistinguishable from globally circulating logics of data-driven growth and entrepreneurialism. Epistemic injustice refers to the forms of unfair treatment that relate to issues of knowledge, understanding, and participation in communicative practices (Kidd, Medina, and Pohlhaus Jr 2017). Epistemic injustice happens as a result of how epistemic institutions are structured – for example, when unequal access to epistemic goods creates both marginalized and privileged communities, making it difficult for them to know what is in their interest to know. Epistemic injustice also takes place when some groups and their issues are not understood or are not believed, distorting understanding and stymieing enquiry (Coady 2010; Dotson 2011; Fricker 2007; Pohlhaus Jr 2017).

Although the notion of epistemic injustice predates the emergence of the smart city, so-called smart technologies transform structures of meaning-making and knowledge-producing practices in the city. The data infrastructures associated with the smart city movement easily might introduce new forms of epistemic injustice owing to their tendency to create a narrative of their use of urban data in terms of what Donna Haraway (Haraway 1991, 195) describes as "the view from above, from nowhere, from simplicity," claiming neutral and omniscient knowledge of people's lives. Complex and contradictory situated knowledges are disembodied, simplified, and potentially subjected to epistemic violence as data collection, analysis, communication, and decision-making are performed by strangers in the dataset – actors one or more steps removed from the local context of the data, places, and bodies that constitute the city (Dotson 2011; Spivak 1998).

The epistemic injustices that are often embedded in the design of data infrastructures and data-driven algorithmic solutions to urban problems have a tendency wilfully or neglectfully to reinforce existing structural power differentials, because the perspectives of the dominators are readily storied as the unbiased default – unmediated and disembodied (D'Ignazio and Klein 2020; Haraway 1991). Distant data centres and omnipresent but invisible interfaces between the human and the computational mask the people, methods, questions, and rationales that lie behind the claims of seemingly neutral and objective data-driven knowledges. The supposedly objective and non-political knowledges so produced conveniently happen to align with the perspectives and agendas of the people and institutions in positions of power (D'Ignazio and Klein 2020, 76), whose worldviews and value systems might not always align with those of the communities where data-driven solutions are deployed. Marginalized groups thus become epistemically disadvantaged as data that would let them make sense of their experiences or make justice claims about them are filtered out or considered irrelevant.

In the following section we present a case study revealing the politics and epistemic conflicts that shaped a series of smart and proto-smart data infrastructures in Milton Keynes. We collated and analysed a variety of qualitative data from primary and secondary sources following a clustering and coding method to trace the history of the MKiO and the policy decisions regarding its development, maintenance, and eventual closure as it was rendered obsolete by newer, smarter infrastructures. We analysed a selection of fifty policy and corporate documents produced between 1999 and 2021 to investigate the stated goals of local and national actors regarding observatories, smart cities, and data infrastructures and practices. We then contrasted these somewhat idealized accounts with primary data in the form of twenty-eight interviews collected between 2017 and 2019 for the "Smart Cities in the Making" research project, funded by the Economic and Social Research Council, which explored how social difference affects participation in the smart city and is affected by it. Interviewees included members of the coalition responsible for the smart city program in Milton Keynes, as well as former users, designers, or administrators of the MKiO.

The Case Study

Background: Early Data Infrastructures

Milton Keynes is a new town located approximately 100 kilometres north of London, founded in 1967 as part of the third wave of a new town program that was initiated to replace bombed-out housing stock after the Second World War, later expanded to accommodate the "overspill" from densely populated areas of deprivation. Prior to its designation, the area had a population of less than 50,000, but it was expected to house 250,000 residents and was the subject of careful planning to allow for such rapid growth.

As a planned city, Milton Keynes had two salient characteristics that had an enduring effect on its data practices. First, the masterplan called for a framework for the city based upon a grid of fast roads designed to facilitate fluid transport. The grid effectively atomized the city into 106 estates measuring one kilometre by one kilometre each (Figure 8.1). Although the estates were connected by pedestrian and cycling paths, the high-speed roads that separated them encouraged a certain insularity that resulted in each estate's having its own distinct personality, demography, and social issues, a fact that, over time, would call for place sensitivity in the data practices of city managers.

The second feature was the production of a large body of planning data in a relatively brief period by the development corporation responsible for building the new town, right before the use of computers in planning became widespread. As planning and management responsibilities were transferred from the development corporation to the Commission for New Towns, then to English Partnerships, and later on to the city council, the original body of planning data was transferred from paper archives to spreadsheets, then to databases, and later to a geographic information system (GIS) with mapping and data analysis capabilities.

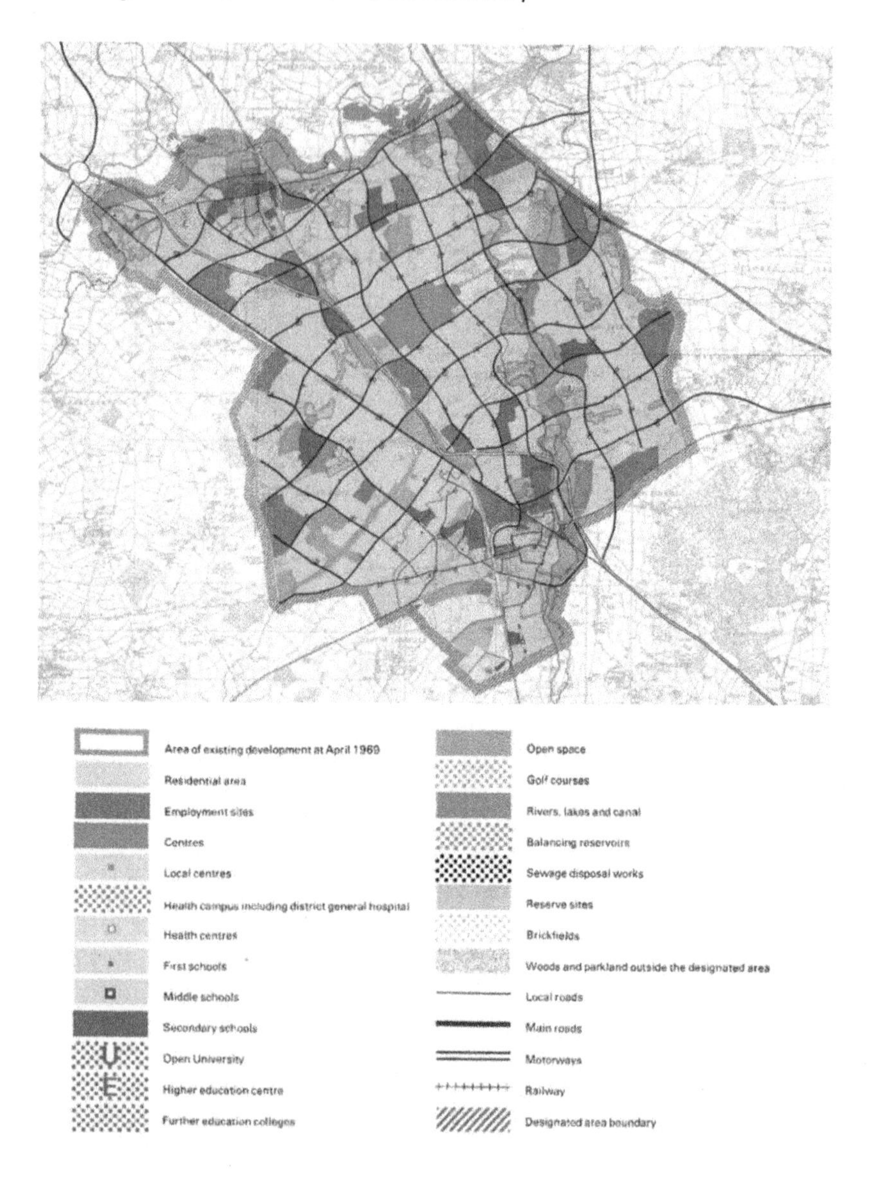

Figure 8.1. The Master Plan for Milton Keynes

The mapping and analytical capabilities of the GIS were used at first to inform matters of physical geography – for example, by drawing data from historic and planning archives to mark sites such as former railyards, tanneries, landfills, and scrap yards in order to identify contaminated land within the projected boundaries of the city. Intelligence officers gradually found themselves using the system to address matters of social geography – for example, using the GIS to identify school catchment areas. Over time, police, health, and social services found further uses for the data – for instance, by mapping the high incidence of diabetes and obesity or areas where children were registered with child protection.

As intelligence officers began to apply the mapping system to address social issues, their work found a user base and support network in the various agencies and volunteer organizations that worked closely with the communities they served and that could see the value of place-centric urban data. The same approach faced resistance, however, in the political climate of the late 1980s and early 1990s, which framed social problems in terms of individual responsibility. Making underlying patterns of inequality visible and drawing attention to places or communities left behind was seen as implicitly placing responsibility on government, and thus countering ideologies and policies advocated by national government at the time.

Proto-Smart Data Infrastructures: The Milton Keynes Intelligence Observatory

Political change in the early 2000s turned national and local authorities more amenable to the use of urban data to make patterns of urban inequality visible. The Milton Keynes Intelligence Observatory was commissioned by the local council in 2003 and launched in 2004, following engagement and consultation with Milton Keynes stakeholders to assess needs and to draw together the demographic and analytic capacity of a network of partners that included the council, the primary care trust, English Partnerships, the National Health Service, local partnerships, and emergency services. The observatory's primary objective would be to explore the leading causes of health inequality in Milton Keynes, thus informing strategies for improving health outcomes (MKC 2003, 2004). As the project was funded through the South East England Development Agency, which provided £1.9 million through its social inclusion program, the observatory was also expected to address economic disadvantage in the most deprived areas of the city (MKC 2006). The observatory was therefore tasked with the provision of actionable data to "reduce health inequalities as measured by an increase in life expectancy and decrease in infant mortality" and to "reduce social exclusion as measured by the number of children in poverty, people in low-income housing and workless households, particularly focusing on disadvantaged groups such alone parents and people with disabilities" (MKC 2004, 3).

The observatory was accessible via the Internet, but it was recognized that an online presence was necessary but not sufficient to achieve the desired outcomes. The brief of the intelligence team was expanded, encouraging them to work and collaborate with partners to collect, process, and disseminate information about Milton Keynes and the surrounding area to all interested parties. Thus, the practices and networks related to community-oriented urban data, which had been previously resisted, were now supported and fully funded. Intelligence officers facilitated a dialogue between the various partners of the observatory, so that each organization would provide an inventory of the data they could contribute as well as the data they needed from other organizations. Urban data revealing patterns of disadvantage were sough and correlated with the Index of Multiple Deprivation to identify areas that would benefit from special initiatives or programs and to determine their eligibility for specific funding streams. The information provided ranged from population and housing projections, fire and crime statistics, census data, and social and economic indicators, to statistics about excess winter deaths. When possible, data were collected with a high spatial resolution, making it possible to visualize social difference with a resolution of at least one kilometre by one kilometre, roughly the size of each of the estates created by the road grid.

As soon as the observatory was launched, it became apparent that it would be of interest to many users outside of the partner organizations, including schoolchildren, the general public, and the business community. Besides using data that were already in the observatory, anyone could contact the observatory and request information, including members of the public, councillors, the police force, and volunteer organizations. Additionally, the observatory team established quarterly meetings of the users' group to provide information and context about new datasets and to ask for feedback about any data they needed. When possible, the team tried to address users' requests, as they noticed that even for data that seemed to be of interest to only a few users, once it was uploaded to the observatory other groups and organizations found it interesting and useful as well. The user base for the observatory grew steadily. By 2006 there were over 300 registered users, and by August 2007 over 500 (Table 8.1)

Although at first the observatory team focused on collecting data and making them available, they gradually paid more attention to analysis, signposting, and human mediation as they realized that data without context could be misunderstood or used incorrectly. For example, this could mean explaining the difference between two different ways to measure unemployment and the impact that differences in frequency of updates, people included or excluded from the count, and spatial resolution of the datasets would have. In the words of one of the intelligence officers, "it'll be a case of trying to understand what the user requesting it would need it for and then apply whichever one is needed and make them aware of what it includes and what it doesn't include."

Table 8.1. Users of the Milton Keynes Intelligence Observatory, Clustered by Sector and Organization

Sector/Organization	Number of Users
Milton Keynes Council	151
Private companies/organizations/businesses	78
Milton Keynes Primary Care Trust	42
Other councils/government agencies/health bodies	39
Universities/schools/colleges	20
Voluntary organizations/charities	17
Open University	9
Fire departments	8
Thames Valley Police	7
English Partnerships	6
Councillors	5
MK Chamber	3
Local Enterprise Partnerships	3
Non-work, personal email address supplied	156

Source: MKC 2007.

The observatory was used by social services and community organizations to identify areas where intervention was needed and to deliver Milton Keynes's joint strategic needs assessment (JSNA). The JSNA drew on quantitative and qualitative data to provide information about the health and well-being of the population and about its current and future needs, thus informing the delivery of health services in the area as well as the local plan. After urban data were used to identify deprived areas, community development officers would be sent to listen to the people in these areas and provide support. By the early 2010s the MKiO was a known and reliable tool in the data practices of a wide network of public and private actors in Milton Keynes. Nonetheless, a combination of local, national, and global factors eventually resulted in its closure. One local issue was the expiration of the licence for key GIS software. In addition to the cost of renewal, there were data security considerations, as the commercial company providing the software decided to transform its product into a platform. In consequence, data storage and processing would be handled remotely, potentially creating a security risk. Nationally, politics once again moved away from government action and sought to frame societal issues as either matters of personal responsibility or issues best addressed by market forces or by "big society" (Fenwick and Gibbon 2017).

Smart Infrastructures: MK:Smart

The closure of the observatory was also related to the rise of globally circulating smart city visions. Smart city narratives are largely predicated on the use of urban data to synchronize urban processes algorithmically, with the ambition to improve resource efficiency, distribution of services, and urban participation (Gabrys 2014). As such, they found fertile ground in Milton Keynes, the population of which had grown from 50,000 to 250,000, becoming the fastest-growing UK city. The smart city seemed to provide a data-driven tool for addressing the tensions that arose as rapid growth encountered limits of physical infrastructure (particularly transport and energy), financial constraints, and environmental concerns.

In 2014 Milton Keynes initiated the MK:Smart program, a £16 million smart city initiative whose main deliverable would be the MK Data Hub. The hub, jointly developed by BT and data scientists from the Open University, was designed to draw together and make available information relevant to how the city functions, including data from key infrastructure networks (energy, water) and sensor networks providing real-time information on weather, pollution, and traffic; satellite data; and data crowdsourced from citizens and volunteer organizations. Access to anonymized data in the Data Hub, under protocols governed by Milton Keynes Council, was expected to encourage the development of new approaches and applications to address the city's challenges by project partners and independent and commercial developers.

The decision to end the MKiO's activities, therefore, resulted from a reduction in funding associated with political changes, the impossibility of maintaining the status quo as software providers migrated to a more profitable platform model, and the availability of a privately owned smart city platform that appeared to provide the same functionality. Minutes of the budget scrutiny committee of Milton Keynes Council, held on 12 January 2017 (MKC 2017a), identify various areas where spending cuts would be required due to the economic austerity policies prevalent at the time, and indicate that, "[t]he Council, in association with the Open University, had developed the MK Smart Data Hub which would continue to supply, via its website, the statistical data currently produced by the Research Team" (MKC 2017a). Owing to the new (and privately owned) infrastructure, data previously available via MKiO would still be freely available, only not presented through the "observatory" interface and the Social Atlas. The budget scrutiny committee also reported that the council had been finding it difficult to recruit suitable people to the intelligence team owing to the high demand in industry for employees with that area of expertise. While the committee raised concerns about "the risk to the Council, both to its reputation and financially, which could arise from poor decision-making due to Service Groups no longer being able to access the data interpretation and analysis skills currently provided by MKi Observatory" (MKC 2017b), the decision to cease the activities of the intelligence unit and the observatory was confirmed shortly afterwards, in the council meeting of 15 February 2017.

Following the decision to close the observatory, data were readily transferred to the MK Data Hub, but the logic of the smart city is very different to that of the observatory. The dominant smart city strategy at the time was largely oriented towards technology development that supported the increasingly efficient use of infrastructure, as well as economic growth resulting from the transition to a knowledge economy (United Kingdom 2013, 2016). As Torin Monahan discusses in Chapter 5 of this book, platforms have the capacity to "extend into and poach upon existing *public* infrastructures, converting them, in the process, into resources for capitalist expansion." Although in the case of Milton Keynes the smart city platform attracted inward investment and ultimately might support a transition to more environmentally sustainable ways of living in the city, it has not engaged with social change and inequality as its predecessors did. The expected smart successor to the MKiO was side-lined:

> Since the launch of the portal, there are still not new contents published or new users from the MK Council departments, yet to realise the replacement of the centralised management of city data with the distributed technology-supported approach ... a complex service of data analysis and tailoring provided by a group of data experts interacting with local stakeholders and with a technological platform supporting their activities, had been progressively replaced by an open data portal to make data available for public uses. This transition did not meet the expectations resulting in a limited use of the current portal by both end-users and Council officers ... highlighting the importance of human support to understanding and contextualization of data to make them actionable. (Luppi and Antonini 2018, 7)

Discussion and Conclusions

The case study suggests that smart city practices and platforms as they were applied in Milton Keynes did not readily facilitate the use of urban data to make inequality visible. The case is relevant to a broader discussion for several reasons. First, the data-driven smart city in the form of MK:Smart represents the implementation of a globally circulating narrative that has found fertile ground in countless other cities. Second, the smart city as an empty signifier has a chameleonic logic that drives it to assimilate existing initiatives in the places where it is implemented. There is a risk that, as was the case in Milton Keynes, other smart cities will assimilate existing urban data infrastructures and practices. As smart cities largely follow an entrepreneurial logic (Shelton, Wiig, and Zook 2015; United Kingdom 2013; see also Tavmen, in this volume), human mediation is seen as an unjustifiable source of inefficiency that should be replaced, with ubiquitous sensors taking the place of humans seeing the reality on the ground and engaging in dialogue with the people whose lives are affected by data, and with algorithms taking the place of decisionmakers. Consequently, the outcome of such smart transformations might indeed be more efficient but also potentially less just. Smart cities offer compelling benefits, and indeed in the case of Milton Keynes, the smart data hub still fulfils other useful social functions, but something meaningful was clearly lost.

The notion of epistemic injustice as discussed at the beginning of this chapter might provide a useful framework for understanding what was lost in the transition from one urban data infrastructure to another. If, as the case study suggests, smart cities become agents of widespread epistemic injustice, we might at least envision alternatives and learn from other urban data infrastructures such as the MKiO, which arguably were more just, more attuned to the bodies and the places they were expected to improve. The notion of epistemic injustice then can be applied to explore how one infrastructure might have been more just than another, and also might shed light on the ways in which the smart city can be rendered more just.

The case study reveals that, prior to the foundation of the MKiO in 2004, data infrastructures in Milton Keynes were held back by a mode of epistemic injustice that can be described as "wilful hermeneutic ignorance" (Pohlhaus Jr 2017, 17), as inconvenient aspects of the world – such as structural responsibility for what authorities would prefer to frame as individual failures – were dismissed by authorities who refused to become proficient in the epistemic resources required to attend to those parts of the world well. Another form of epistemic injustice takes place when unequal access to epistemic goods creates both marginalized and privileged communities. In the case of the MKiO, intelligence officers undertook a constant labour of engagement to ensure data reached those who needed them, not only in government or business, but also volunteer organizations, community mobilizers, and the general public. In the case of the MK Data Hub, urban data were open to the public, but human mediators were not there to promote them, explain them, and provide their context. Whereas the observatory organized user groups, took calls from schoolchildren, and presented its data in colour-coded maps (Figure 8.2), the data hub offered online access to a catalogue of raw data (Figure 8.3). Or, as a retired officer from the observatory said regarding its smart successor: "I pick up that there's actually a lot of cynicism that it's not been properly resourced and it's remote and people don't have individual human beings to relate to within that system. What they have is quite sort of a remote, online, techie-competent system. There are no names behind it anymore. There are no real people that you can talk to and you're not sure what you"re looking for or where to look."
Epistemic injustice also takes place when some bodies and their issues are maliciously or neglectfully not understood. Here, D'Ignazio's notion of "missing

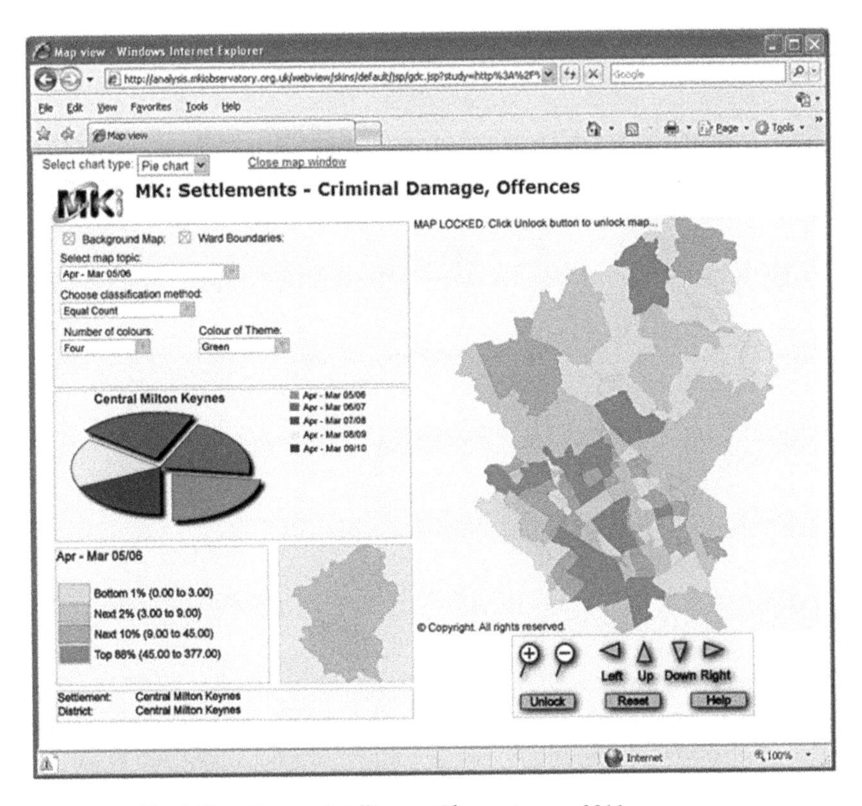

Figure 8.2. The Milton Keynes Intelligence Observatory, c. 2011
Source: Oakford and Williams 2011.

body problems," as explained in her interview with Thylstrup and Veel (2017, 69) might be usefully applied to understand one form of epistemic injustice in the case study. Particularly, data practices might be unjust when specific bodies go uncounted – the issue of what people in power decide is worthy of allocating scarce resources towards quantifying and, once data are collected, disaggregating them into categories and analysed to determine patterns. The case study suggests that, before the observatory was established, there was active opposition to making patterns of inequality visible. After the end of the observatory, social difference became invisible again, not necessarily because of active opposition, but because it simply did not fit the narrative of efficiency and economic growth associated with smart cities. Here, it is important to draw attention to the way in which social difference fell off the agenda apparently through the lack of action by any body. D'Ignazio suggests that another problem that might make data practices unjust is that bodies, specifically the bodies of those setting the agenda,

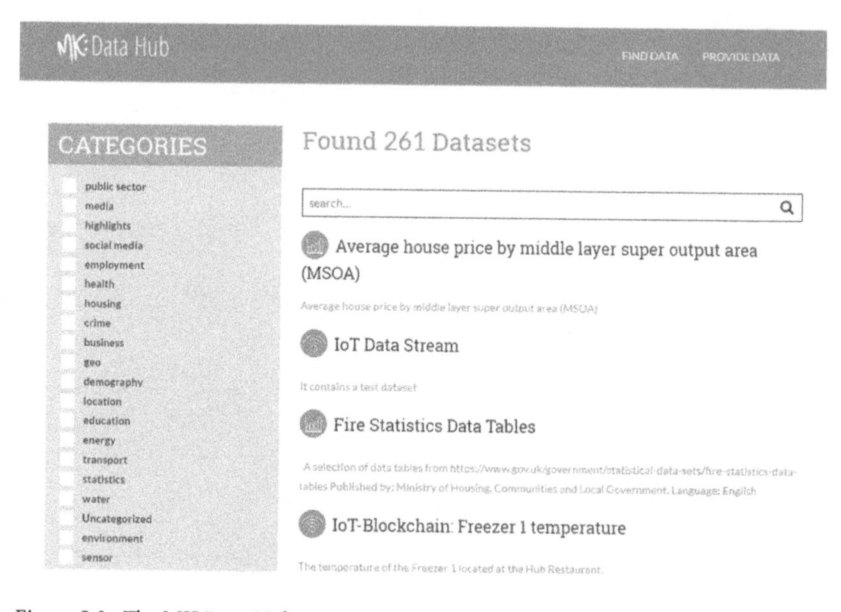

Figure 8.3. The MK Data Hub, c. 2019
Source: MK:Smart 2019.

are rendered invisible. Data infrastructures, visualizations, and algorithms wield a tremendous amount of power – rhetorical power as well as the power to decide how scarce resources will be allocated. In smart cities projects such as MK:Smart, it is often the case that distant data centres and omnipresent but invisible interfaces between the human and the computational act as an invisibility cloak, masking the people, methods, questions, and rationales that lie behind the claims of seemingly neutral and objective data-driven knowledges. There is therefore a risk that epistemic violence will take place as data collection, analysis, communication, and decision-making are performed by strangers in the dataset, actors one or more steps removed from the local context of the data.

Milton Keynes, as are countless other cities across the world, is transitioning towards a smart city paradigm. As the editors discuss in the Introduction to this book, smart city projects are frequently defined in vague, optimistic language so that they may claim a broad legitimacy for guiding stakeholders and constitute an attractive reference for actors at all levels and across sectors. It remains rather open, however, what the actual pursuit of a smart city is and, therefore, which winners and losers we are to expect from its deployment. Smart cities are often framed as empowering, with policymakers claiming that the data-driven city will be open to tinkering by citizen-scientists and citizen-innovators and that smart initiatives therefore will empower the community (Townsend 2013). Avelino (2011) suggests that the success of empowerment in transitions can be evaluated only at the moment of physical materialization, possession, and profit: who has materialized the new institutional arrangements, who possesses the new resources, who profits from them, and whose goals have been realized? In the case of Milton Keynes, at least initially, the interests of health care workers, community organizers, and people in deprived areas seem to be largely absent from the smart city agenda. The mechanism of epistemic injustice through which this is achieved is that of rendering decisionmakers invisible. The authorship of algorithms intended to guide the distribution of civic resources is itself an inherently political act (Del Casino Jr, et al. 2020; Kitchin 2017), but smart city narratives in Milton Keynes and elsewhere are largely framed as non-political, bypassing mechanisms of democratic accountability. Smart city logics are presented as objective and rational, and therefore they face little or no opposition. The research discussed in this chapter has sought to draw attention to smart cities as arenas of epistemic conflict, since urban data infrastructures not only frame solutions to urban problems; they also seek to define the nature and definition of the problem itself. Smart cities largely frame problems in terms of inefficiencies that can be addressed algorithmically, potentially drawing attention and resources away from other uses of urban data that might challenge the status quo by making injustice and inequality visible.

ACKNOWLEDGMENTS

This chapter results from a research project funded by the Economic and Social Research Council grant reference ES/N014421/1, "Smart Cities in the Making: Learning from Milton Keynes." The research team members were Professor Gillian Rose (Principal Investigator), Dr Nick Bingham, Professor Matthew Cook, Professor Parvati Raghuram, Dr Alan-Miguel Valdez, Professor Sophie Watson, Dr Edward Wigley, and Dr Oliver Zanetti.

REFERENCES

Avelino, F. 2011. *Power in Transition: Empowering Discourses on Sustainability Transitions*. Rotterdam: Erasmus Universiteit Rotterdam.

Coady, D. 2010. "Two Concepts of Epistemic Injustice." *Episteme* 7 (2): 101–13. https://doi.org/10.3366/epi.2010.0001

Del Casino Jr, V.J., L. House-Peters, J.W. Crampton, and H. Gerhardt. 2020. "The Social Life of Robots: The Politics of Algorithms, Governance, and Sovereignty." *Antipode* 52 (3): 605–18. https://doi.org/10.1111/anti.12616

D'Ignazio, C., and L.F. Klein. 2020. *Data Feminism*. Cambridge, MA: MIT Press.

Dotson, K. 2011. "Tracking Epistemic Violence, Tracking Practices of Silencing." *Hypatia* 26 (2): 236–57. https://doi.org/10.1111/j.1527-2001.2011.01177.x

Fenwick, J., and J. Gibbon. 2017. "The Rise and Fall of the Big Society in the UK." *Public Money & Management* 37 (2): 126–30. https://doi.org/10.1080/09540962.2016.1266162

Fricker, M. 2007. *Epistemic Injustice: Power and the Ethics of Knowing*. Oxford: Oxford University Press.

Gabrys, J. 2014. "Programming Environments: Environmentality and Citizen Sensing in the Smart City." *Environment and Planning D: Society and Space* 32 (1): 30–48. https://doi.org/10.1068/d16812

Haraway, D. 1991. *Simians, Cyborgs and Women*. New York: Routledge.

Kidd, I.J., J. Medina, and G. Pohlhaus Jr. 2017. *The Routledge Handbook of Epistemic Injustice*. New York: Tailor and Francis.

Kitchin, R. 2017. "Thinking critically about and Researching Algorithms." *Information, Communication & Society* 20 (1): 14–29. https://doi.org/10.1080/1369118X.2016.1154087

Luppi, L., and A. Antonini. 2018. "From Service to Data Infrastructure – The Transition from MK Intelligence Observatory to MK:Insight." Milton Keynes, UK: Open Research Online. http://oro.open.ac.uk/62253/1/From%20MKiO%20to%20MKI%20-%20Technical%20Report%20%281%29.pdf, accessed 25 June 2021.

MKC (Milton Keynes Council). 2003. "Developing a Programme for Improving Health and Reducing Health Inequalities in Milton Keynes." Online at https://milton-keynes.cmis.uk.com/milton-keynes/Search.aspx, accessed 25 June 2021.

MKC. 2004. "Milton Keynes Council – Corporate Priority 8: Creating Social Inclusion." Online at https://milton-keynes.cmis.uk.com/milton-keynes/Search.aspx, accessed 25 Jume 2021.

MKC. 2006. "Milton Keynes Council Report on Corporate Priorities." Online at https://milton-keynes.cmis.uk.com/milton-keynes/Search.aspx, accessed 25 June 2021.

MKC. 2007. "MKi Observatory – Reasons for Having the Observatory." Online at https://milton-keynes.cmis.uk.com/milton-keynes/Search.aspx, accessed 25 June 2021.

MKC. 2017a. "Minutes of the Meeting of the Budget Scrutiny Committee Held on Thursday 12 January 2017." Online at https://milton-keynes.cmis.uk.com/milton-keynes/Search.aspx, accessed 25 June 2021.

MKC. 2017b. "Responses to Budget Scrutiny Committee Recommendations 2017/2018." Online at https://milton-keynes.cmis.uk.com/milton-keynes/Search.aspx, accessed 25 June 2021.

MK:Smart. 2019. "Data Hub – Datasets." Online at https://datahub.mksmart.org/, accessed 1 September 2020.

Oakford, A., and P. Williams. 2011. "The Use and Value of Local Information Systems: A Case Study of the Milton Keynes Intelligence (MKi) Observatory." Aslib Proceedings 63 (5): 533–48. https://doi.org/10.1108/00012531111165003

Origgi, G., and S. Ciranna. 2017. "The Case of Digital Environments." In *The Routledge Handbook of Epistemic Injustice*, ed. I.J. Kidd, J. Medina, and G. Pohlhaus Jr, 303–12. New York: Routledge.

Pohlhaus Jr, G. 2017. "Varieties of Epistemic Injustice." In *The Routledge Handbook of Epistemic Injustice*, ed. I.J. Kidd, J. Medina, and G. Pohlhaus Jr, 13–26. New York: Routledge.

Sen, A. 1999. *Development as Freedom*. New York: Anchor Books.

Shelton, T., M. Zook, and A. Wiig. 2015. "The Actually Existing Smart City." *Cambridge Journal of Regions, Economy and Society* 8 (1): 12–25. https://doi.org/10.1093/cjres/rsu026

Spivak, G. 1998. "Can the Subaltern Speak?" In *Marxism and the Interpretation of Culture*, ed. C. Nelson and L. Grossberg, 271–315. Urbana: University of Illinois Press.

Taylor, L. 2017. "What Is Data Justice? The Case for Connecting Digital Rights and Freedoms Globally." *Big Data & Society* 4 (2): 1–14. https://doi.org/10.1177/2053951717736335

Thylstrup, N., and K. Veel. 2017. "Data Visualization from a Feminist Perspective – Interview with Catherine D'Ignazio." *Kvinder, Køn & Forskning* 1: 67–71. https://doi.org/10.7146/kkf.v26i1.109785

Townsend, A.M. 2013. *Smart Cities: Big Data, Civic Hackers, and the Quest for a New Utopia*. New York: W.W. Norton.

United Kingdom. 2013. "Smart Cities Background Paper." London: Department for Business Innovation & Skills. Online at https://assets.publishing.service.gov.uk/, accessed 26 June 2021.

United Kingdom. 2016. "Smart Cities Pitchbook." London: UK Trade & Investment. Online at https://assets.publishing.service.gov.uk/, accessed 26 June 2021.

9 The Politics of Re-membering: Inequity, Governance, and Biodegradable Data in the Smart City

NATHAN A. OLMSTEAD AND ZACHARY SPICER

For many of us, the idea that our behaviour is being observed, measured, and managed by the cities we live in is no longer surprising. In municipalities around the world, an expanding network of data-based technologies determines everything from traffic and policing patterns to snow removal, pothole repair, and parking. While much of this data-gathering technology is hidden and imperceptible, the debate about how data are used in projects such as these has received a high degree of visibility in recent years (Srnicek 2017; Zuboff 2015). More specifically, the growing prevalence of data-based technologies has raised important questions about privacy, ownership, and transparency (Murakami Wood and Mackinnon 2019; Scassa 2017).

Highly visible projects with largely ambiguous data governance policies, such as the Quayside project in Toronto, have highlighted this debate (see Scassa 2017; Wylie 2017). In fact, a torrent of criticism has been directed at policymakers, technologists, and strategists who have promoted and adopted such data-intensive infrastructure projects. Much of this criticism centres on the collection, storage, and use of data (Valverde and Flynn 2018). Technologists often claim this data collection is ubiquitous, anonymized, and harmless to the general public (Murakami Wood and Mackinnon 2019). Critics counter by arguing that data are a valuable commodity that, once collected, can generate value for data firms at the expense of individual privacy (Tusikov 2019).

There is thus an apparent tension between the rights and protections expected by city residents and the enthusiastic pursuit of data-based solutions to municipal challenges (Taylor 2017; Townsend 2013). Although large-scale, private surveillance and data collection can lead to negative ends once commercialized by increasingly powerful multinational data firms, data collection by public bodies offers the hope of gaining important insights into the equitable and fair allocation of public resources and aiding in better policy and program creation (Robinson and Coutts 2019; Tusikov 2019). Simply put, intelligence about city life often comes from the same large-scale surveillance efforts decried by critics of smart city projects (Kitchin 2014).

This chapter aims to reimagine the relationship between cities and their data. Drawing on posthuman notions of "biodegradability," we position data as a communal resource and prioritize their role in government accountability and responsivity. We argue that data-based technologies provide an opportunity for municipal governments and their residents to reflect on, and participate in, the past, present, and future of their communities. In contrast to the contemporary emphasis on speed and progress, we propose a more deliberate "politics of re-membering," which balances accountability to residents, including the previously unaccounted for, with the need to protect individuals.

Smart Cities, Data Ownership, and Privacy

Making cities *smart* through the incorporation of information and communication technologies has been part of a normative discourse that has made technology synonymous with efficiency and resilience (Hollands 2008; Johnson 2014). Smart cities have an operating system that requires a constant stream of data to be responsive to individual users. The *smart* vision of cities and communities is an automated urban setting, where cameras and sensors are installed on nearly every surface, from traffic lights to garbage bins to living spaces (Kim, Ramos, and Mohammed 2017). In doing so, smart city technology is intended to allow cities to deliver increased convenience for residents, while also providing more efficient management of urban programs and services.

To accomplish these ends, smart city technology requires data (Kim, Ramos, and Mohammed 2017). Purveyors of smart city technology contend that the mass collection of data was needed to create a reactive environment (Batty 2013). To solve urban problems and alleviate the nagging consequences of urban life, we are told that smart city technology requires a stream of intelligence about the world with which it interacts (Batty 2013). Similarly, resources could not be deployed efficiently if not monitored. Data were touted as the fuel for the smart city: a necessary ingredient in the embrace of a digital urban future (Ahvenniemi et al. 2017).

The data collected by smart city technology are the public – the movement, habits, and predilections of those occupying urban space – but they are often held in private hands and, so, frequently commercialized. Although smart city developers contend the data are required for the community's operating system, there is an immense value to the mass collection of data (Tusikov 2019). Data are valuable and increasingly a leading commodity in the world economy. Data, unlike oil or other natural resources, are not finite and cannot be exhausted (Johnson 2014). Combined with other data sources, data can be used to build predictive systems for commercial uses or leased or sold to third parties for their own uses. Those controlling or owning more data inevitably create a monopolistic position for themselves, which creates competition for data generation and collection (Scassa 2017).

Since much of the data collected in smart city developments would be used to operate public infrastructure, these types of developments introduce a challenge to traditional municipal governance, along with a host of privacy concerns. Residents could very well enter a smart community without knowing their movements were being monitored. They would have no opportunity to accept a data collection agreement. Even those who understand how much data are being collected might not know how their data are stored or used. Given this scenario, many have called for the halt of smart city projects until proper data governance guidelines can be developed or until private smart developers submit to a litany of public demands about data collection, storage, and use (Goodman and Powles 2019).

A focus on privacy and data ownership has followed the increased development of data-intensive public infrastructure projects and new data-based technologies. This is understandable, given the mass amount of data collected by smart city projects. However, there is some danger that curtailing data collection and failing to recognize the important distinction between private and public sector data use might harm our ability to understand vital nuances about our urban communities. Data collection by public bodies is not necessarily a nefarious act. Governing requires intelligence gathering about the public that might use a particular program or service. Developing good public policy with precision cannot be done in the dark. A more equitable approach to data collection, storage, and use needs to emerge for the public to realize the benefits of good data in policymaking and ultimately serving the community's needs, while adequately protecting privacy from commercial interests. A new dialogue on data is needed, as the debate about data use in smart cities increasingly becomes polarized. Data collection is not inherently bad; data collection for commercial purposes is not necessarily passive or neutral. We need to refine the conversation to understand fully the versatile place of data in modern urban governance. In the remainder of the chapter, we suggest that the foundations for an intriguing alternative approach to data governance and data policy can be found in Jacques Derrida's engagement with notions of "biodegradability."

Lessons in Biodegradability

Derrida's interest in biodegradability begins as an extension of his more general concern for the survivability, inheritance, and influence of literature, art, and other forms of publication. Just as an ever-growing selection of products is designed to dissipate into the earth gradually, so Derrida (1989, 816) contends that the ongoing influence and memorability of a text depends on the extent to which it is absorbed and turned over into a wider discursive context. Survivability, in other words, requires that a text be revisited, reimagined, and repeated over time.

Decomposition is not neutral, however. To be taken up is to be reimagined and reinterpreted in contexts that differ from the original in which a piece was created. Thus, while biodegradability is indicative of the way a text survives across time, it also reflects the ways in which this survival depends on a corruption of, or distancing from, the original. As Michael Naas (2018, 193) reflects, "biodegradable is the name given to things destined and designed to become other than the things they are." Survival comes at a cost, then: the conversion of an original into particulates that can be translated and understood in new, alternative contexts, existing in a co-constitutive reciprocity with the surrounding soil.

For Derrida, such decomposition is a prerequisite to survivability; it is the necessary mechanism by which individual products and persons exist in co-constitutive relation to existing systems of signification, categorization, and evaluation (see Frazer and Hutchings 2011, 135). Yet total biodegradability also risks a different type of erasure. Texts (and persons) can be forgotten not only by remaining untranslated or inaccessible, but also by dissolving so much into the cultural milieu that they become unidentifiable. There is, as Derrida (1989, 863) says, "the possibility of a radical destruction without displacement, of a forgetting without remainder." Resisting this total annihilation requires a "singular impropriety" that enables a text to be revisited without ever being fully absorbed or exhausted by interpretation, something infinitely unique which escapes closed significatory systems (846).

To survive, then, is to strike a balance between absorption and resistance, repetition and singularity. "The difference," explains Naas (2018, 193), "is between *two kinds of survival,* the one anonymous, its singularity or its event indistinguishable from its context or milieu, a sort of biodegradable survival, and the other identifiable and identified ... a sort of nonbiodegradable survival."

In sum, biodegradability can be thought of as a question of time, of the dislocating temporalities of inheritance, legacy, and survivability. In this sense, there is harmony between Derrida's work on biodegradability and his work on the spectre as that which threatens to return from the past in only partial and corrupted forms. Texts survive but only at the expense of an original that has since passed away – an original that remains alluded to but never seen. "The house will always be haunted," Derrida (2006, 25) writes, "rather than inhabited by the meaning of the original."

Biodegradable Data

This connection between notions of biodegradability and the survivability of a text is significant because, despite frequent claims to the contrary, data are not a naturally occurring resource for cities and companies to extract and refine for value. Instead, they are the artificial translation of cities into a binary vocabulary that can be understood, transmitted, and modelled using increasingly complex software (Loukissas 2019, 169). Data, in other words, are the tracing through, taking up, and turning over of unique cityscapes into webs of categorization, labelling, and intelligibility (boyd and Crawford 2012, 670). In short, they are textual.

In the case of smart cities, this inscription is frequently interwoven with technocratic and managerial notions of governmentality, such that collecting data about city patterns is meant to render these patterns "visible, readable, and thereby governable" (Chandler 2015, 836). There are no data prior to encoding, however; no decipherability prior to artificiality. Data are not something waiting to be discovered, but rather something produced through a series of inherently political choices. "Raw data," writes Gitelman (2013, 1), "is an oxymoron."

While this inscription renders cities legible, it also obfuscates the unique complexities of cities behind decisions regarding what type of information to collect, how to collect it, and, of course, how to analyse it. As Ben Green (2019) writes, data-based technologies provide "myopic models that do not – for they cannot – capture the full complexity of society" (85). Data remain partial and particulate references to events, individuals, and spaces that remain elsewhere – a pixelated photograph that marks the withdrawal of the very thing it captures (Anderson 2006, 8). Put differently, there remains a spacing between the city *as such* and its appearance in the form of data.

Even in the case of seemingly instantaneous technologies, there is a deferral, a latency between input and output that ensures that cities are only ever encountered on delay (Kitchin 2017, 28). Rather than an immediate and objective view of the city, then, datafication is the inheritance of a city that is already passing away. It is a matter of conjuring: the creation of mirror worlds (Gelernter 1993, 5) and data doubles (Haggerty and Ericson 2000, 606) that only ever point to a world that has already disappeared. Data are "the phantom of a friend returning" (Derrida 2005, 75). In contrast to idealizations of instantaneous governance and the real-time city, then, notions of biodegradability hint at "the more corruptible image of the spectral city – an incomplete city haunted by the ghosts (and composts) of the past" (Olmstead 2019, 3).

Like a bottle seeping its contents into the ground, this translation of cities into data depends on the sociopolitical and sociotechnical contexts in which datafication takes place. There are no data without context (boyd and Crawford 2012, 670). On the one hand, this means that decisions about collection, storage, and analysis are often imbued with the assumptions of those in positions of power. These decisions are further constrained by the limitations of contemporary computation and hardware, which frequently force designers to decide between detail and efficiency (Kitchin 2017, 28).

On the other hand, processes of datafication take place in an urban soil already saturated with other systems of signification, evaluation, and power. As a result, data-based technologies tend to reflect and reproduce existing inequalities. Based on data accumulated in the context of overpolicing and systemic racism, for example, predictive policing algorithms have tended to allocate disproportionate resources to already marginalized communities and to reinforce the disproportionate incarceration of Black and Indigenous residents (Rieland 2018). Somewhat less egregiously, yet no less interwoven with histories of racism and colonialism, data collected through 311 applications have tended to privilege not only the technologically savvy, but also those most comfortable interacting with their local government – namely, affluent white residents. "The public might have a new tool to inform municipal operations," writes Green (2019, 43), "but that tool is employed within the framework of traditional relationships."

Data nevertheless remain an important way for individuals to exert influence over their increasingly technological cities. Like decomposing particulates turning over in the soil, data gathered about cities influence the very context in which they were created. This is especially true in the context of data-based technologies, wherein contemporary datafication not only influences outputs, but, in doing so, influences future renderings of those same technologies. Loukissas (2019, 52) describes this process as an "ontological looping effect whereby [data] help to shape the practices and institutions that create them."

Within these "recursive processes," there are risks to remaining unaccounted (Beer 2017, 4). In particular, those excluded from datasets risk being victimized by the decisions made based on those data. In our own Canadian context, for example, responses to COVID-19 in many cities failed to capture the disproportionate impact of the virus on the lives and well-being of racialized and lower-income communities. As a result, cities failed to introduce early interventions to counter these trends. These oversights and erasures manifest even more directly in the case of machine-learning algorithms, which depend on data for their training and implementation. To name a popular example, facial recognition technology trained on exclusively white faces tends to misidentify people of colour, especially Black and Indigenous women, at disproportionately higher rates than white faces (Vincent 2019).

Even if datafication represents a type of decomposition, then, there are risks to resisting this erosion outright. In short, to borrow a phrase from Karen Barad (2018, 226), "erasure is a material practice that leaves its trace in the very worlding of the world." Given the proliferation of smart city initiatives, concerns over privacy must therefore be balanced with a recognition that datafication is an increasingly common antecedent to being accounted for within the city. That is, rather than a tool for optimization and coordination, urban data are repositioned as the "re-turning" through which delayed renderings of cities and their residents come to "nourish the infinitely rich ground of possibilities for living and dying otherwise" (242). Urban data are the means by which individuals exert influence over digital space – not in real time but on a delay that always implies futurity. It is through data, in other words, that unique cityscapes come to influence the "ongoing re-worlding of the world" (241).

With this in mind, notions of biodegradability orientate digital governance towards what Barad calls a politics of "re-membering" (2018, 229). That is, before speed, optimization, and efficiency, responsible governance is ensuring that all are accounted for – that nobody gets left behind. This includes the work of making previous erasures known and recognizing the discourses and textualities that have been marginalized by the focus on neoliberal and capitalistic processes of encoding and evaluation. As smart cities come to depend on data generated entirely through digital interfaces, critics must also be wary of the potential alienation of those without access to, or appropriate knowledge of, the technologies on which these infrastructures depend. More generally, re-membering requires opening discursive space to those whose unique positionalities have slipped through the cracks of hegemonic systems of categorization.

Such a project is nevertheless haunted by the fact that any account is doomed to be incomplete and insufficient. Each account is unavoidably a decomposition of the thing it hopes to represent – an infinitely unique and complex individual that will be lost in translation. And of course, there are past erasures that cannot be rectified. There are those "who can no longer be counted, whose cultures, livelihoods, peoples, or species are gone forever" (Olmstead 2019, 15). A politics of re-membering is thus also a "work of mourning," always cognizant of what is (and has been) lost along the way (Derrida 2006, 9). Instead of viewing datasets as totalizing and enclosed captures of an objective reality, then, practices of re-membering are never complete. There is always more work to do. There is more turning-over yet to come.

Admittedly, there are risks to biodegradable data in the technocratic and managerial landscape of the contemporary smart city movement. In ways that stretch the limits of Derrida's metaphor, unprecedented growth in computational power means that cities and companies now wield the power to reassemble particulates in ways that, even if removed from a phenomenological "as such," nevertheless make it possible to identify particular individuals even from seemingly innocuous datasets (Green 2019, 93). Insofar as decomposition takes place in the context of existing power structures and inequalities, this degree of identifiability-through-data can prove disastrous for those most vulnerable. Calls for more re-membering run headlong into the realities of surveillance capitalism.

To this understandable challenge, we offer two disclaimers. First, though notions of biodegradability certainly lend themselves to a framing of data-as-accountability, they are also a rejection of the top-down approaches to digital governance that have thus far determined data-based technologies. Data are not an objective measurement from above, but the product of an iterative relationship between text and context that ensures individuals are only ever rendered through processes of differentiation and, consequently, relationality (Matzner 2016, 206). That is, like subjectivity more generality, the textuality of data ensures that residents are only ever rendered as subjects in the form of self-referential categories and co-constitutive discourses. Somewhat paradoxically, data thus de-centre the very subjects they aim to pinpoint, decomposing individuals into a soil that, like the ecology of cities themselves, is interconnected and tentacular. Data are, to borrow Haraway's (2016, 55) phrase, "partnered all the way down" – even so far as a simple binary.

On the one hand, this means that data tend to reflect, and provide insight into, the patterns of relationality, power, and inequality which encase the very subjects they aim to identify. On the other, and in contrast to the top-down managerialism that has characterized the smart city movement, notions of biodegradability recast data-based technologies in terms of a democratic and participatory vision of urban governance. Data are not a pre-existent by-product to be collected and exploited by managerial governments or private corporations. They are public from the very beginning. Decomposition is the responsibility of the soil.

In the context of urban data, this entails access not just to data themselves, but also to the preliminary decisions made about data collection. As Derrida (1996, 4) writes, "effective democratization can always be measured by this essential criterion … the participation in and access to the archive, its constitution, and its interpretation." Re-membering is therefore something of a procedural question before it is ever about who makes it into the data. It is a matter of ensuring that decisions over how data are collected, stored, and used remain in the soil, made by those most affected by those decisions. Indeed, if biodegradability aims to make sure all are accounted for in the data, it also works to ensure that all are accounted for in the policymaking process. Paradoxically, control over data can influence the types of data residents might be comfortable sharing in the first place (Van Zoonen 2016, 472).

Second, and relatedly, data-as-biodegradable implies a recognition that data-based technologies are only one form of textuality among many through which individuals are rendered present within cities. Rather than technology for technology's sake, in other words, notions of biodegradability point cities towards the question of whether digital technology is the most effective means of responding to *all* residents. In some cases, responsibility might even entail more direct, less private encounters with those positioned to be affected by certain policies. In this context, data should be thought of as nothing more than a "point of contact" that points to more complicated persons and processes that remain out of view

(Loukissas 2019, 196). As Derrida (2006, 126) puts it, "the non-presence of the specter demands that one take its time and its history into consideration, the singularity of its temporality or of its historicity." Data thus point towards the opening of a space in which particulate individuals might speak for themselves, even in ways that threaten to undo the models and assumptions that have thus far dominated the discourses. Through data, cities are directed towards the muck – the tentacular hummus of their unique cityscapes. As Loukissas (2019, 196) writes, "do not mistake the availability of data as permission to remain at a distance."

Conclusion

Data-based technologies are becoming an important part of urban governance. As municipalities around the world embrace digital infrastructure, there are increasing risks to remaining unaccounted for in the datasets with which these technologies operate. At the same time, the tendency of smart cities towards technocratic, managerial, and privatized forms of governance has raised important concerns about privacy and data ownership.

In this chapter, we have proposed an alternative understanding of the relationship between cities and their data based on poststructuralist notions of "biodegradability." In particular, we have suggested that data represent the decomposition of unique cityscapes into co-implicated and self-referential networks of differentiation, deferral, and accountability. Rather than a resource with which cities can be coordinated and optimized, data are the means by which cities might come to understand and respond to the networks of power and relationality that determine urban space.

In turn, notions of biodegradability implicate digital governance in a "politics of re-membering." Responsibility entails ensuring that all are accounted for, a process we have suggested is always, already, a work in progress. Although this might seem an unconditionally pro-data position, we have suggested that a politics of re-membering raises broader procedural questions and replaces the top-down managerial approach of the smart city with a more tentacular and democratic politics. Cities require a new relationship with data, one that avoids a patchwork of policy responses that centralize decision-making about the purposes of data and data collection in urban life. In doing so, they will find a more robust equilibrium that provides intelligence for policymaking, a central spine of community life and a protection for necessary privacy.

REFERENCES

Ahvenniemi, H., A. Huovila, I. Pinto-Seppä, and M. Airaksinen. 2017. "What Are the Differences between Sustainable and Smart Cities?" *Cities* 60 (February): 234–45.

Anderson, N. 2006. "(De)constructing Technologies of Subjectivity." *Journal of Media Arts and Culture* 3 (3): 1–9. Online at https://research-management.mq.edu.au/ws/portalfiles/portal/16985370/mq-1139-Publisher+version+%28open+access%29.pdf

Barad, K. 2018. "Troubling Times and Ecologies of Nothingness: Re-turning, Re-membering, and Facing the Incalculable." In *Eco-Deconstruction: Derrida and Environmental Philosophy*, ed. M. Fritsch, P. Lynes, D. Wood, F. Clingerman, and B. Treanor, 160–86. New York: Fordham University Press.

Batty, M. 2013. "Big Data, Smart Cities and City Planning." *Dialogues in Human Geography* 3 (3): 274–9. https://doi.org/10.1177/2043820613513390

Beer, D. 2017. "The Social Power of Algorithms." *Information, Communication & Society* 20 (1): 1–13. https://doi.org/10.1080/1369118X.2016.1216147

boyd, d., and K. Crawford. 2012. "Critical Questions for Big Data: Provocations for a Cultural, Technological, and Scholarly Phenomenon." *Information, Communication & Society* 15 (5): 662–79. https://doi.org/10.1080/1369118X.2012.678878

Chandler, D. 2015. "A World without Causation: Big Data and the Coming Age of Posthumanism." *Millennium* 43 (3): 833–51. https://doi.org/10.1177/0305829815576817

Derrida, J. 2006. *Spectres of Marx: The State of the Debt, the Work of Mourning and the New International*. Trans. P. Kamuf. New York: Routledge Classics.

Derrida, J. 2005. *The Politics of Friendship*. Trans. G. Collins. London: Verso.

Derrida, J. 1996. *Archive Fever: A Freudian Impression*. Trans. E. Prenowitz. Chicago: University of Chicago Press.

Derrida, J. 1989. "Biodegradables: Seven Diary Fragments." Trans. P. Kamuf. *Critical Inquiry* 15 (4): 812–73. https://doi.org/10.1086/448522

Frazer, E., and K. Hutchings. 2011. "Remnants and Revenants: Politics and Violence in the Work of Agamben and Derrida." *British Journal of Politics and International Relations* 13 (2): 127–44. https://doi.org/10.1111/j.1467-856X.2010.00428.x

Gelernter, D. 1993. *Mirror Worlds: Or the Day Software Puts the Universe in a Shoebox… How It Will Happen and What It Will Mean*. Oxford: Oxford University Press.

Gitelman, L. 2013. *Raw Data Is an Oxymoron*. Cambridge, MA: MIT Press.

Goodman, E.P., and J. Powles. 2019. "Urbanism under Google: Lessons from Sidewalk Toronto." *Fordham Law Review* 88 (2): 457–98. https://doi.org/10.2139/ssrn.3390610

Green, B. 2019. *The Smart Enough City: Putting Technology in Its Place to Reclaim Our Urban Future*. Cambridge, MA: MIT Press.

Haggerty, K., and R. Ericson. 2000. "The Surveillant Assemblage." *British Journal of Sociology* 51 (4): 605–22. https://doi.org/10.1080/00071310020015280

Haraway, D. 2016. *Staying with the Trouble: Making Kin in the Chthulucene*. Durham, NC: Duke University Press.

Hollands, R.G. 2008. "Will the Real Smart City Please Stand Up?" *City* 12 (3): 303–20. https://doi.org/10.1080/13604810802479126

Johnson, J.A. 2014. "From Open Data to Information Justice." *Ethics and Information Technology* 16 (4): 263–74. https://doi.org/10.1007/s10676-014-9351-8

Kim, T.-H., C. Ramos, and S. Mohammed. 2017. "Smart City and IoT." *Future Generation Computer Systems* 76 (November): 159–62. https://doi.org/10.1016/j.future.2017.03.034

Kitchin, R. 2014. "The Real Time City? Big Data and Smart Urbanism." *GeoJournal* 79: 1–14. https://doi.org/10.1007/s10708-013-9516-8

Kitchin, R. 2017. "The Realtimeness of Smart Cities." *Technoscienza: Italian Journal of Science & Technology Studies* 8 (2): 19–42.

Loukissas, Y. 2019. *All Data Are Local: Thinking Critically in a Data-driven Society.* Cambridge, MA: MIT Press.

Matzner, T. 2016. "Beyond Data as Representation: The Performativity of Big Data Surveillance." *Surveillance & Society* 14 (2): 197–210. https://doi.org/10.24908/ss.v14i2.5831

Murakami Wood, D., and D. Mackinnon. 2019. "Partial Platforms and Oligoptic Surveillance in the Smart City." Surveillance & Society 17 (1/2): 176–82. https://doi.org/10.24908/ss.v17i1/2.13116

Naas, M. 2018. "E-Phemera: Of Deconstruction, Biodegradability, and Nuclear War." In *Eco-Deconstruction: Derrida and Environmental Philosophy*, ed. M. Fritsch, P. Lynes, D. Wood, F. Clingerman, and B. Treanor, 187–205. New York: Fordham University Press.

Olmstead, N. 2019. "Data and Temporality in the Spectral City." *Philosophy & Technology* 34 (December): 1–21. https://doi.org/10.1007/s13347-019-00381-8

Rieland, R. 2018. "Artificial Iintelligence is now used to predict crime. But is it biased?" *Smithsonian Magazine*, 5 March. Online at https://www.smithsonianmag.com/innovation/artificial-intelligence-is-now-used-predict-crime-is-it-biased-180968337/, accessed 13 January 2021.

Robinson, P., and S. Coutts. 2019. "The Case of Quayside, Toronto, Canada." In *Smart City Emergence: Cases from Around the World*, ed. L. Anthopoulos, 333–50. New York: Elsevier.

Scassa, T. 2017. "Sharing Data in the Platform Economy: A Public Interest Argument for Access to Platform Data." *UBC Law Review* 50 (4): 1017–71.

Srnicek, N. 2017. *Platform Capitalism*. Cambridge, UK: Polity.

Taylor, L. 2017. "What Is Data Justice? The Case for Connecting Digital Rights and Freedoms Globally." *Big Data & Society* 4 (2): 1–14. https://doi.org/10.1177/2053951717736335

Townsend, A. 2013. *Smart Cities: Big Data, Civic Hackers and the Quest for a New Utopia*. New York: W.W. Norton.

Tusikov, N. 2019. "Precarious Ownership of the Internet of Things in the Age of Data." In *Information, Technology and Control in a Changing World: Understanding Power Structures in the 21st Century*, ed. B. Haggart, K. Henne, and N. Tusikov, 121–48. Cham, UK: Palgrave Macmillan.

Valverde, M., and A. Flynn. 2018. "'More Buzzwords than Answers' – To Sidewalk Labs in Toronto." *Landscape Architecture Frontiers* 6 (2): 115–23. https://doi.org/10.15302/J-LAF-20180212

Van Zoonen, L. 2016. "Privacy Concerns in Smart Cities." *Government Information Quarterly* 33 (1): 472–80. https://doi.org/10.1016/j.giq.2016.06.004

Vincent, J. 2019. "Gender and racial bias found in Amazon's facial recognition technology (again)." *Verge*, 25 January. Online at https://www.theverge.com/2019/1/25/18197137/amazon-rekognition-facial-recognition-bias-race-gender, accessed 13 January 2021.

Wylie, B. 2017. "Think Hard Before Handing Tech Firms the Rights to Our Cities' Data." *Information Policy*, 13 November. Online at https://www.i-policy.org/2017/11/think-hard-before-handing-tech-firms-the-rights-to-our-cities-data.html.

Zuboff, S. 2015. "Big Other: Surveillance Capitalism and the Prospects of an Information Civilization." *Journal of Information Technology* 30: 75–89. https://doi.org/10.1057/jit.2015.5

PART THREE

Infrastructures of Injustice

A Dialogue with Vincent Mosco

Part Three of the book foregrounds empirical cases of smart cities, focusing on the often-vanishing smart infrastructures that increasingly (in)script urban life. Exploring a range of technologies – mobile applications, artificial intelligence (AI), transportation systems, parking, and supranational policy – authors in this part grapple with enduring questions of privatization, corporatization, and automation, as well as the implications these infrastructures have for manifold forms of justice.

To contextualize the vanishing, mutating, and mundane infrastructures of the smart city, here we dialogue with Dr Vincent Mosco, Professor Emeritus in the Department of Sociology at Queen's University and Distinguished Professor, New Media Centre, School of Journalism and Communication, Fudan University, Shanghai. Dr Mosco is former Canada Research Chair in Communication and Society, and his work focuses on the political economy and cultural analysis of media and communications technologies.

Editors: In your 2005 book, The Digital Sublime, *you discuss the power of myths. Specifically, you explore how technological transformations around "the cyber" have brought with them promises of transcendence – offering ways to lift us out of the monotony of everyday life. Many authors in this part – in particular, Alberto Vanolo, Liam Heaphy, and Karol Kurnicki – explore mundane and everyday infrastructures of the smart city. Heaphy stresses that "much of the digital revolution is low key, gradualist, and developed through broad assemblages of companies, city engineers, and research with little fanfare." With an eye towards the emergence of autonomous technologies, Federico Cugurullo reminds us of their "smart" inheritance and ethical implications. Considering remediation and precursory technologies, could you explain some of the pressing connections between the digital sublime and the smart city movement? How did we get here?*

VM: In almost any project I take on, whether it's examining cloud computing, the Internet of Things, or, in this case, smart cities, it is incumbent to look at the ways in which the social whole is constituted in the specific subject under investigation. In this case, having described some of the broadly political, economic, and technological contours of the smart city, I then moved on to a discussion of the culture of the smart city; primarily through the entry point that I used in *The Digital Sublime*: the concept of myth.

I examine myth as a transcendent narrative that people use to make sense of the world in which they live. We make myths whenever we make technology, but I think it is also the case that we make myths whenever we make smart cities. In *The Smart City in a Digital World*, I decided to look back at some of the central mythologies around cities and how they originated, and then examine some of the questions they were trying to address or narratives they were interested in developing.

I looked at Ebenezer Howard's "garden city," Le Corbusier's "radiant city," and Jane Jacobs's "urban village." From there I moved on to Richard Florida's "creative city" and then to the smart city of today. These are not just concepts one uses to analyse the material reality of a city; they are also myths and stories, narratives and imaginaries that provide people with a sense of the culture of the city.

In this case, the smart city is in itself a mythology, a story about how technology can transform the massive materiality of the city into a well-functioning machine. Today's smart city advocates see the city as a computer, as a rational instrument for making life – specifically, communication, transportation, and energy use – more efficient and effective. By building what amounts to the city as a computer or a platform, as some are describing it, one can make major improvements in the challenges facing urban life. In essence, the smart city is a cultural object and not just a set of technologies.

Editors: Similar to your point, Alberto Vanolo argues that "technologies become smart city technologies by their association with certain narratives, logics, practices, symbols, and ideologies." Focusing on the Italian COVID-19 context, Vanolo highlights the numerous ways the digital sector has been reconfigured. In your view, is there a city that exists outside the myth? Is there a city that underlies the smartness myth to which the cultural myth then attaches?

VM: The answer to that is embedded in my notion of mutual constitution. That is, there is never a material world and a mythic world that are separate from each other. They are constantly in the process of making and remaking each other. They are, in a real sense, mutually determinant or, if you choose, mutually constituted: myth exists outside the realm of the city as a material object, but the city as a material object also exists outside the realm of myth, and they work on one another to constitute the city.

Editors: In the pursuit of justice, do we need to abandon the smartness framework and, subsequently, the smartness mythology?

VM: I tend not to go as far as to argue for conceptual abandonments because that means taking on board something that we may or may not be in a position to accomplish. I think that what we need to do is make use of dominant narratives to understand how people bring meaning to the city they're in.

We are constantly challenging the mythos, as I demonstrated in my book. It is one thing to understand Ebenezer Howard's myth of the garden city, but, at the same time, it's important to comprehend how that fed into debates about eugenics, so that we both understand a dominant way of imagining a city and critique it. On the way to justice, we absolutely need to critique the notion of the smart city. In my view, the concept of the smart city is not ours to abandon. Rather, it is ours to understand and make use of for what it is we want to achieve.

In this sense, what you are asking about is more of a political question. That is, if you want to achieve justice in the smart city, do you abandon the notion of smart? In my view, you accept it as the grounding of current debates, and you join in those debates. It does not mean fully accepting the notion of the smart city. Yes, I want to create a smart city. And no, it does not start with a surveillance-driven Operations Centre run by IBM, Cisco, or Siemens. Instead, my conception of smartness is a city that is democratic, that is just, and that respects the rights of its residents to make determinations about their own lives in the city. That is precisely what smart should mean.

I think it is fine to talk about the smart city because that is on the agenda today. Abandoning the concept of the smart city makes it more difficult to be a part of an ongoing debate or discussion. Would I use the term if I had the opportunity to start the debate? Probably not. But that is what I have been given, and I would be satisfied to take over the concept and make it what I think it should be. It is important for us to be part of the debate, not just within the scholarly community, but also in the communities where we live and where we need to be activists.

Editors: Throughout your work, you have highlighted the social justice consequences of post-industrial thinking and consumption. In The Smart City in a Digital World, *you question "how smart it is to entrust the future of our cities to billionaires who are fuelling greater inequalities."*

Authors in this part of the book investigate the intersections of digitization, automation, responsibilization, privatization, and marketization. For instance, Torin Monahan demonstrates how "private platforms are rapidly becoming the interfaces through which people experience the city and themselves in it." Kurnicki cautions that our reliance on "digitalization and automation is set to maintain the inequalities and injustices that have roots in wider urban issues, such as ownership of land, planning, or the financialization of every possible urban resource, including space itself."

Therefore, as spaces are captured and privatized, governance is handed over to tech companies, and citizenship is recast through user agreements, what are new or existing questions that we need to be asking about linkages between justice, smartness, and the city? What are the social justice consequences of digital smart thinking and consumption?

VM: One of the central points *The Smart City in the Digital World* makes is that governance is of utmost significance. It's not about hiring a company to install technologies, as if these were neutral instruments to produce efficiencies. Any decision about a city involves governance, and so I chose to cut into the many questions surrounding smart cities by looking at different forms of governance and different ways of thinking about governance. In essence, to think about governance is a step on the way to building a genuinely democratic and thereby genuinely smart city.

There are many ways to interpret justice, but I get to justice through democracy, which is the fundamental right to control your own life and join with fellow citizens in the construction of community. It means advancing the fullest possible public participation in the decisions that affect our lives. Making an organization, a nation-state, or a city more democratic seems to me to be a central goal in helping people to realize their own humanity, giving them greater control over their own lives and over their own communities. When I think about smart cities, I imagine places, with or without advanced technology, that adhere to some vision of giving people the power they need to govern themselves.

My critique of the smart city, in a broadly political sense, looks at how technology as an index of smartness has become a Trojan horse for advancing undemocratic ideas about a fundamental restructuring of the city around control by private enterprise over the public interest that turns cities, including streets, sidewalks, and other public places, into the private property of global companies – increasingly Big Tech companies. The notion here is that, if you support democracy, then you oppose, in most cases, public-private partnerships and so-called business improvement districts because these have been instruments to support private interests and restrict citizens' rights.

Editors: You have characterized the smart city movement as a distraction. Connected to your work on the environmental effects of cloud computing, smart cities pose massive environmental implications (while simultaneously promising to overcome them). Citing Chris Hume, you note: "the smart city industry is a Trojan horse for technology companies. They come in under the guise of environmentalism and improving quality of life, but they are here for the money" (Mosco 2019, 232). In this part of the volume, Lorena Melgaço and Lígia Milagres also challenge these forms of greenwashing, arguing that smart cities' rationalities reinforce colonial practices and inequalities between the global North and South. Can you say more about the intersections of environmental and digital justice?

VM: Consider cloud computing, a major technology in smart city discussions. There is now, thankfully, a strong and growing literature about the environmental impacts of the power requirements of cloud systems. The sheer power needed to keep computers, servers, and data centres going is massive. Nevertheless, in many cases, the technological infrastructure is not taken into account in examining environmental impacts.

Environmental issues are very serious, but are, by and large, ignored in the public relations campaigns touting energy efficiency that come from discussions about smart cities. The notion that we can make our cities more energy efficient by deploying advanced technologies, without considering their own power requirements and the extent of environmental pollution, including e-waste, that grows out of these technologies, is seriously problematic.

We need to account for the environment in a much broader way than we have when it comes to smart cities. We're beginning to see some of this now in the challenges brought against, for example, Google's Sidewalk Labs project in Toronto, where opponents raised serious environmental questions. It is encouraging to observe that social movement organizations – and there are many of them that have arisen in the context of debates about smart cities – are beginning to account more thoroughly for environmental issues.

More broadly, it is important to account for popular resistance to technology-driven smart cities, which has been growing in recent years. This includes, for example, in New York City, where activists forced Amazon to back away from a major development project because they concluded it would provide few meaningful jobs. The same happened in Toronto, where resistance led Google to end its Sidewalk Labs project. In each of these cases, environmental consequences played a significant role in resistance to the Big Tech version of the smart city.

Editors: At the end of The Smart City in a Digital World, *you offer a manifesto for the smart city. Given that smart and its mutations are perhaps here to stay, you argue that we must reconsider "what is smart." Specifically, you posit that: (1) people make cities smart; (2) smart cities are democratic cities; (3) smart cities value public space; (4) smart cities share data; (5) smart cities defend privacy; (6) smart cities do not discriminate; (7) smart cities preserve the right to communicate; (8) smart cities protect the environment; (9) smart cities and their streets are about people; and (10) smart cities deliver services.*

Other contributors in the volume expand this list to suggest that the smart city embraces informality, or idiocy (see Alan Smart and Alison Powell, in particular). Clearly you see a purpose in reclaiming "smart"; can you talk more about this and why? Is justice possible when smart qualifies cities? What sort of work should the commons or utopia do in our conceptions and visions of a just smart city, or perhaps simply a just city?

VM: I think both the concepts of a "just smart city" and a "just city" are important. Depending on the venue, one might use one or the other in making arguments, but we want just cities that are governed not by private companies, or oppressive governments, but as fully as possible by citizens themselves. Technology plays a role, but the most important role is played by people living in communities who are entitled to a set of basic rights, including the right to the democratic operation of their community and their city, and rights to fundamental social services, including housing, education, and health care. All discussion of what makes a city smart and just must begin with the fundamental needs of its citizens.

What does it take to be a just smart city? Well, what does it take to create democratic communities that work on behalf of the people who participate in those communities and on behalf of the land they occupy, and in ways that advance their fundamental rights?

I would challenge fundamentally any discussion of how to make a city better place solely through the use of technology. I insist that, as a starting point to achieve smartness, we need to mobilize our communities; we need to bring people together to understand how to make those communities better; we need fair housing; we need a better environment; we need to assure that our schools are better; we need to build better public infrastructure. It is not a question of how to get people to accept smart technology. Rather, it means remaking the concept of smart itself, in terms of what we want to achieve. That's part of the process of critiquing and debunking technological determinism: a recognition that technology short-circuits all of the major questions we've ever asked about communities.

Editors: In your work you have traced practices of digitization and their implications Kurnicki contends that "[d]igitalization, while disruptive in some ways, entrenches existing spatial relations where immobility transformations are critically important for the survival of urban life. In this context, digital justice has to go hand-in-hand with mobility justice." From thinking about sites of care-full justice (see Teresa Abbruzzese and Brandon Hillier) to actually-existing social justice (Inka Santala and Pauline McGuirk) authors across this volume emphasize the importance of urban agency. Can you make explicit or articulate your approach to social justice?

VM: Any approach to social justice depends on the conditions that we face and the problem we aim to address. Today, social justice grows out of a recognition of people's fundamental rights to a set of services, including housing, education, vibrant communities, a good environment, and the opportunity to control the decisions that affect our lives. So, justice, for me, combines both rights and democracy and is a set of guarantees of the conditions that can make one a full and effective citizen, broadly understood, and provide opportunities to control the decisions that affect one's life. In essence, we fight against whatever gets in the way of achieving human rights and against what blocks our ability to control the decisions that affect our lives. Our struggles are against the erosion of those values and for expanding those values to all of humanity and to the non-human world that we inherit and that we require for a better world.

Practically speaking, if we're talking about smart cities, it means that people have the right to the data collected about them, whether in an analogue or in a digital form. This is fundamental. So yes, we might use technologies to make cities more efficient, but people need to be involved in the construction of the data and need to understand the systems that generate data about them.

We are now finally opening the black box of algorithms and recognizing the inherent biases that are involved in systems created by a narrow elite, whether of engineers or policy makers. We are also recognizing that better results come from a more democratic development of our instrumentation. Resisting racist, sexist, and classist algorithms is an act of social justice today and central to the construction of genuinely democratic (and smart) cities.

FURTHER READING

Florida, R. 2018. *The New Urban Crisis*. New York: Basic Books.

Harvey, D. 2019 *Rebel Cities*. London: Verso.

Howard, E. 1946. *Garden Cities of Tomorrow*. London: Faber.

Jacobs, J. 2011. *The Death and Life of Great American Cities, 50th Anniversary Edition*. New York: Modern Library.

Le Corbusier. 1987. *The City of Tomorrow and Its Planning*. New York: Dover Books.

Mosco, V. 2005. *The Digital Sublime: Myth, Power, and Cyberspace*. Cambridge, MA: MIT Press.

Mosco, V. 2015. *To the Cloud: Big Data in a Turbulent World*. New York: Routledge.

Mosco, V. 2017. *Becoming Digital: Toward a Post-Internet Society*. London: Emerald Group.

Mosco, V. 2019. *The Smart City in a Digital World*. London: Emerald Group.

Newitz, A. 2021. *Four Lost Cities*. New York: Norton.

Ratti, C., and M. Claudel. 2016. *The City of Tomorrow: Sensors, Networks, Hackers, and the Future of Urban Life*. New Haven, CT: Yale University Press.

Rossi, U. 2017. *Cities in Global Capitalism*. London: Polity.

Rudofsky, B. 1987. *Architecture Without Architects: A Short Introduction to Non-Pedigreed Architecture*. Repr. Albuquerque: University of New Mexico Press.

10 Good and Evil in the Autonomous City

FEDERICO CUGURULLO

Everyday Trolley Problems

Artificial intelligence (AI) is becoming a prominent element in the governance, planning, and everyday experience of cities (Batty 2018; Cugurullo 2020). AI operates across different scales and assumes heterogeneous shapes, thereby impacting multiple spheres of urban sustainability (Yigitcanlar and Cugurullo 2020). AI is in urban transport as an autonomous car, in foodservice as a service robot, and in urban governance as an autonomous digital platform. Yet, despite such diversity of urban forms and roles, AI technology presents the same crucial ethical challenge: understanding what is *right* and *wrong* in the city and acting accordingly. In abstract terms, this challenge can be understood as a trolley problem, a classic thought experiment in ethics asking the participant to visualize a situation in which a runaway trolley is heading fast towards a group of people unable to move. You are a bystander and have the option to pull a lever that would divert the trolley onto a side track where it would hit and kill only one person. What would you do?

In AI studies, the question becomes: *what would an AI do?* AI is not a mere spectator. It is a sentient technology capable of choosing and, above all, in the position of having to choose in urban contexts where harm could be unavoidable: autonomous cars traversing roads where a jaywalker might be around the corner; algorithms choosing which customer will get a mortgage and be able to buy a flat; intelligent digital platforms deciding whether to build more premium office space or focus on social housing. These are all trolley problems that AIs face every day in the city. The aim of this chapter is to examine critically how diverse artificial intelligences process issues of ethics in the city. *Justice*, a key ethical concern since antiquity, is one of the main dimensions through which I explore the ethics of AI, specifically in connection with urban development. Given the thematic focus of the chapter, I approach the concept of justice in a multidisciplinary manner, by drawing upon the insights of moral philosophers, urbanists, and AI scholars. I evoke notions of *justice* and *injustice* in relation to different AIs operating in the city to illuminate a wide spectrum of ethical issues, ranging from unwanted bias and uneven distribution to unavoidable harm and the need for public political debate.

The structure of this chapter is fourfold. First, I illustrate the emergence of AI in cities during the passage from *smart* to *autonomous* urbanism. Second, I explain the ethics of AI, and, third, how broad ethical questions, such as justice and non-maleficence, relate specifically to urban artificial intelligences. In the conclusion, I question the fairness of an urban future shaped by a non-biological intelligence, and stress the need to politicize AI and its ethical choices.

From the Smart City to the Autonomous City

The rise of AI is changing the dynamics and very essence of smart city initiatives, thereby forming a new type of urbanism called *autonomous urbanism* (Cugurullo 2020). Since artificial intelligence is crucial to the formation and understanding of autonomous urbanism, it is important to clarify the nature and capabilities of artificially intelligent entities. Although a single and universal definition of AI does not exist, decades of studies in the fields of computer science, engineering, and, more recently, social sciences and humanities can help us understand what artificial intelligences are and do in general.

First, AI is *artificial* in the sense that it is not the product of a natural process of evolution, as in the case of the human brain, for example. AI is human made or it can be made by machines, and normally resides in an artifact whose shape can range from a personal computer to a car. Second, the *intelligence* of AI is expressed by means of five interconnected key skills: learning, extracting concepts, handling uncertainty, behaving rationally, and acting autonomously. AIs are capable of acquiring information on the surrounding environment (Russell and Norvig 2016), both directly through sensors such as cameras and microphones and indirectly through large datasets installed by the developers. AIs make sense of the acquired information by extracting concepts from it (Bostrom 2017). This is how an AI, for instance, can understand that the colour red in traffic lights means *stop* and green signifies *go ahead*. AIs should use the information that has been acquired and comprehended, to act in uncertain situations (Kanal and Lemmer 2014). In cases where some information might be missing, incomplete, or unclear, AIs would act rationally according to predefined performance measures and goals supposed to clarify what is right or wrong (Russell and Norvig 2016). Finally, AIs would exercise all these skills in an autonomous manner. Being *autonomous* means that AIs are unsupervised (Levesque 2017). They can learn about their environment and act within it, with humans out of the loop.

These core skills and capabilities denote what makes *AI* different from *smart*, but also what AI tech has in common with traditional smart tech. On the one hand, AI tech relies on smart tech. For example, AIs need sensors in order to learn directly about the surrounding environment, and they need to be fed with large datasets, the so-called *Big Data*, to learn indirectly about any possible subject whose understanding has been digitalized. AIs also need the Internet of Things (IoT) to transfer and circulate in *real time* knowledge among themselves. There is nothing new about sensors, Big Data, and IoT. These technologies and techniques of data collection have been around for decades and represent the essence of smart urbanism (Karvonen, Cugurullo, and Caprotti 2018; Kitchin 2014). On the other hand, AI tech goes beyond smart tech in terms of functions, capabilities, and related urban challenges. Smart sensors, Big Data, and IoT are points of intersection between smart urbanism and emerging autonomous urbanism, presenting the city with a new set of technologies and issues (Cugurullo 2020, 2021).

Autonomous urbanism is characterized by three main technologies that constitute the material embodiment of AI in cities: autonomous cars, robots, and city brains (Cugurullo 2021). *Autonomous cars* are automobiles driven by artificial intelligence, and their employment in cities is rapidly increasing (Cugurullo et al. 2020; Milakis, Van Aren, and Van Wen 2017). An autonomous car is capable of sensing the surrounding urban spaces by means of cameras, radars, and lidar systems, and it can learn about the city through downloadable datasets such as roadmaps and weather forecasts. At the highest level of autonomy (level 5), no human input is required, and the autonomous car theoretically should be able to handle uncertain situations by itself. *Robots* are a multiform category that, unlike autonomous cars, operate in many different urban domains. They range from unmanned air vehicles (drones) to humanoid machines, and from minuscule nanobots to androids (Russell and Norvig 2016). Their presence in cities is substantial and visible in retail, customer service, education, hospitality, security, and in the maintenance of urban infrastructure (Macrorie, Marvin, and While 2021; Tiddi et al. 2019; While, Marvin, and Kovacic 2020). Like most artificial intelligences, robots sense, learn, and ultimately act in situations full of uncertainty. On these terms, service robots in particular are emblematic, given that they often operate on the front line and have to accommodate the unknown requests of unknown customers without the help of human operators (Wirtz et al. 2019). Finally, *city brains* are large-scale artificial intelligences located in a digital platform that controls entire parts of the governance of cities, such as transport, security, health, and planning (Zhang et al. 2019). Alibaba's City Brain, for instance, is an AI that autonomously manages traffic in several Chinese and Malaysian cities (Alibaba 2020; Caprotti and Liu 2020; Curran and Smart 2021). It acquires information from hundreds of cameras placed in different urban areas and controls traffic lights, thereby controlling the mobility of most vehicles in the city.

Autonomous cars, robots, and city brains can be understood as *urban artificial intelligences* (Cugurullo 2020). They are artefacts operating in cities that, in theory, are capable of learning about the surrounding urban environment, extracting concepts, behaving rationally, and acting autonomously without humans. Some of these skills and capabilities originate from smart tech, and resonate with technologies and practices that are common in smart urbanism. Others are new, and it is from them that a new set of urban problems is coming. In terms of analogies and correspondences with smart urbanism, examples abound. Autonomous cars rely on voluminous and detailed urban maps that are updated in real time and fall within the Big Data category. Service robots require *sensors* such as microphones and cameras to perceive customers and understand their requests. In order to collect information, city brains need numerous cameras that manage to communicate in real time with the central digital platform because of the Internet of Things. The Big Data-sensors-IoT trinity functioning in real time is, then, the technological foundation of smart urbanism, and it is also what allows urban artificial intelligences to operate.

Urban artificial intelligences, however, and so autonomous urbanism, go beyond the now-standard Big Data-sensors-IoT trinity. Autonomous cars, robots, and city brains are supposed to make sense of the acquired data, to understand what is right or wrong, and to act autonomously in real-life situations in which benefits and harm inevitably will have to be distributed. These urban AIs will have to make choices that might even determine somebody's life or death, and whatever the choice will be, it will be *their* choice. Discerning good from evil and choosing how to distribute pros and cons are the quintessential ethical questions, and it is to ethics that I now turn.

The Ethics of Artificial Intelligence

Ethical questions have become prominent in both the study and the development of artificial intelligences. There is a quasi-universal agreement that AI should be ethical, but "what constitutes ethical AI" is highly debatable (Jobin, Ienca, and Vayena 2019, 389). The debate about AI ethics is vast and complex, primarily for two reasons. First, many heterogeneous actors are involved. National and international organizations, non-governmental organizations, and private companies have produced numerous documents in attempting to provide ethics guidelines that are supposed to discipline the creation of AI tech (Jobin, Ienca, and Vayena 2019). Prominent examples include the *White Paper on Artificial Intelligence*, published by the European Commission (2020), Google's (2020) *AI Principles*, and the *Toronto Declaration* launched by Amnesty International (2018). Further, several scholars, primarily and unsurprisingly from the fields of philosophy and ethics, have written articles, commentaries, and books meant to inspire the design and genesis of ethical AIs (see, for instance, Floridi et al. 2020). To further complicate the landscape of AI ethics, sometimes all these different actors come together in the production of a single document, such as in the case of the *Beijing AI Principles*, created by a coalition of representatives from the Beijing Academy of Artificial Intelligence (BAAI), Peking University, Tsinghua University, the Chinese Academy of Sciences, Baidu, Alibaba, and Tencent (BAAI 2019).

Second, questions of AI ethics involve concepts and themes that are rarely self-evident and universal, and are part of a long-standing tradition of debates in humanities and the social sciences. Typical ethical questions investigate what is *right* or *wrong*, *fair* or *unjust*, *true* or *false*, *good* or *bad*. There is no universal and detailed definition of *good* and *evil*, *justice* and *injustice*, *truth* and *lie*. These are complex notions whose understanding has repeatedly changed across space and time, thus assuming different meanings. For example, fundamental ethical questions were explored by Aristotle in his *Nicomachean Ethics* in the fourth century BC. Although Aristotle's philosophy is still employed today to comprehend concepts such as *justice* and *generosity*, there are acts and establishments that were ethically sound in ancient Greece – *slavery*, for instance – that would not be tolerated in most contemporary societies. This is to stress that ethics can be a very unstable ground when it comes to building the foundations of normative principles meant to make explicit what activities should be avoided or pursued.

On the positive side, despite the complexity and heterogeneity of the landscape of AI ethics, the literature shows a significant agreement about what general principles should guide the development of ethical AIs. Recent documents on AI ethics demonstrate that, despite geographical and cultural differences, recurring ethical themes resonate with the conceptual perspectives of contemporary philosophers (Floridi et al. 2018, 2020; Hagendorff 2020; Jobin, Ienca, and Vayena 2019). As of this writing, the most recurring AI-related ethical themes are *transparency*, *justice*, *non-maleficence*, *responsibility*, and *privacy*. It is beyond the scope of this chapter to discuss in depth all these different themes, so here I provide only an overview of their meaning and related key references for those who wish to dig deeper.

Transparency is about making clear how AI functions and operates. For example, a transparent AI should make decisions in a way that people can understand, thus preventing any doubt about how a given decision was reached. On these terms, transparency relates to the issue of *explainability* and to the fact that, as Greenfield (2018) laments, autonomous technologies are arcane technologies whose very basic mechanics are hard to grasp by whoever does not have a background in computer science or engineering (Jobin, Ienca, and Vayena 2019). *Justice* concerns issues of fairness and equity and, more specifically, the "prevention, monitoring or mitigation of unwanted bias and discrimination" (Jobin, Ienca, and Vayena 2019, 394; see also O'Neil 2016) that AI might be causing. The question of justice is also a matter of evenly distributing AI tech so that its benefits are equally accessible (Yigitcanlar and Cugurullo

2020). *Non-maleficence* as an ethical principle means that AI must not cause intentional or unintentional harm, where *harm* is intended in both psychological and bodily terms (Floridi et al. 2018; Jobin, Ienca, and Vayena 2019). The spectrum of possibilities is vast and scary, and ranges from autonomous cars running over pedestrians to a chatbot *à la* Replika (2020) saying something emotionally painful to its human interlocutor. *Responsibility* is normally about *accountability* and making sure that someone can be "named as being responsible and accountable for AI's actions and decisions" (Jobin, Ienca, and Vayena 2019, 395). *Someone* could be the designer of an AI, the company that produced it, or, potentially, even the AI itself if, by law, it was considered to be an intelligent and independent agent: this is the ethical matter on which there is consensus the least. Finally, *privacy* deals with data protection and security (Jobin, Ienca, and Vayena 2019), a recurring ethical theme that puts emphasis on the concerns that individuals, companies, and institutions have about sharing personal and confidential information with a non-biological intelligence (see, for instance, Lobera, Fernández Rodríguez, and Torres-Albero 2020).

Some of these ethical themes have been discussed extensively in smart city studies. Questions of justice, transparency, and privacy, in particular, have been unpacked both theoretically and empirically in the critical analysis of smart urbanism (Cugurullo 2018a; Cugurullo and Ponzini 2018; Cardullo, Di Feliciantonio, and Kitchin 2019; Kitchin 2016; Robinson and Franklin 2020). As an evolution of smart urbanism, autonomous urbanism is inheriting the ethical issues of its predecessor. Issues such as the uneven and unjust geographical distribution of smart technologies, the violation of data privacy, and the lack of transparency in smart city operations are well known, which this volume makes it evident – see, in particular, Chapters 7 (Monahan), 9 (Olmstead and Spicer), 12 (Heaphy), and 19 (Dierwechter). These issues will not disappear, but will hunt autonomous technologies, urban artificial intelligences, and, more generally, cities that are employing AI in their governance and infrastructure. Furthermore, AI is generating unprecedented and less-known urban issues, on which I now focus.

The Ethics of Urban Artificial Intelligence

With the rise of AI, "we are entering an age in which machines are tasked not only to promote well-being and minimize harm, but also to distribute the well-being they create, and the harm they cannot eliminate" (Awad et al. 2018, 59). This age is an urban age since artificially intelligent machines operate primarily in cities, and it is in cities that they will have to make complex ethical choices. Here, I shift the focus back to *urban artificial intelligences* (Cugurullo 2020) and, building upon the ethical theories discussed above, I relate them specifically to urban AIs: autonomous cars, robots, and city brains.

Autonomous cars are the primary example of an ethical dilemma in motion. It is a materially substantial and heavy piece of technology whose volume is enough, even in a low-speed impact, to seriously harm or even kill humans. This is not a hypothetical scenario, as evidenced by the killing of Elaine Herzberg in March 2018 (see Stilgoe 2019), and it is an *urban* scenario. Autonomous cars are operating outside testing facilities. They have entered ordinary cities and they are here to stay (Acheampong et al. 2021; McCarroll and Cugurullo 2022). In common urban spaces, shared with pedestrians, cyclists, and motorists, they will have to avoid harming humans and distribute harm when accidents cannot be prevented. In ethical terms, an autonomous car should embody the principle of *non-maleficence*. In practice, this is an unprecedented urban ethical issue because, although many technologies have recurringly harmed humans in cities (normal automobiles, in particular), autonomous cars are conscious of the act that is causing harm. They are intelligent machines whose decisions and actions might result in bodily harm. They are not just the medium: they are the agent. When harm is unavoidable, another ethical principle comes into play: *justice*. In an influential study on the ethics of autonomous cars, Awad and his colleagues ask us to "think of an autonomous vehicle that is about to crash, and cannot find a trajectory that would spare everyone. Should it swerve onto one jaywalking teenager to spare its three elderly passengers?" (Awad et al. 2018, 59). This is an uncomfortable and yet necessary ethical question, based on a real-life urban context, which exemplifies how difficult it is to determine what is fair when something bad inevitably is going to happen.

When it comes to robots, it is almost an instinctive thought to correlate the ethical principle of non-maleficence to the image of cruel machines physically harming humans. This is because decades of science fiction have saturated pop culture with images of killer robots – a myth that has been debunked by academics working in the field of AI and dismissed as an unlikely scenario (Tegmark 2017), since most of the harm that robots will cause will not be bodily and immediately visible. Service robots, for example, are essentially an extra workforce penetrating cities and, as Bissell and Del Casino (2017, 436) point out, "few employment fields are immune." A direct consequence and form of harm that this phenomenon might be causing is mass unemployment, with many people brought below the poverty line and rendered incapable of fulfilling their basic needs. A subtler consequence could be triggered by those human workers who, in order to escape from the robotization of labour, seek jobs in professional areas that require the kind of creative, critical, and lateral thinking that AI is not yet capable of delivering. This, in turn, opens up a thorny question of justice, since forcing people to migrate to professions that are not in synch with their vocation might be considered unfair (Loi 2015).

City brains, due to their vast scale and large scope, escalate the ethical dilemmas of AI to levels that were inconceivable until recently. These AIs are used to govern entire urban sectors, ranging from transport to urban planning to health and safety. While in the field of transport *traffic* is a relatively uncontroversial matter, since ethical concepts such as *good* and *just* can be interpreted by the AI as *reducing traffic congestion* and *avoiding car accidents*, the same cannot be said for fields such as safety and planning. Governing safety in cities, for example, is about determining what actions can be dangerous and for whom, and identifying who is dangerous and why. Ultimately, the urban governance of safety depends upon ethical assumptions and norms about what is good or evil and who is a good citizen and who is not. Historically, these assumptions and norms have been developed (and contested) by human intelligences. City brains would interrupt this tradition by placing a non-biological intelligence behind the curtain, where vital ethical decisions are being made. Similar ethical issues can be observed in the field of planning, which de facto is a large-scale trolley problem where unavoidable harm is constantly distributed in the city. Planning for *housing* is an emblematic instance of uneven urban development. Planning the construction of premium real estate implies, almost certainly, the exclusion of low-income workers and students. Planning for social housing means there will be more affordable flats and houses in the city, but not necessarily enough for all those in need of accommodation. Planning the development of business districts and office spaces results in fewer and more expensive domestic spaces. These planning choices are actually complex ethical choices whose results, unlike a classic trolley problem, do not correspond to the life or death of somebody, but instead produce better or worse *lives*: the lives of people whose urban future is now being planned, in some cities, by city brains.

Conclusions: Autonomous Urban Futures

The autonomous city is inheriting the baggage of ethical issues from the smart city. These issues have been around for over two decades, since the early smart city initiatives were developed (Cugurullo 2018b; Vanolo 2014). Issues of scarce privacy, uneven geographical distribution, and lack of transparency, repeatedly reported in critical urban studies, have rarely been addressed in practice by policy makers and urban developers and, unsurprisingly, they have not disappeared. Apart from a few progressive exceptions (see, for instance, Smith and Martín 2020; Trencher and Karvonen 2019), smart urbanism continues to be ethically problematic, and its evolution into autonomous urbanism is feeding long-standing problems of injustice into emerging urban artificial intelligences. With the advent of AI, the technological portfolio of smart cities has changed, but fundamental ethical questions have stayed the same. We cannot expect technology to evolve autonomously into ethically sound machinery. Technology will not purify itself of its ethical flaws. Matters of injustice have to be raised proactively and tackled by the designers of smart and AI tech, its producers, its users, and, above all, by the public institutions that allow the use of AI in cities and societies.

The ominous ethical inheritance of smart urbanism is now adding to the recent ethical challenges of autonomous urbanism. Ethical issues of intelligent machines that can consciously harm humans and find themselves in the position of having to distribute unavoidable harm are unprecedented and pose new questions that go beyond the smart city. We are being surrounded by sentient machines such as autonomous cars that, in the worst-case scenario, will have to choose whom to kill, non-biological intelligences such as service robots that are replacing humans and hindering their self-realization, and large-scale artificial intelligences that are being stored in digital platforms where the future of the city and that of its inhabitants is being planned according to unknown algorithms. Is this the urban future we want?

Before concluding, I want to consider one final ethical question: who are *we* in the autonomous city? Urban artificial intelligences are largely and enthusiastically being promoted by the private sector as commodities meant to be sold in order to realize economic ambitions. Several studies show that people's opinions about AI are much less adamant and optimistic (Acheampong and Cugurullo 2019; Cugurullo et al. 2020; Hulse, Xie, and Galea 2018). Many are scared of simply sharing urban spaces with an AI. Some do not want AI at all in their city. It is, therefore, a crucial matter of justice to open the future of the autonomous city to public political debate.

REFERENCES

Acheampong, R.A., F. Cugurullo, M. Gueriau, and I. Dusparic. 2021. "Can Autonomous Vehicles Enable Sustainable Mobility in Future Cities? Insights and Policy Challenges from User Preferences over Different Urban Transport Options." *Cities* 112: 103134. https://doi.org/10.1016/j.cities.2021.103134

Acheampong, R.A., and F. Cugurullo. 2019. "Capturing the Behavioural Determinants Behind the Adoption of Autonomous Vehicles: Conceptual Frameworks and Measurement Models to Predict Public Transport, Sharing and Ownership Trends of Self-Driving Cars." *Transportation Research Part F: Traffic Psychology and Behaviour* 62 (April): 349–75. https://doi.org/10.1016/j.trf.2019.01.009

Alibaba. 2020. "City Brain Overview." Online at https://www.alibabacloud.com/et/city, accessed 11 October 2020.

Amnesty International. 2018. "The Toronto Declaration: Protecting the Right to Equality and Non-discrimination in Machine Learning Systems." Online at https://www.amnesty.org/en/documents/pol30/8447/2018/en/, accessed 11 October 2020.

Awad, E., S. Dsouza, R. Kim, J. Schulz, J. Henrich, A. Shariff, J. Bonnefon, and I. Rahwan. 2018. "The Moral Machine Experiment." *Nature* 563 (7729): 59–64. https://doi.org/10.1038/s41586-018-0637-6

BAAI (Beijing Academy of Artificial Intelligence) 2019. "Beijing AI Principles." Online at https://ceedia.org/wp-content/uploads/2021/05/BEINJING-AI-PRINCIPLES.pdf, accessed 24 March 2022.

Batty, M. 2018. "Artificial Intelligence and Smart Cities." *Environment and Planning B* 15: 3–6. https://doi.org/10.1177/2399808317751169

Bissell, D., and V.J. Del Casino. 2017. "Whither Labor Geography and the Rise of the Robots?" *Social & Cultural Geography* 18 (3): 435–42. https://doi.org/10.1080/14649365.2016.1273380

Bostrom, N. 2017. *Superintelligence: Paths, Dangers, Strategies.* Oxford: Oxford University Press.

Caprotti, F., and D. Liu. 2020. "Platform Urbanism and the Chinese Smart City: The Co-Production and Territorialisation of Hangzhou City Brain." *GeoJournal* 1–15. https://doi.org/10.1007/s10708-020-10320-2

Cardullo, P., C. Di Feliciantonio, and R. Kitchin, eds. 2019. *The Right to the Smart City.* Bingley, UK: Emerald Publishing.

Cugurullo, F. 2018a. "Exposing Smart Cities and Eco-Cities: Frankenstein Urbanism and the Sustainability Challenges of the Experimental City." *Environment and Planning A: Economy and Space* 50 (1): 73–92. https://doi.org/10.1177/0308518X17738535

Cugurullo, F. 2018b. "The Origin of the Smart City Imaginary: From the Dawn of Modernity to the Eclipse of Reason." In *The Routledge Companion to Urban Imaginaries*, ed. C. Linder and M. Meissner, 113–24. London: Routledge.

Cugurullo, F. 2020. "Urban Artificial Intelligence: From Automation to Autonomy in the Smart City." *Frontiers in Sustainable Cities* 2 (38): 1–14. https://doi.org/10.3389/frsc.2020.00038

Cugurullo, F. 2021. *Frankenstein Urbanism: Eco, Smart and Autonomous Cities, Artificial Intelligence and the End of the City.* London: Routledge.

Cugurullo, F., R.A. Acheampong, M. Gueriau, and I. Dusparic. 2020. "The Transition to Autonomous Cars, the Redesign of Cities and the Future of Urban Sustainability." *Urban Geography*: 1–27. https://doi.org/10.1080/02723638.2020.1746096

Cugurullo, F., and D. Ponzini. 2018. "The Transnational Smart City as Urban Eco-modernisation." In *Inside Smart Cities: Place, Politics and Urban Innovation*, ed. A. Karvonen, F. Cugurullo, and F. Caprotti, 149–62. London: Routledge.

Curran, D., and A. Smart. 2021. "Data-driven Governance, Smart Urbanism and Risk-Class Inequalities: Security and Social Credit in China." *Urban Studies* 58 (3): 487–506. http://dx.doi.org/10.1177/0042098020927855

European Commission. 2020. "White Paper on Artificial Intelligence: A European Approach to Excellence and Trust." Online at https://ec.europa.eu/info/sites/default/files/commission-white-paper-artificial-intelligence-feb2020_en.pdf, accessed 24 March 2022.

Floridi, L., J. Cowls, M. Beltrametti, R. Chatila, P. Chazerand, V. Dignum, et al. 2018. "AI4People – An Ethical Framework for a Good AI Society: Opportunities, Risks,

Principles, and Recommendations." *Minds and Machines* 28 (4): 689–707. https://doi.org/10.1007/s11023-018-9482-5

Floridi, L., J. Cowls, T.C. King, and M. Taddeo. 2020. "How to Design AI for Social Good: Seven Essential Factors." *Science and Engineering Ethics* 26: 1771–96 https://doi.org/10.1007/s11948-020-00213-5

Google. 2020. "Artificial Intelligence at Google: Our Principles." Online at https://ai.google/principles/, accessed 11 October 2020.

Greenfield, A. 2018. *Radical Technologies: The Design of Everyday Life*. London: Verso Books.

Hagendorff, T. 2020. "The Ethics of AI Ethics: An Evaluation of Guidelines." *Minds and Machines* 30: 99–120. https://doi.org/10.1007/s11023-020-09517-8

Hulse, L.M., H. Xie, and E.R. Galea. 2018. "Perceptions of Autonomous Vehicles: Relationships with Road Users, Risk, Gender and Age." *Safety Science* 102: 1–13. https://doi.org/10.1016/j.ssci.2017.10.001

Jobin, A., M. Ienca, and E. Vayena. 2019. "The Global Landscape of AI Ethics Guidelines." *Nature Machine Intelligence* 1 (9): 389–99. https://doi.org/10.1038/s42256-019-0088-2

Kanal, L.N., and J.F. Lemmer, eds. 2014. *Uncertainty in Artificial Intelligence*. Amsterdam: Elsevier.

Karvonen, A., F. Cugurullo, and F. Caprotti, eds. 2018. *Inside Smart Cities: Place, Politics and Urban Innovation*. London: Routledge.

Kitchin, R. 2014. "The Real-Time City? Big Data and Smart Urbanism." *GeoJournal* 79 (1): 1–14. https://doi.org/10.1007/s10708-013-9516-8

Kitchin, R. 2016. "The Ethics of Smart Cities and Urban Science." *Philosophical Transactions of the Royal Society A: Mathematical, Physical and Engineering Sciences* 374 (2083). https://doi.org/10.1098/rsta.2016.0115

Levesque, H.J. 2017. *Common Sense, the Turing Test, and the Quest for Real AI: Reflections on Natural and Artificial Intelligence*. Cambridge, MA: MIT Press.

Lobera, J., C. J. Fernández Rodríguez, and C. Torres-Albero. 2020. "Privacy, Values and Machines: Predicting Opposition to Artificial Intelligence." *Communication Studies* 71 (3): 448–65. https://doi.org/10.1080/10510974.2020.1736114

Loi, M. 2015. "Technological Unemployment and Human Disenhancement." *Ethics and Information Technology* 17 (3): 201–10. https://doi.org/10.1007/s10676-015-9375-8

Macrorie, R., S. Marvin, and A. While. 2021. "Robotics and Automation in the City: A Research Agenda." *Urban Geography*: 42 (2): 197–217. https://doi.org/10.1080/02723638.2019.1698868

McCarroll, C., and F. Cugurullo. 2022. "Social Implications of Autonomous Vehicles: A Focus on Time." *AI & Society* 37: 791–800. https://doi.org/10.1007/s00146-021-01334-6

Milakis, D., B. Van Arem, and B. Van Wee. 2017. "Policy and Society Related Implications of Automated Driving: A Review of Literature and Directions for

Future Research." *Journal of Intelligent Transportation Systems* 21 (4): 324–48. https://doi.org/10.1080/15472450.2017.1291351

O'Neil, C. 2016. *Weapons of Math Destruction: How Big Data Increases Inequality and Threatens Democracy*. London: Penguin.

Replika. 2020. "About Replika." Online at https://replika.ai/about/story, accessed 11 October 2020.

Robinson, C., and R.S. Franklin. 2020. "The Sensor Desert Quandary: What Does It Mean (Not) to Count in the Smart City?" *Transactions of the Institute of British Geographers*. https://doi.org/10.1111/tran.12415

Russell, S.J., and P. Norvig. 2016. *Artificial Intelligence: A Modern Approach*. Harlow, UK: Pearson Education.

Smith, A., and P.P. Martín. 2020. "Going Beyond the Smart City? Implementing Technopolitical Platforms for Urban Democracy in Madrid and Barcelona." *Journal of Urban Technology*: 1–20. https://doi.org/10.1080/10630732.2020.1786337

Stilgoe, J. 2019. *Who's Driving Innovation? New Technologies and the Collaborative State*. Berlin: Springer Nature.

Tegmark, M. 2017. *Life 3.0: Being Human in the Age of Artificial Intelligence*. London: Penguin.

Tiddi, I., E. Bastianelli, E. Daga, M. d'Aquin, and E. Motta. 2020. "Robot–City Interaction: Mapping the Research Landscape – A Survey of the Interactions between Robots and Modern Cities." *International Journal of Social Robotics* 12 (2): 299–324. https://doi.org/10.1007/s12369-019-00534-x

Trencher, G., and A. Karvonen. 2019. "Stretching 'Smart': Advancing Health and Well-being through the Smart City Agenda." *Local Environment* 24 (7): 610–27. https://doi.org/10.1080/13549839.2017.1360264

Vanolo, A. 2014. "Smartmentality: The Smart City as Disciplinary Strategy." *Urban Studies* 51 (5): 883–98. https://doi.org/10.1177/0042098013494427

While, A.H., S. Marvin, and M. Kovacic. 2020. "Urban Robotic Experimentation: San Francisco, Tokyo and Dubai." *Urban Studies*. https://doi.org/10.1177/0042098020917790

Wirtz, J., P.G. Patterson, W.H. Kunz, T. Gruber, V.N. Lu, S. Paluch, and A. Martins. 2019. "Brave New World: Service Robots in the Frontline." *Journal of Service Management* 29 (5): 907–31. https://doi.org/10.1108/JOSM-04-2018-0119

Yigitcanlar, T., and F. Cugurullo. 2020. "The Sustainability of Artificial Intelligence: An Urbanistic Viewpoint from the Lens of Smart and Sustainable Cities." *Sustainability* 12 (20): 8548. https://doi.org/10.3390/su12208548

Zhang, J., X.S. Hua, J. Huang, X. Shen, J. Chen, Q. Zhou, et al. 2019. City Brain: Practice of Large-Scale Artificial Intelligence in the Real World." *IET Smart Cities* 1 (1): 28–37. https://doi.org/10.1049/iet-smc.2019.0034

11 Pornhub Helps: Digital Corporations in Italian Pandemic Cities

ALBERTO VANOLO

At the beginning of March 2020, Pornhub offered Italian users a free month of Pornhub Premium as part of a program to "keep Italians at home" and to "keep company to Italians." It was sufficient to register. The company also donated part of its profits to Italian social initiatives. This approach was not new in the pornography sector, as xHamster had proposed a similar one a few weeks before. But there is more. In the same month, the Italian national social security system, the Istituto Nazionale Previdenza Sociale (INPS), introduced a public policy supporting certain categories of citizens with a €600 bonus. In order to access the bonus, beneficiaries had to fill out a form on the INPS website. On the first day, the website collapsed because of the high number of requests (about 1.5 million), and it publicly displayed private data of those who applied, causing a national scandal. The following day, Pornhub offered its help for the improvement of the INPS website and for the management of its huge amounts of data (Pornhub is visited by about 120 million users every day, without any crash). Pornhub did not receive a reply from the INPS, but the news gained tremendous popularity, boosting Pornhub's corporate image in the country.

Smart City, Solutionism, and Injustices

Over the past decade, critical scholars in urban studies have built a solid literature on the perils and limits of the smart city. Although critical discussions were proposed almost twenty years ago by authors such as Graham and Marvin (2001), the first to analyse fully and explicitly the politics and discursive strategies of smart cities was Hollands (2008), who focused on the tendency to reproduce celebrative, superficial, elitist, and business-led approaches to technology and to the management of cities. Building on his original contribution, a number of authors explored further dimensions of the problem, discussing the ambiguity of the meaning itself of the expression "smart city," which seems to include very different ideas and imaginaries of technologies and urban life (see, for example,

Greenfield 2013; Kitchin 2015; Townsend 2013; Vanolo 2014). The bulk of the first generation of critical literature in the field analysed ideologies of solutionism and neoliberalism, particularly by focusing on issues of participation, citizenship, surveillance, injustice, and the new forms of capital extraction and work exploitation allowed by digital technologies (for reviews, see Cardulo, Di Feliciantonio, and Kitchin 2019; Karvonen, Cugurullo, and Caprotti 2019).

Of course, the aim of this critical literature is not to demonize new technologies or to suggest any kind of Luddite vision. Obviously, innovations may determine dramatic improvements in the quality of life, they may sustain empowerment and participation, and they may help in fixing current urban problems and injustices. But the smart city does not strictly coincide with its technical infrastructure, nor does it correspond to any single technology or set of technologies: sensors, networks, and algorithms generally associated with the smart city could be developed and related to other settings, contexts, and discourses (Sadowski and Bendor 2018). Smart technologies are conceived as such not because of technical or institutional reasons – they can differ dramatically from each other from the technological point of view, and they are not necessarily produced by similar actors or implemented by specific institutions. Rather, technologies become smart city technologies by their association with certain narratives, logics, practices, symbols, and ideologies. In this sense, most of the critical literature, – particularly that developed during the first part of the past decade – partly developed as a reaction to a massively diffused framework of celebrative accounts of digital technologies, excessive hype, and uncritical support for everything in the field, from "e-whatever" public policies to market enthusiasm for private startups, from branding new apps and digital services as tools for "social innovation" to the supposed general belief that digital infrastructure is the core ingredient for fixing a variety of urban problems (see the edited collection by Marvin, Luque-Ayala, and McFarlane 2016; see also Wiig and Wyly 2016). More recent strands of the literature have not only pushed critical understandings of the smart city; they have also proposed specific and situated analysis of the way people coexist with smart technologies, sensors, and codes; how places are shaped, mediated, and co-constituted by the digital; how subjectivities, rationales, and identities take form in the urban digital space; and how citizens might resist and reclaim an informational right to the city, or a right to the smart city (see, for example, Cardulo 2021; Kuecker and Hartley 2020; Safransky 2020; Shaw and Graham 2017; Zandebergen and Uitermark 2020).

This chapter aims to contribute to the literature by situating a reflection on the provision of digital "solutions" in the very specific and contingent time-space of Italy during the COVID-19 pandemic, from February 2020 to the actual time of writing of this chapter (August 2020). In the dramatic situation of sufferance, fear, and lockdown, politicians, technological gurus, and a number of other urban experts looked at digital technologies and smart city discourses in search of solutions and knowledge to manage the crisis. As I discuss, the pandemic has been a testbed for new technologies with ongoing effects in the country that still have to be analysed fully.

The chapter focuses on a particular aspect of this scenario: the reconfiguration of major corporations working in the digital sector as fair and responsible subjects with the power, capability, and ethical will to help citizens and, ultimately, to operate for the general benefit of society. The state of emergency caused by the pandemic helped overcome many critical discourses about the pervasiveness of smart technologies, surveillance systems, the gig economy, and platform capitalism, contributing to reframing discursively these elements as resources and opportunities in the struggles of the crisis (see Newell 2021; Taylor et al. 2020). This trend is coherent with the general tendency to *solutionism* that characterizes the ideology of the smart city. Solutionism is basically a variant of the modernist ideology, based on the idea that the right technology or the right app might fix every kind of problem, including problems "invented" by the technology itself, without the need for general readjustments in society (Cardulo 2021; Krivý 2018; March 2018; Morozov 2013; Vanolo 2016).

The critical perspective proposed in this chapter, although not strictly focused on cities, builds on ideas developed in the urban studies literature, and looks mostly at discourses enacted through journalistic representations and political interventions. The analysis is based on empirical materials, including journal articles from the main Italian newspapers, fragments of different kinds of news and public discourses circulating on the web, and the direct experience of living in a locked-down Italian city.

The Italian Pandemic: Basic Facts

Italy was hit severely by the COVID-19 pandemic, particularly in the first phase of the diffusion of the virus in Europe. In March 2020, it was the country with the most coronavirus deaths in the world, overcoming China.[1] In the weeks following 21 February 2020, a growing number of "non-essential" services and activities were closed in order to limit the diffusion of the virus. At the beginning of the crisis, regulation was mostly provided at the urban level, particularly by mayors of cities severely hit in the northern region of Lombardy (the Milan region). Regulation was extended to the entire country on 9 March and progressively tightened, with the fining of people leaving home without "good reasons." Schools and universities were closed; basically, only supermarkets, banks, pharmacies, hospitals, and post offices kept working, with limited opening hours. Travel inside the country was banned. To leave home, citizens had to fill self-declaration forms providing meaningful justification, and the police were given permission to use drones to patrol streets.

This severe lockdown phase ended on 4 May, with the beginning of a "phase 2," characterized by the progressive reopening of activities and slowing down of containment measures. For the first two weeks, it meant permission to visit family members within the same region, or to take away food from restaurants and bars. On 25 May, most urban services were reopened, including, for example, gyms. As this chapter was written (August), Italy was in a period of progressive lightening of containment measures.

Italian Platforms and Tracking Apps: Some Facts and Examples

In this section, I present some evidence and anecdotes concerning the growing legitimation of companies operating in the digital sphere during the months of the COVID-19 crisis. I begin by presenting speculations connected to digital corporations and the digital infrastructure; I then focus on the specific case of Immuni, the Italian mobile tracking app.

Companies Care a Lot

With the progressive lockdown of urban activities, digital relations and digital services acquired a prominent role: in Italy, web traffic in February 2020 was 90 per cent higher than in February 2019.[2] The digital dimension of urban life and the digital right to the city quickly became crucial elements in everyday life, with the huge diffusion of *smart working* and online courses as two of the most evident effects. It became clear that the possibility and capability of accessing Internet, web-based services and digital resources was going to mean a fundamental change in times of isolation and social distance. The scenario obviously widened the divide between those who were able to get online and those who were not. At the same time, it also revealed the limits of Italy's digital infrastructure: put simply, several website and digital services crashed or became unusable due to an excessive number of users. As a personal account, I can mention the case of the university where I work: for some weeks, due to excessive traffic, its website was de facto unusable for streaming lectures, at least during the daytime, and several students suggested that I upload my lectures to the more reliable and efficient YouTube platform.

In general, major Italian telecommunication companies (such as Telecom Italia) denied the risk of a collapse of their digital infrastructure. Institutional bodies, however – particularly the Autorità per le Garanzie nelle Comunicazioni (AGCOM, the Italian authority for communications guarantees) – opened a series of roundtables to discuss how to expand the digital infrastructure given the temporary suspension of existing laws and regulations.[3] In Italy, there is a structural divide in the provision of hardwired broadband connections: most private operators have concentrated their investments in densely populated areas with great demand and where there are better possibilities for returns on investments – a phenomenon known in urban studies as "splintering urbanism" (see Graham and Marvin 2001). About 17 per cent of the Italian population, however, lives in small towns with fewer than 5,000 inhabitants, and almost half lives in towns with fewer than 20,000. According to the Ministry for Economic Development, the pandemic made it clear that Italy had to improve its digital infrastructure, possibly by sustaining the formation of a single operator scenario, in order to avoid the duplication of investment and infrastructure. Debates on this topic, and on the potential role of the state in regulating this crucial and inefficient market, are ongoing.[4]

In the meantime, in March 2020, Netflix decided to reduce temporarily the quality of its streamed videos in Europe by a ratio of 25 per cent, in order to reduce data consumption and hence to avoid the collapse of digital infrastructure. The decision was reached after a discussion between a European Commissioner and a Netflix executive in relation to the challenges of the coronavirus. After a few days, similar interventions were declared by Amazon Prime, Disney+, Facebook, Instagram, and others. Several Italian newspapers described Netflix's decision as "sensitive" and "responsible" in order to maintain the "security" of the digital infrastructure (see, for example, Bottin 2020). Netflix also donated €1 million to sustain Italian workers in the TV and cinema industries.

While Netflix took action to preserve digital infrastructure, the Italian government's ability to deal with social and spatial divides in access to digital resources has been limited. To be clear, Internet connections are continuing to be provided by a small number of Italian commercial information and communications technology companies. Discount rates have been applied and advertised during the pandemic, particularly for those living in cities severely hit by the virus, but the commercial basis of those services remains untouched. The main official intervention introduced by the Italian government has been the project (i.e., a website) named Solidarietà Digitale, an initiative by the Ministry for Technological Innovation and Digitalization aimed at "reducing the social and economic impact of the coronavirus thanks to innovative solutions and services."[5] The ministry intends the portal to create a network between the private and the social in times of crisis. In fact, however, the website simply provides a heterogeneous list of online services and contents delivered by private entities, ranging from free apps for smart working to free newspaper and libraries to e-learning resources. In many cases, simple "free trials" have been advertised through the instructional webpage of the initiative – for example, in the form of two weeks of free software for those registering, a common commercial strategy in the field. Beyond the good intentions of public administration, companies, and users, the initiative turned, to a certain degree, into a mere window for the advertising of services and brands.

Overall, a number of digital companies clearly have had the opportunity to expand their global role and credibility during the pandemic. To quote an example, on 13 March 2020, US President Donald Trump declared: "I want to thank

Google. Google is helping to develop a website, it's gonna be very quickly done ... to determine whether a test is warranted and to facilitate testing at a nearby convenient location. Google has 1,700 engineers working on this right now, made tremendous progress." Shares of Google parent company, Alphabet Inc., rose immediately by more than 9 per cent, regardless of the fact that Trump's account was not accurate.[6] Similar facts suggest that, at the global level, the technocratic elite has been increasingly framed as the ultimate solution to problems, from developing the right app to finding an effective medical treatment.

As a matter of fact, global digital corporations – and other categories, such as pharmaceutical companies – generally have increased their market positions during the crisis. In the first half of 2020, many of them experienced huge increases in their global stock exchange values, as in the cases of Amazon (+43.8 per cent), Microsoft (+22.4 per cent), PayPal (+51.5 per cent), Netflix (+38.9 per cent), and Zoom (+255.1 per cent).[7] These global figures are coherent with Italian trends. Zoom, for example, has been by far the most downloaded app in the country during the months of lockdown, being downloaded 2.5 million times in the month of March 2020 alone. Of course, Zoom is formally free, but it is also a commercial product developed by a Californian private company, founded by a former Cisco engineer, and has expanded its worldwide presence significantly due to the pandemic.

In Italy the pandemic has allowed several discursive moral reconfigurations about high-tech companies and smart initiatives. Surely, the pandemic has offered meaningful opportunities for enterprises, operating in very different economic sectors, to take a mindful approach to corporate social responsibility policies: beyond humanitarian reasons, consumers are expected to be proud of companies and brands that support their employees, donate money and equipment, and help with the crisis (He and Harris 2020). As an example, at the end of March 2020, Amazon decided to stop selling non-essential products. Many advertising features were also removed from its Italian website. This looked like radical and responsible behaviour for a commercial company. Despite its huge gains, Amazon apparently decided to limit commercial transactions in order to diminish the number of deliveries and, arguably, to follow a reasonable growth strategy for the company. Amazon also donated €2.5 million to the Italian government and €1 million to local non-government organizations, and it shared for free some digital contents (Prime videos, Kindle, and Audible books). It also actively engaged against speculation on masks and sanitizer gel prices, and it introduced an Alexa skill for donating money (Cosimi 2020). Such initiatives surely helped boosting the company' image, which was suffering in Italy due to debates on tax avoidance and unfair working conditions.

Another example of improvement in the image of companies accused of exploiting workers might be related to the case of food deliveries. During the entire lockdown period, food delivery platforms remained fully operational in Italian cities. Restaurants were closed, and during the most severe weeks of the lockdown, riders on bicycles and scooters were the main visible human presence in the streets. Delivery services were considered "fundamental services" by public institutions, allowing both the provision of food to people and the survival of many restaurants that were closed to the public. According to a sectorial research report by a private food delivery company, 90 per cent of Italians (out of a sample of 30,000 citizens) considered food delivery "an essential service."[8] Clear data are still unavailable, and it is quite difficult to say whether the sector expanded during the lockdown or suffered because of the general economic crisis and the reduced number of active restaurants. Surely, the situation allowed weakening hostile sentiments against food delivery: riders have been often described in Italian newspapers as "heroes" keeping cities alive (see, for example, Crippa 2020). Major companies have been praised because they enacted voluntary welfare policies, as in the case of Deliveroo, which activated for free a health insurance for riders contracting COVID-19. Still, riders have kept on mobilizing in Italian cities, asking for more decent working conditions in the gig economy.

In Search of a Tracking App

Technological corporations immediately considered the possibility of developing smart apps that might help limit the pandemic's spread by tracing everyday contacts and signalling contact with infected people, originating what Taylor et al. (2020, 11) called an "epidemiological turn in digital surveillance" (see also French and Monahan 2020; Newell 2021). In some cases, individual companies proposed their own apps, experiments, and protocols, but the most relevant examples have been developed at the national level, with different kinds of support from government. Different countries have also proposed very different apps, characterized by individual approaches to technology and ethics (Newell 2021), posing relevant questions about the degree to which civil liberties may be sacrificed for public health (Kitchin 2020). The Chinese health code system, introduced in hundreds of cities, or the South Korean Corona 100m, are commonly considered as negative benchmarks in terms of intrusive surveillance and lack of transparency, allowing public authorities to gather data about citizens' identities, positions, movements, and commercial transactions, among other details, and posing alarming threats to freedom and privacy. Still, the Chinese state, which managed to arrest the spread of the virus by using extreme techniques of surveillance-based control and containment, has also been praised by Western observers who previously might have denounced such tactics as abuses of human rights (French and Monahan 2020).

At the other end of the spectrum, apps such as the Danish Smittestop, Austrian Stopp Corona, Canadian Covid Alert, Singaporean Trace Together, and Italian Immuni are widely considered positive examples (O'Neill, Ryan-Mosley, and Johnson 2020). Their use is voluntary, unlike the cases of Qatar and the United Arab Emirates, where the COVID app is mandatory and citizens refusing to install it or to register can be fined. The more positive examples set limitations on the amount and use of collected data, unlike in Turkey, for example, where data are used for law enforcement. Finally, transparency in these examples has been preserved by diffusing open-source code bases and making policies and designs publicly available.

Immuni has been promoted by the Italian Ministry of Public Health, and realized for free by the Italian company Bending Spoons. The idea of developing a mobile app for the purpose was first discussed in Italy in March 2020, causing severe reactions from political quarters worried about issues of surveillance and privacy. Despite this criticism, the shocking situation of emergency, risk, and fear ultimately made it possible to develop the idea, which arguably would have been considered unacceptable in different times (see Kitchin 2020). Still, newspapers mostly discussed the potential benefits of a tracking app, keeping in mind crucial and problematic examples of invasive surveillance technologies such as the South Korean Corona 100m (Sonn, Kang, and Choi 2020); see also Soave n.d.). Despite resistance, on 23 March 2020, a "fast call" for applications was launched by the Ministry for Technological Innovation and Digitalization, and on 16 April a task force of seventy-four national experts decided to opt for Immuni, which was quite optimistically expected to be made available before the end of the month. The following week, as a consequence of unexpected cooperation between Google and Apple to develop a common technological protocol, Bending Spoons announced the use of decentralized technology, which basically means that data are kept and processed in smartphones, and not uploaded to a central server (Taylor et al. 2020). This solution reassured part of the public about concerns of privacy, transparency, and data protection. Immuni uses Bluetooth technology in order to allow communication between mobile phones, and it notifies about potential risks. In case of a contact warning about the virus, it is up to citizens to inform the medical services and alert recent contacts, giving form to specific types of "lateral," "social," and "self"-surveillance (French and Monahan 2020). The software code was made available on 25 May, and the app was available to the public on 8 June, in a period of meaningful decline in the spread of the virus all over the country. Two months later, the app had been downloaded 4.5 million times.

There are very different opinions and discourses about Immuni. On the one hand, positive accounts emphasize the overall quality of the app, which has been considered internationally as an example of simplicity, transparency, and privacy protection (O'Neill, Ryan-Mosley, and Johnson 2020). On the other hand, critical voices describe the project as a total failure, especially if compared with initial intentions: policy makers expected two-thirds of the Italian population to use the app, and many actively supported the "sharing for common good" ethos.[9] Current figures (4.5 million downloads out of a population of more than 60 million, August 2020) show that the app has been ineffective, especially considering that the figures include users who are not actually using it. By the end of July 2020, the app had detected fewer than 50 positive cases in the entire country, out of a total about 13.000 infected citizens. According to Italian commentators, the limited diffusion of the app is due to a mixture of laziness, difficulties in understanding the mechanics of the app (how devices communicate, how privacy is protected), and generic lack of enthusiasm for surveillance technologies, which apparently are of limited use at the individual level (Capitanio 2020). Political enthusiasm is clearly weak, and a number of national policy makers have ridden the momentum by explicitly declaring that they were not going to download the app (Bozza 2020). Overall, it seems that this "technology theatre" at the intersection of technology and politics has added little, or even harmed, public trust in health policies (McDonald 2020). Surely, a winner is Bending Spoons, whose name has acquired huge popularity – the names and faces of the four young, talented founders, managers, and technological heroes of the company are regularly displayed in the news.

Confronting the Virus, the Digital Economy, and Smartness

According to Sadowski and Bendor (2018), the imaginary of the smart city strictly relates to the idea that society, cities, and urban leaders are confronted with critical problems and crises threatening our very life, forcing us to discover new ways to deal with them. Crises, and the perception of an imminent catastrophe, are key elements at the core of the entire smart city narrative: the future of the planet depends on our ability to fix urban problems of actually existing cities. Specifically, the authors detect three types of crises: rapid urbanization, fiscal austerity, and climatic catastrophe. All of them suggest the need for smart infrastructure, smart urban governance, data-driven systems, public-private partnerships, and adaptive technologies to produce what IBM calls a "smarter planet." In this framework, the smart city may be conceptualized as both a reactionary narrative of technological modernization – that is, maintaining stability and controlling uncertainty – and the ultimate technological utopia, grounded in optimistic visions of social and technological progress. In the smart city, the unknowable and the uncontrollable, which are supposed to be at the basis of most urban problems, are tamed through data, code, and algorithms.

The global pandemic crisis was not really unpredictable. Scholars and experts warned for years about the potential impact of coming global epidemics (see, for example, Žižek 2020). Still, the pandemic overwhelmed and shocked people, governments, institutions, and companies, threatening the very basis of urban and economic life. If, on the one hand, the crisis pushed many to rethink old rationales and to consider seriously different forms of change, at both the individual and collective levels, on the other hand it offered further possibilities for the consolidation of solutionism and the concentration of power, wealth, and credibility in the hands of a limited number of interests, including global digital corporations. In fact, the pandemic offered new opportunities for the technology and digital surveillance industries to reassure, comfort, and please citizens, and companies from Amazon to Pornhub, Deliveroo, and Netflix have had numerous opportunities to boost and humanize their image.

In the case of Italy, the crisis implied not only a further acceleration of the outsourcing of public functions, but also the framing of information technology companies and service providers as heroes that help. For example, the entire educational infrastructure essentially has been left in the hands of users and teachers, who rely on Zoom, Classroom, WebEx, and similar "free" private e-learning technological solutions. Surely, this has forced technological upgrading in schools, but it evidently has also raised technological divides, leaving it to families to cover the costs of PCs, connections, and technical skills for e-learning, exacerbating existing inequalities and fuelling the notion that digital technology will relieve us from in-person interactions and thus reduce coronavirus spread (see Burns 2020; Manzo and Minello 2020).

On a psychological level, it can be also argued that the diffusion of a climate of fear has grown side by side with the need to control, tame, and turn the unpredictable into the predictable (Žižek 2020). In Italy, the diffusion of everyday statistics, facts, maps, and models about the spread of the virus surely might be connected to this psychological need for control and for the reduction of uncertainty in a more and more unstable (and liquid) society, in a way not much different from the desire obsessively to control one's bodyweight or the number of calories consumed every day (see Bauman 1999). Complex problems surely are at play, but the perceived need to limit and control bit-rate usage or compulsive commodity consumption, while sustaining new programs of digital surveillance, also might be explained partly by such psychological dynamics. Certainly, the perception of digital insecurity has not been contained by the Italian government, which has only limited credibility as a regulator, at least in the digital realm – what we might call "smart governance." Fears of a collapse of the digital infrastructure, the ironic case of Pornhub's offering help to the crashing website of the Italian social security system, or the low number of downloads for Immuni, are somehow symptomatic of a situation – at the time of writing of this chapter – where citizens do not perceive the state as a credible player in the field. Moreover, although the Italian government has had to introduce severe and unpopular lockdown measures, global corporations have had the opportunity to play a different role by offering – in many cases apparently for free – solutions, recreation, possibilities for social relations, imaginaries of control, and a number of other tools and resources. Immuni fits in this framework: although the app has not been that successful – the fate of similar software all over the world (see Taylor et al. 2020) – it is coherent with the logic of taming and controlling the world through the extraction and management of data, which is at the core of smart city ideology and contemporary surveillance capitalism. For experts, it was clear from the start that digital tracking systems would be insufficient to limit the pandemic, as traditional tracking techniques, which include fieldwork, are crucial (Newell 2021; Kitchin 2020; Sonn, Kang, and Choi 2020). Still, Immuni is a politically relevant experience in the building of Italian smart city culture. The shocking state of emergency caused by COVID-19 allowed the introduction of a form of digital surveillance that had been politically unthinkable a few months before, opening the way to future experiments in the field. Immuni is also coherent with the idea that technology is the ultimate weapon to tackle urban problems and that the state cannot really master technical tools. The controversial declarations of several Italian policy makers made it evident that the political elite have had quite confused ideas about tracking technologies. Instead, grounded knowledge has had to be provided by an external, young, and growing company. The credibility and popularity of Bending Spoons, which offered skills, knowledge, and talent "for free," is currently high in the country.

Overall, the still-evolving case of Italy confirms a further acceleration in both the diffusion and the legitimation of smart city logics and digital platforms. The problems and injustices related to technologies and to the ideology of solutionism have not been reduced, but probably exacerbated by an uneven situation of crisis and responsibilization of individuals in accessing resources, including digital ones: the case of smart work is crucial in this sense. Still, the smart and digital narrative has gained momentum, meaningfully shaping our experience and perception of the pandemic, and opening new spaces of credibility and opportunity for corporations.

NOTES

1 "Italy becomes country with most coronavirus deaths," *Guardian*, 19 March 2020. Online at https://www.theguardian.com/world/2020/mar/19/france-may-refuse-entry-to-britons-if-no-strict-lockdown-is-imposed-in-uk-coronavirus, accessed August 2020.

2 "Emergenza Covid-19 e internet: c'è il rischio collasso?" *Teknoring*, 30 March 2020. Online at https://www.teknoring.com/news/ingegneria-informatica/emergenza-covid-19-internet-collassera/, accessed August 2020.

3 See Autorità per le Garanzie nelle Comunicazioni, "Emergenza COVID-19 – Tavoli tecnici con gli Operatori," n.d., online at https://www.agcom.it/emergenza-covid-19-tavoli-tecnici-con-gli-operatori, accessed August 2020.

4 See Agenzia Italia, "Che cos'è la rete unica nazionale e quali sono gli interessi in gioco," 14 August 2020, online at https://www.agi.it/politica/news/2020-08-14/rete-unica-nazionale-tim-enel-open-fiber-9417051/, accessed August 2020.

5 See the website at https://solidarietadigitale.agid.gov.it

6 The quote has been widely reported in the media – for example, "Google to develop website to help with coronavirus test – Trump," *CNBC*, 13 March 2020, online at https://www.reuters.com/article/health-coronavirus-google-idUKL4N2B654T, accessed August 2020. The account was basically inaccurate because the project was more limited and was developed by another distinct Google-related company, Verily.
7 See the website at https://www.ft.com/markets
8 See "Coronavirus: food delivery essenziale per 90% italiani, boom gelato, +133%," *ADNKronos*, 20 April 2020, online at https://www.adnkronos.com/sostenibilita/tendenze/2020/04/20/non-solo-pizza-boom-gelato-anche-cocktail-food-delivery-tempi-del-covid_eQs6o0YhN7Y7ArAPVyrGnJ.html, accessed August 2020.
9 See "Lo spettacolare fallimento delle app contro il coronavirus," *Repubblica*, 9 July, 2020, online at https://www.repubblica.it/dossier/stazione-futuro-riccardo-luna/2020/07/09/news/lo_spettacolare_fallimento_delle_app_contro_il_coronavirus-261374292/, accessed August 2020.

REFERENCES

Bauman, Z. 1999. *In Search of Politics*. Cambridge, UK: Polity Press.

Bottin, M. 2020. "Netflix e Prime Video ridurranno la qualità dello streaming per un mese in Europa (aggiornato)." *Smartworld*, 20 March. Online at https://www.smartworld.it/streaming/netflix-riduce-bitrate-ue.html

Bozza, C. 2020. "Immuni, dal renziano di ferro ai big di centrodestra: i politici che non vogliono scaricare l'app anti-contagio." *Corriere Della Sera*, 3 July. Online at https://www.corriere.it/politica/20_giugno_03/immuni-renziano-ferro-big-centrodestra-politici-che-non-vogliono-scaricare-l-app-anti-contagio-a6a515d4-a574-11ea-9dea-fe0c662b4b9d.shtml, accessed August 2020.

Burns, R. 2020. "A COVID-19 Panacea in Digital Technologies? Challenges for Democracy and Higher Education." *Dialogues in Human Geography* 10 (2): 246–9 https://doi.org/10.1177/2043820620930832

Capitanio, M.E. 2020. "Immuni, che flop! Solo 4,5 milioni (su 60) hanno scaricato l'app." *Huffington Post*, 5 August. Online at https://www.huffingtonpost.it/entry/immuni-che-flop-solo-45-milioni-su-60-hanno-scaricato-lapp_it_5f2aafa4c5b64d7a55ed0383, accessed August 2020.

Cardullo, P. 2021. *Citizens in the "Smart City": Participation, Co-Production, Governance*. New York: Routledge.

Cardullo, P., C. Di Feliciantonio, and R. Kitchin, eds. 2019. *The Right to the Smart City*. Bingley, UK: Emerald Publishing.

Cosimi, S. 2020. "Coronavirus, prezzi alle stelle per gel e mascherine: Amazon bacchetta gli speculatori in Italia e all'estero." *Repubblica*, 26 February. Online at https://www.repubblica.it/tecnologia/prodotti/2020/02/25/news/prezzi_alle_stelle_amazon_bacchetta_gli_speculatori_in_italia_e_all_estero-249620527/?ref=search, accessed August 2020.

Crippa, M. 2020. "Un treno per i rider eroi." *Il Foglio*, 19 June. Online at https://www.ilfoglio.it/contro-mastro-ciliegia/2020/06/19/news/un-treno-per-i-rider-eroi-321197/, accessed August 2020.

French, M., and T. Monahan. 2020. "Dis-Ease Surveillance: How Might Surveillance Studies Address COVID-19?" *Surveillance & Society* 18 (1): 1–11. https://doi.org/10.24908/ss.v18i1.13985

Graham, S., and S. Marvin. 2001. *Splintering Urbanism. Networked Infrastructures, Technological Mobilities and the Urban Condition*. London: Routledge.

Greenfield, A. 2013. *Against the Smart City*. New York: Do Publications.

He, H., and L. Harris. 2020. "The Impact of COVID-19 Pandemic on Corporate Social Responsibility and Marketing Philosophy." *Journal of Business Research* 116 (August): 176–82. https://doi.org/10.1016/j.jbusres.2020.05.030

Hollands, R.G. 2008. "Will the Real Smart City Please Stand Up? Intelligent, Progressive or Entrepreneurial?" *City* 12 (3): 303–20. https://doi.org/10.1080/13604810802479126

Karvonen, A., F. Cugurullo, and F. Caprotti, eds. 2019. *Inside Smart Cities: Place, Politics and Urban Innovation*. London: Routledge.

Kitchin, R. 2015. "Making Sense of Smart Cities: Addressing Present Shortcomings." *Cambridge Journal of Regions, Economy and Society* 8 (1): 131–6. https://doi.org/10.1093/cjres/rsu027

Kitchin, R. 2020. "Civil Liberties *or* Public Health, or Civil Liberties *and* Public Health? Using Surveillance Technologies to Tackle the Spread Of COVID-19." *Space and Polity* 24 (3): 362–81. https://doi.org/10.1080/13562576.2020.1770587

Krivý, M. 2018. "Towards a Critique of Cybernetic Urbanism: The Smart City and the Society of Control." *Planning Theory* 17 (1): 8–30. https://doi.org/10.1177/1473095216645631

Kuecker, G.D., and K. Hartley. 2020. "How Smart Cities Became the Urban Norm: Power and Knowledge in New Songdo City." *Annals of the American Association of Geographers* 110 (2): 516–24. https://doi.org/10.1080/24694452.2019.1617102

Manzo, L.K.C., and A. Minello. 2020. "Mothers, Childcare Duties, and Remote Working under COVID-19 Lockdown in Italy: Cultivating Communities of Care." *Dialogues in Human Geography* 10 (2): 120–3. https://doi.org/10.1177/2043820620934268

March, H. 2018. "The Smart City and other ICT-Led Techno-Imaginaries: Any Room for Dialogue with Degrowth?" *Journal of Cleaner Production* 197: 1694–703. https://doi.org/10.1016/j.jclepro.2016.09.154

Marvin, S., A. Luque-Ayala, and C. McFarlane, eds. 2016. *Smart Urbanism: Utopian Vision or False Dawn?* New York: Routledge.

McDonald, S.M. 2020. "Technology Theatre and Seizure." In *Data Justice and COVID-19: Global Perspectives*, ed. L. Taylor, G. Sharma, A. Martin, and S. Jameson, 20–7. London: Meatspace.

Morozov, E. 2013. *To Save Everything, Click Here: Technology, Solutionism and the Urge to Fix Problems That Don't Exist*. London: Allen Lane.

Newell, B. 2021. "Surveillance and the COVID-19 Pandemic: Views from around the World." *Surveillance & Society* 19 (1): 81–4. https://doi.org/10.24908/ss.v19i1.14606

O'Neill, P.H., T. Ryan-Mosley, and B. Johnson. 2020. "Covid Tracing Tracker." *MIT Technology Review*, 7 May. Online at https://www.technologyreview.com/2020/12/16/1014878/covid-tracing-tracker/

Sadowski, J., and R. Bendor. 2018. "Selling Smartness: Corporate Narratives and the Smart City as a Sociotechnical Imaginary." *Science, Technology, & Human Values* 44 (3): 540–63. https://doi.org/10.1177/0162243918806061

Safransky, S. 2020. "Geographies of Algorithmic Violence: Redlining the Smart City." *International Journal of Urban and Regional Research* 44 (2): 200–18. https://doi.org/10.1111/1468-2427.12833

Shaw, J., and M. Graham. 2017. "An Informational Right to the City? Code, Content, Control, and the Urbanization of Information." *Antipode* 49 (4): 907–27. https://doi.org/10.1111/anti.12312

Soave, I. n.d. "Il 'caso Corea': zero nuovi contagi, ma la app è da 'Grande Fratello.'" *Corriere Della Sera*. Online at https://www.corriere.it/esteri/20_maggio_02/caso-corea-zero-nuovi-contagi-ma-app-grande-fratello-e2a450d0-8c8e-11ea-9e0f-452c0463a855.shtml, accessed Auguat 2020.

Sonn, J.W., M. Kang, and Y. Choi. 2020. "Smart City Technologies for Pandemic Control without Lockdown." *International Journal of Urban Sciences* 24 (2): 149–51. https://doi.org/10.1080/12265934.2020.1764207

Taylor, L., G. Sharma, A. Martin, and S. Jameson. 2020. "Global Data Justice?" In *Data Justice and COVID-19: Global Perspective*, ed. L. Taylor, G. Sharma, A. Martin, and S. Jameson, 9–17. London: Meatspace.

Townsend, A. 2013. *Smart Cities: Big Data, Civic Hackers, and the Quest for a New Utopia*. New York: Norton.

Vanolo, A. 2014. "Smartmentality: The Smart City as Disciplinary Strategy." *Urban Studies* 51 (5): 883–98. https://doi.org/10.1177/0042098013494427

Vanolo, A. 2016. Is There Anybody Out There? The Place and Role of Citizens in Tomorrow's Smart Cities." *Futures* 82 (September): 26–36. https://doi.org/10.1016/j.futures.2016.05.010

Wiig, A., and E. Wyly. 2016. "Introduction: Thinking through the Politics of the Smart City." *Urban Geography* 37 (4): 485–93. https://doi.org/10.1080/02723638.2016.1178479

Zandbergen, D., and J. Uitermark. 2020. "In Search of the Smart Citizen: Republican and Cybernetic Citizenship in the Smart City." *Urban Studies* 57 (8): 1733–48. https://doi.org/10.1177/0042098019847410

Žižek, S. 2020. *Pandemic! COVID-19 Shakes the World*. New York: Policy.

12 Trajectories of Data-Driven Urbanism and the Case of Intelligent Transport Systems

LIAM HEAPHY

The expansive social literature on smart cities has revived debates on the "right to the (digital) city" (Foth, Brynskov, and Ojala 2015) and new concepts of digital citizenship (Cardullo and Kitchin 2019a) in the wake of concerns about the loss of political autonomy as Big Tech corporations impinge on citizens' sovereignty (Carr and Hesse 2020). Dystopian accounts of corporate control over the smart city are frequently stirred by the ambitious horizontal integration of urban functions and services as proposed by technology giants such as IBM, Cisco, and Sidewalk Labs (Goodman and Powles 2019; Sadowski and Bendor 2019; Shapiro 2018). However, as Cullen (2016, 134) demonstrates in her inside study of IBM, this is often a case of mistakenly conflating the placeholder tech portfolios of marketing divisions in corporations with the operational divisions that actually develop, implement, and support urban technologies. The problematic communication and ethical frameworks of these superlative urban projects is at least partly a product of unknowable experimental outcomes and unproven business models. This is evident in the eventual failure of Sidewalk Lab's Toronto Quayside project (Mitanis 2020) and in how a consortium of companies (including Sidewalk Labs) fell short of its ambition to profitably distribute 4,500 connectivity/advertising booths across New York (Rubinstein and Anuta 2020).

In contrast to these experimental megaprojects, case studies on more mundane (Star and Ruhleder 2001) or arcane urban functions such as transport, drainage, waste, and fleet logistics might not lend themselves to oft-repeated dystopian visions of neoliberalization or technocratic depoliticization. While such forewarnings are prescient, as weakened public actors lose negotiating power vis-à-vis more powerful corporate leaders through platform capture (see Monahan and Kurnicki, both in this volume), it should not be overlooked that much of the digital revolution is low key, gradualist, and developed through broad assemblages of companies, city engineers, and research with little fanfare. Indeed, as Ylipulli and Luusua (2020, 7) argue, it is appropriate that "the

SC [smart city] be seen as a more politically pre-defined subcategory of urban digitalization."

Grounded studies on digital infrastructure, whether the spatial logics of Github (Straube 2016) or the temporalities of traffic management (Coletta and Kitchin 2017), might provide novel insights and "middle-range theory" (Wyatt and Balmer 2007) for comparing and charting the evolution of digital technologies and data analytics. Kitchin and Lauriault (2014, 1) advance the concept of a "data assemblage that encompasses all of the technological, political, social and economic apparatuses and elements that constitutes and frames the generation, circulation and deployment of data." Data assemblages can function as a means of expanding discussion to include how data reshape society and co-develop with society (Flyverbom and Murray 2018). This chapter builds on a framework for understanding data-driven urban change, formulated as three interrelated translations. The first concerns *data expansionism*, as new data infrastructures are established and shared with an expansive public of interested parties, including communities and corporations. The second is *data experimentalism*, as this public crafts new functions for these data, such as consolidating multiple datasets or providing new informational services. Lastly, *operational consolidation* describes how new procedures and their institutional settings are created or adapted to integrate new data-driven functionalities emerging from the foregoing experimentation (Heaphy 2019). The role of citizens is mediated through the succession of phases, which admit greater participation during periods of expansionism and experimentalism, such as open data movements, artistic and intellectual exploration, and commercial product development. The role of citizens, however, is frequently circumscribed by positioning them as "users" within a largely neoliberal framing of the smart city (Cardullo and Kitchin 2019b; Vanolo 2014) and by a technical complexity that limits participation to those with resources and motivation.

This case study considers the deployment of intelligent transportation systems (ITS) technologies in Dublin, Ireland, to outline a framework for understanding urban digitalization. It follows the rollout of Automatic Vehicle Location Systems (AVLS) and Real-Time Passenger Information (RTPI) on bus services in Dublin. It examines how data generated through vehicle locational technologies are being integrated into automated traffic control systems, and how they feed into new powers of organizational control and coordination through visualization and dashboards.

The present generation of RTPI was implemented in 2011–12 on Bus Átha Cliath (henceforth Dublin Bus), and has been integrated with RTPI from tram services, national coach companies, train services, and other operators that now fall under the jurisdiction of Ireland's National Transport Authority (NTA). The empirical work consists of twelve semi-structured interviews with fourteen participants from two phases of research: the first (2015–16) was on Dublin's efforts to coordinate and promote smart city technologies and initiatives, and was closely linked with the strong presence of leading information technology (IT) companies and universities (see, for example, Coletta, Heaphy, and Kitchin 2019; Kitchin, Coletta, and Heaphy 2019); the second phase (2016–17) expands on the first phase to chart the rollout of RTPI and traffic management technologies. Participants comprised ten transport engineers, two front-line traffic operators in control rooms, a consultant, and a marketing representative.

All interviews were conducted *in situ* in back offices (NTA, Luas, Bus Éireann) and control rooms (Dublin Bus, Dublin City Council, South Dublin County Council). Technologies such as graphical user interfaces, remote monitoring technologies, and dashboards were demonstrated to the researcher, and additional perspectives were gleaned from impromptu conversations with operators and other staff in the control rooms. Interviewees were asked about the deployment of RTPI in their organization and about coordination with other transport providers and regulators. The empirical work was supplemented by reports and secondary literature on the longer timeline of ITS in Dublin.

Participation in the design and rollout of these technologies was driven largely through the political system and the state, which retains power over providers and regulatory structures and whose vacillations shaped the final outcome. Although Ireland is a liberal market economy, its rollout of ITS technologies has not resulted in the removal of agency from the public sphere. By fortune, and possibly through weariness of the austerity drive that followed the 2008 recession, the transport sector has remained closer to the European social democratic model (see Ylipulli and Luusua 2020). The relationship between wavering state support for transport provision and digital advancement is evident in preceding pilots of AVLS and RTPI, as covered in the next section, with the eventual full rollout followed by data experimentation as state companies developed new data-driven functions. The section thereafter explores their integration with control rooms and city-wide decision-making. Finally, I discuss the future of dashboards and data-driven urbanism prior to the conclusion.

Digital Infrastructures for Transport in Dublin

ITS originates in the context of managing and coordinating road use, but interfaces with air, rail, and water systems (Williams 2008). It relies on complex assemblages of information networks, human engineers, transport policies, roadside sensors, in-vehicle computers, and global positioning systems (GPS) to choreograph an increasingly wide array of objects in space.

The digital infrastructure for scheduling bus services and tracking their locations in real time is laid onto an existing network with a long-standing terminology of stages, duties, rosters, and depots. This digital overlay allows for, first, greater control over internal management (scheduling, planning, and performance monitoring) and, second, user-oriented real-time information that can be consulted when planning journeys. Moreover, transport data can be combined with other systems and services to create new datasets and new functionalities. As such, the power of accurate, near real-time data is reworking and retooling public transport, morphing into data-driven transport planning and design as it integrates further into everyday practice.

AVLS and Data-Driven Rationalization of Services

Radio-based locational devices on buses were tested in the 1970s to coordinate services and improve punctuality (Roth 1977), and tested successfully in Dublin in the late 1970s (World Bank 2011). This triangulation-based technology reported the positions of each bus back to a central computer every 45 seconds. Each bus depot was responsible for the coordination of its bus services, rather than there being a centralized control room. The technology was revolutionary from an operations perspective, as it allowed human operators in each depot to focus their attention on routes for which there were identifiable or anticipated issues, such as a bus running ahead or behind schedule due to traffic or driver behaviour (World Bank 2011). Furthermore, it was known from the 1970s that AVLS could be used to provide real-time information to bus passengers (Symes 1980), but this service was not scaled up in Dublin until rapid economic growth had replaced decades of subpar economic performance and stagnation. The original AVLS service remained in place until 1994, and when decommissioned was not replaced.

In the mid-1990s, Dublin Bus faced uncertainty about whether it would follow the fate of UK public transport into wholesale privatization, given the presence of political messages signalling the arrival of competition in an open market.[1] In this prolonged state of uncertainty, investment in a second trial of GPS-based AVLS with RTPI on selected services from 2001 to 2004 was not upscaled and operationalized across the network. Instead, the next and definitive implementation of RTPI occurred in tandem with large organizational changes as transport services were modernized in line with international best practices in the late 2000s and 2010s. These changes led to the creation of the NTA as the national regulator, the preservation of Dublin Bus as the primary bus operator in the city region, and a small market share for new competing operators under strict conditions stipulated by the NTA.

The new AVLS and RTPI system was implemented during the 2009–11 period in partnership with a German company, INIT. GPS and odometer readings are used for discerning location, and they report back to a central system every 30 seconds. Given the maturity of transport technologies at time of rollout, there were several market-leading options for scheduling and monitoring software. Dublin Bus uses Microbus for scheduling and INIT's system (MOBILE-ITCS) for dispatcher control, while national coach company Bus Éireann uses Hastus and Trapeze, respectively, for these functions. Transport operators from around Europe and beyond meet regularly at conferences and expositions to discuss new developments, implementations, and support systems. As a result, they generally use standardized systems of information exchange for timetabling and RTPI data. Germany is a world leader in ITS and its national standard, VDV452, is the favourite in Europe for sharing scheduling data, while a second European standard called SIRI (Service Interface for Real-time Information) is preferred for RTPI.

In common with other modern implementations, dispatchers at both Dublin Bus and Bus Éireann have visual interfaces that highlight services running ahead or behind schedule, so that dispatchers can intervene or request further information directly from drivers using voice communications. The three monitors typically show a map view, a timetable display, and a line display showing punctuality of services. For example, in Figure 12.1, Bus Éireann services are monitored in real time with a map view on the left (Cork City in this case), a table view on the right, and a general performance monitoring dashboard to the rear.

During fieldwork, there were large-scale public works to facilitate a new tram line through the city centre, creating frequent interruptions to bus services on which the city is heavily reliant. This was observed as a dispatcher spotted a bus shown in red that was running heavily behind schedule near the end of its route to South Dublin. The controller, an ex-driver like most of the controllers, contacted the driver by radio and requested the driver to curtail the service early. Controllers and drivers alike in Dublin Bus are trained to use "Press-It" devices on buses that signal drivers when stationary whether they are running ahead of schedule (less desired than running behind). Thus, a suite of technologies are coercing bus services to become more punctual, efficient, and reliable. This is in line with both rising demands from patrons and a regulator that now sets demanding minimum standards for operators, including Dublin Bus, on pain of losing or diminishing their almost exclusive public service obligation contract for services in the city.

Such technologies initially incurred mistrust among drivers but were quickly incorporated. Former drivers and existing controllers were trained to use the new software at Dublin Bus and Bus Éireann. This minimized resistance and allowed both organizations to instate a new culture driven by quality control. Furthermore, the technologies increased the sense of security, particularly welcomed by drivers in remote areas or where the possibility of theft is higher:

> So certainly it is unusual, though because when we were starting off there was a lot of concern, maybe paranoia, about being monitored in real time and there was certainly concern about the system. But it has kind of gone full circle now where,

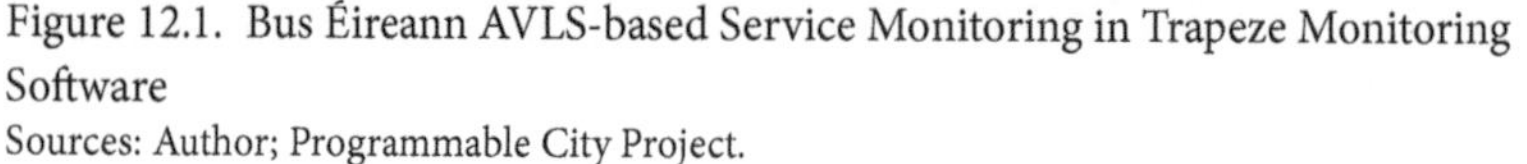

Figure 12.1. Bus Éireann AVLS-based Service Monitoring in Trapeze Monitoring Software
Sources: Author; Programmable City Project.

> if we don't have AVL working, drivers will complain because they feel isolated if they are out in a vehicle. So it has gone full circle now that they do actually rely on it now and look for it to be fully operational. (Interview RTPI08, Bus Éireann)

The technologies are now largely normalized and increasingly integrated with broader transport systems. Experimentation with these systems, as discussed below, involves tinkering around the edges by linking to other transport functions and through innovative use of data analytics to wrest better efficiency from existing services.

Data Analytics and Performance Monitoring

Thus far I have discussed technologies that have been imported and grafted onto local organizations, with little innovation or novelty of interest to practitioners elsewhere. It is with data analytics, however, that some high-end capacities come to light. A phase of data experimentation follows AVLS as engineers use data analytics to provide a belated, but effective, data-driven rationalization of schedules and routes, driven by the more demanding performance monitoring of the new transport regulator.

Dublin Bus technicians with statistical expertise used both INIT's Mobile Statistics application and a further general-purpose statistical analysis application called Crystal Reports to identify route interruptions and improve punctuality. This took the form of eliminating unscheduled stops (typically resulting from heavy traffic), recalibrating schedules based on historical data, and implementing AVLS-based prioritization at traffic lights. Technicians, provided with ample digital resources, could now experiment with functions that would highlight least-performing routes and their pinch points:

> And that is where we started getting into using Crystal [Reports] properly that we are able to run a full week's data, or a full month's data or whatever was required and do a proper ranking based on the trips that … would go through that area against the number of unscheduled stops that we actually create a percentage … I mean there were some cases there where it was going through certain areas and [junction saturation] was 110 per cent, so that means that no matter what you did that bus was always going to get stopped there. Or where people perceive an area as bad it might be down at 20 to 30 per cent. (Interview RTPI04, Dublin Bus)

The statistician achieves a privileged viewpoint through progressively and analytically refining the report by, for example, eliminating incidents of buses remaining stationary for longer than six minutes, and distinguishing between perceived delays versus those reported frequently and consistently by the AVL system. This was rapidly upscaled to improve longitudinal corridors, rather than just individual junctions:

> The junctions were getting great results on individual junctions, but then when you went outside of that and went a kilometre or two kilometres either side of that one junction, benefits are lost. So we were kind of saying, okay, we need to kind of stretch this out and do maybe two junctions around that. So that is where it went, and that was good, we were getting great results … But last year we said … we would like to move into looking at corridors and over a full corridor. And that would be from a point of view of looking at priority at junctions but also looking at efficiency around bus stops, how easy it is to get in and out of bus stops. Like, could we have the yellow box in the right place or is the yellow box in the wrong place? Just have a look at the whole layout of different things, and not just for the bus. The bus may be the priority that we are looking at, but it is also consideration for cyclists, consideration for pedestrians, for everyone. (Interview RTPI04, Dublin Bus)

Having made piecemeal improvements in areas of highlighted concern, Dublin Bus could then optimize its bus schedules based on measured performance from AVLS data. This is now done on a three-week basis or less using aggregated travel time data over eight to ten days. The improvements feed into RTPI accuracy as reflected in smartphone apps and overhead signs at bus stops. These improvements are driven by the NTA, whose authority is bolstered by the data-driven culture that accompanies RTPI, and shows how, as an assemblage, RTPI forms part of a broad system of improvements beyond arcane technical changes and developments.

Coordinating and Managing Data through Control Rooms

The complex dashboards available to the controllers of Dublin Bus and Bus Éireann represent an organic unfolding of information and metrics made possible through more refined transport technologies. They are now collected in spaces dedicated to the coordination and control of services. For Dublin Bus, this is an arrangement of five pods on one floor of a prefabricated building at its Roadstone depot, with around sixteen controllers at the workstations. Nearby, the Dublin Traffic Management Centre (TMC) is responsible for coordinating the safe circulation of both private and public transport in the city and its orbital motorway. RTPI data are integrated with Dublin's traffic management system to prioritize buses at traffic-light-controlled junctions. Thus, the operations and control rooms of Dublin Bus are linked to the main TMC, and engineers from both organizations work together to decide parameters for prioritization.

Dublin uses SCATS (Sydney Coordinated Adaptive Traffic System) in Dublin City for changing traffic light signals through a series of pre-programmed plans that are alternated according to a schedule (allowing for peak traffic hours) and, dynamically, to real-time detection of traffic and approaching buses and trams (see Figure 12.2). Controllers focus on a pivot phase (Plan A), which responds to prevailing conditions, and alternate this phase with standardized pedestrian phases (automatic, or otherwise initiated by the physical request button). The "degree of saturation" reflects the performance of the junction with respect to the proportion of green-light time that vehicles are using, and is calculated by induction loops installed under the surface of each linking road.

The Dublin Public Transport Interface Module was developed by Dublin City Council to integrate with SCATS and provide prioritization for buses and trams using real-time location data in SIRI format. To allow for the limitations of 30-second polling intervals, a series of trigger conditions were created on the basis of "virtual detectors," based on either the calculation of journey time over a longer segment of road or queuing in areas known to be "bottlenecks," such as junctions where a right turn is required (in a left-hand-drive country) (Kinane and O'Donnell 2013). Once this technology was tested

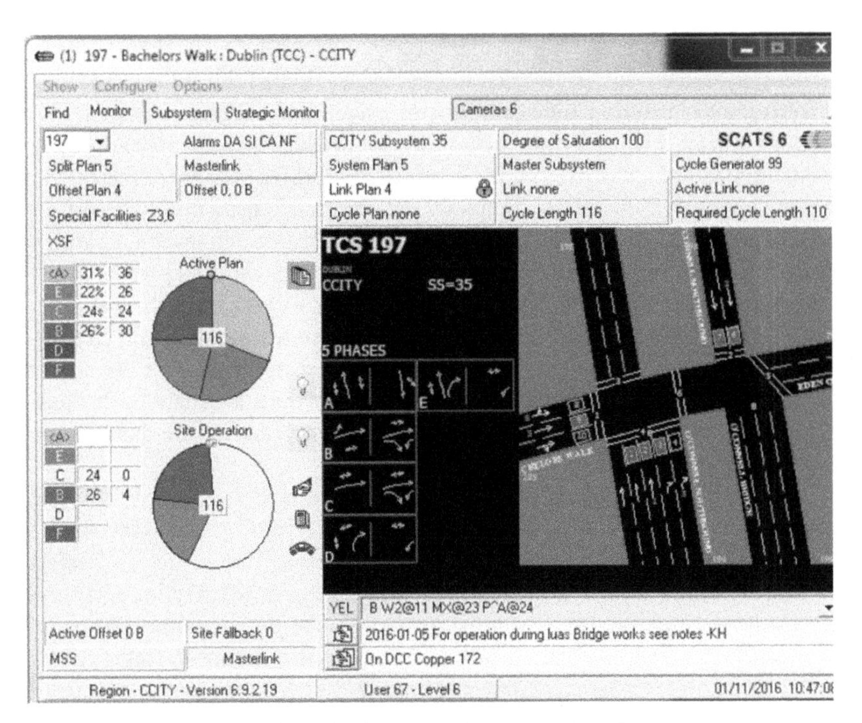

Figure 12.2. Sydney Coordinated Adaptive Traffic System Interface
Note: Figure shows a stylized depiction of the junction and its phases, and information on the proportion assigned to each plan.
Sources: Claudio Coletta; Programmable City Project.

and operational, experiments then involved tweaking sequences of junctions in favour of consistently shorter bus journey times. Initially starting with an individual junction-based bus prioritization, this was later extended to the corridors described above, where a whole series of factors could be considered contiguously.

Bus prioritization plans for junctions are occasionally overridden by human controllers in the TMC when the build-up of traffic as seen on closed-circuit television is deemed excessive. These changes might last for several minutes or longer to "flush out" traffic, but can become permanent changes after discussions between the SCATS controllers and the traffic engineers. The plumbing metaphor is carried through consistently. Rather than being a mellifluous orchestral symphony, it is about ensuring safe and controlled passage of objects through the congested and limited road spaces of an old European city, necessitating human interventions over several timescales: immediate, short-term changes to controllers' voting patterns (or "personalities"); "permanent" alterations for prioritization and general flow; and design changes to the roads themselves in terms of road markings, lanes, and direction, placement of controllers and other infrastructure (including light rail), closures, and road-widening schemes.

During peak-time traffic and occasional events such as concerts and public holidays, the TMC has five or six controllers and three staff from a local radio station and a traffic-alerts company. The spatial arrangement of the control room and the systems on which it depends have gone through several iterations as the logic of centralized coordination imposed itself, with the present iteration dating from 2013. However, unlike the control rooms of cyberneticists, the modern traffic control room does not seek to model, simulate, and control. The control room technologies are anticipatory and reactive, with the actual design decisions made on longer timescales. Its rhythms alternate by day and night, peak and off-peak, with a revolving cast in accordance with daily events (Coletta and Kitchin 2017).

The controllers in Dublin TMC are constantly mediating breakdowns and collisions, occasionally fatal, that are part of the everyday incidents they manage in coordination with emergency services and recovery vehicles. Located adjacent to the room in Figure 12.3 is an incident room where joint decisions are made and modifications to junctions applied using computer terminals. Controllers handle immediate issues, and their rotas reflect peak traffic times, with a skeleton crew covering late evenings and nights. The traffic engineers work standard office hours in physical proximity to controllers, and organize changes to junctions through scheduled meetings:

> So maybe in the case of a bus priority, changing a bus lane, creating a bus lane, extending it, or even just making a junction just a left turn as opposed to, say, it had a left turn or to allow it to be a right-turn-only lane and then to modify the signals to give them more time … We would have a meeting every couple of weeks with other members of the team and then we would raise it up that way and then we would follow them up. We also get a number of queries from residents and users, and they would go through the area engineers, a lot of the time queries would come up that way as suggestions from people as well as to changes they would like to see, and you can evaluate them and see if they are just someone trying to get out of their house quicker or actually valid. (Interview RTPI07, engineer in Dublin City Council)

These changes are conducted in partnership with a broad range of actors, and involve a reprioritization of flows away from private vehicles towards public transport, cyclists, and pedestrians. The engineers, like the human controllers, also use the cameras to validate their changes. The decisions they make mediate between various demands, yet both they and the TMC manager bow to pressures to demote private, low-occupancy vehicles in favour of public

Figure 12.3. Wall of Data, Dublin's Traffic Management Centre
Sources: Claudio Coletta; Programmable City Project.

transport, cycling, and pedestrians, in accordance with the stated objectives of the city plan (Dublin City Council 2016). Several public works in the city have now shifted the balance, including a third cross-city tram line, which has closed private access to one of Dublin's main thoroughfares in favour of restoring the architectural set piece that is the College Green area, enclosed by Trinity College and the old Parliament building.

Such decisions, formerly more problematic in a city where drivers held sway, are supported by the increasing patronage of public transport, itself aided by the provision of higher levels of service through RTPI, a modern fleet, and better adherence to schedule. The implementation and responsibility for monitoring these changes is conferred to the assemblage of engineers, technicians, regulators, controllers, and digital infrastructure that form the urban transport system.

Anticipating the Future of Data-Driven Urbanism

A further conceptual experimentation in data-driven urbanism is the development of city dashboards that show the potential for horizontal integration of urban services and the provision of city-wide benchmarks. Mattern (2015) notes that "the dashboard is an epistemological and methodological pastiche," which represents the many ways a governing entity can decide on important variables (such as key performance indicators) and how they are operationalized into management procedures. Sarikaya et al. (2019) observe the variance between operational dashboards, built around the needs of a clearly defined audience, and communicative advisory dashboards. Examples of the former include those used in the Dublin TMC, while examples of the latter are the city dashboards of London and Dublin (Kitchin, Lauriault, and McArdle 2015; Stehle and Kitchin 2020), functioning as experimental aggregators of open data.

As certain experimental functions developed or discovered by engineers are recognized for their utility, so might they become operationally consolidated into procedural systems such as the dashboards of bus controllers or traffic engineers. Like other infrastructure, they become invisible when operational – black-boxed insofar as the average citizen has little knowledge of their existence. As the case of AVLS and RTPI shows, such changes occur over long timescales that are heavily shaped by a social and political context that might hasten or hinder technological change. These changes are regulated by state and local government actors, encountering citizens only when there are major service alterations or privatization schemes.

At time of writing, the NTA is leading a large-scale, data-driven rationalization of all Dublin Bus services called "BusConnects," with extensive consultation. It favours a hub-and-spoke model with better provision of orbital routes based on a data-driven redesign of routes. It hopes to dispense with accretions from a data-agnostic past, such as an overreliance on services passing through the main thoroughfare of an expanding city. These changes are contentious because, as always, they affect property and land-use values (particularly for road-widening schemes) and add an initial cognitive cost to transport planning at the individual user level. They are therefore as much a program of selling a vision of modernization and progress as they are of convincing people that a given route Y will be superior or equal to route X according to a series of parameters such as journey time, connections, and accessibility. City dashboards, which serve the needs of creating shared vision and data awareness, could play a role in such changes inasmuch as they are communicative and experimental tools, rather than operational devices. As such, they occupy a place between artistic experimentation and policy discourse, stimulating debate on the future of the digital city and its purported benefits to citizens.

Conclusion

ITS technologies fulfil much of the lauded aims of the smart city, integrating multiple rich data sources into algorithmic and informational systems that improve the health and safety of the urban environment. They do so without fanfare or publicity, only retroactively incorporated into the "smart city" (Kitchin, Coletta, and Heaphy 2019). They follow a long timeline that precedes the smart city, as can be seen in the various iterations of RTPI in the case of Dublin Bus.

The rollout of AVLS and RTPI enabled data experimentation by Dublin Bus technicians and city engineers – including a new methodology for traffic prioritization that has been shared with practitioners in international events. The use of AVLS for improving schedule and RTPI accuracy reflects an example of where a previously sporadic or dormant capacity for data-driven cultural change was recognized by those controlling the resources necessary for its actual city-wide implementation after several successful, but disjointed, trials over four decades. Taken together, these complementary technologies provide evidence of a localized realization of the power of data when allied to mature and well-established technologies of management and control. It has also resulted in Dublin City Council's participation in international ITS research projects, including two European Commission–funded projects, Insight-ICT (FP7) and VaVel (Horizon 2020), in partnership with IBM's research centre in Dublin and international universities. This has been a process of collaboration through broad public-private research partnerships, rather than one in which corporations were allowed to capture and reframe the narrative (see Krivý, in this volume).

The roles of control rooms and dashboards need to be demythologized and their development contextualized in relation to technological change. In the case of Dublin, the TMC control room itself is closely bound with the temporal rhythms of existing traffic flows. Control rooms are the nerve centres for the arterial networks of cities and central to the effective coordination of traffic and public transport fleets, exploring a "risk space" (November and Creton-Cazanave 2016) that extends far beyond the room itself. The Dublin TMC is largely responsive and reactive, rather than directing, with each type of intervention requiring different but complementary skill sets and technologies. Decisions on material changes to urban infrastructure such as road layouts and public transport services are informed by technologies of monitoring, modelling, and forecasting, yet ultimately still guided and sanctioned through the planning process, consultation exercises, and traditional representative democracy.

Nevertheless, fears that the architecture of the control room or the selective representation of data in dashboards (Shelton 2017) might facilitate a technocratic, transactionalist form of citizenship (Johnson, Robinson, and Philpot 2019) are not ungrounded, as the case of China shows through its social credits system and integration with transport options (Kostka and Antoine 2018). Such transitions reflect underlying social and political systems, with participants in China broadly supportive of these changes, in contrast to Western liberal democracies that are wary of corporate takeover and authoritarianism. This supports the argument that urban technological development is better understood with reference to extended infrastructural timescales and attention to an assemblage of sociocultural, material, and political factors. These will be manifested in both the gradually evolving technical architectures themselves and in the cultural and experimental practices that facilitate them, including dashboards, art installations, and expositions – hence, the value of attending forthwith to the visions and programs that shape the medium- to long-term future, as these will determine the "intelligent city," "smart city," "resilient city," or whatever appellation commands attention.

ACKNOWLEDGMENTS

The research for this chapter was provided by a European Research Council Advanced Investigator Award, "The Programmable City" (ERC-2012-AdG-323636).

NOTE

1 A notable exception to privatization in the United Kingdom was London, which, after the breaking up of the Greater London Council and the loss of its transport powers, created Transport for London as part of its replacement of the Greater London Authority. This has provided a model for city region devolution in other cities, exerting a strong influence through proximity and familiarity on developments in Ireland, as discerned during the fieldwork.

REFERENCES

Cardullo, P., and R. Kitchin. 2019a. "Being a 'Citizen' in the Smart City: Up and Down the Scaffold of Smart Citizen Participation in Dublin, Ireland." *GeoJournal* 84 (1): 1–13. https://doi.org/10.1007/s10708-018-9845-8

Cardullo, P., and R. Kitchin. 2019b. "Smart Urbanism and Smart Citizenship: The Neoliberal Logic of 'Citizen-Focused' Smart Cities in Europe." *Environment and Planning C: Politics and Space* 37 (5): 813–30. https://doi.org/10.1177/0263774X18806508

Carr, C., and M. Hesse. 2020. "When Alphabet Inc. Plans Toronto's Waterfront: New Post-Political Modes of Urban Governance." *Urban Planning* 5 (1): 69–83. https://doi.org/10.17645/up.v5i1.2519

Coletta, C., L. Heaphy, and R. Kitchin. 2019. "From the Accidental to Articulated Smart City: The Creation and Work of 'Smart Dublin.'" *European Urban and Regional Studies* 26 (4): 349–64. https://doi.org/10.1177/0969776418785214

Coletta, C., and R. Kitchin. 2017. "Algorhythmic Governance: Regulating the 'Heartbeat' of a City Using the Internet of Things." *Big Data & Society* 4 (2). https://doi.org/10.1177/2053951717742418

Cullen, M. 2016. "Cities on the Path to 'Smart': Information Technology Provider Interactions with Urban Governance through Smart City Projects in Dubuque,

Iowa and Portland, Oregon." PhD diss., London School of Economics and Political Science. Online at http://etheses.lse.ac.uk/3392/

Dublin City Council. 2016. "Dublin City Development Plan 2016–22." Online at http://dublincitydevelopmentplan.ie/

Flyverbom, M., and J. Murray. 2018. "Datastructuring – Organizing and Curating Digital Traces into Action." *Big Data & Society* 5 (2). https://doi.org/10.1177/2053951718799114

Foth, M., M. Brynskov, and T. Ojala, eds. 2015. *Citizen's Right to the Digital City: Urban Interfaces, Activism, and Placemaking*. Singapore: Springer.

Goodman, E.P., and J. Powles. 2019. "Urbanism under Google: Lessons from Sidewalk Toronto Symposium: Rise of the Machines: Artificial Intelligence, Robotics, and the Reprogramming of Law." *Fordham Law Review* 88 (2): 457–98. https://doi.org/10.2139/ssrn.3390610

Heaphy, L. 2019. "Data Ratcheting and Data-Driven Organisational Change in Transport." *Big Data & Society* 6 (2): 1–12. https://doi.org/10.1177/2053951719867359

Johnson, P.A., P.J. Robinson, and S. Philpot. 2019. "Type, Tweet, Tap, and Pass: How Smart City Technology Is Creating a Transactional Citizen." *Government Information Quarterly*, October. https://doi.org/10.1016/j.giq.2019.101414

Kinane, D., and M. O'Donnell. 2013. "Dublin Public Transport Interface Module A Pilot Study." In *Proceedings of the ITRN2013*, 19. Trinity College Dublin, Ireland.

Kitchin, R., C. Coletta, and L. Heaphy. 2019. "Actually Existing Smart Dublin: Exploring Smart City Development in History and Context." In *Inside Smart Cities: Place, Politics and Urban Innovation*, ed. A. Karvonen, F. Cugurullo, and F. Caprotti, 83–101. London: Routledge.

Kitchin, R., and T.P. Lauriault. 2014. "Towards Critical Data Studies: Charting and Unpacking Data Assemblages and Their Work." SSRN Scholarly Paper ID 2474112. Rochester, NY: Social Science Research Network. http://papers.ssrn.com/abstract=2474112

Kitchin, R., T.P. Lauriault, and G. McArdle. 2015. "Knowing and Governing Cities through Urban Indicators, City Benchmarking and Real-Time Dashboards." *Regional Studies, Regional Science* 2 (1): 6–28. https://doi.org/10.1080/21681376.2014.983149

Kostka, G., and L. Antoine. 2018. "Fostering Model Citizenship: Behavioral Responses to China's Emerging Social Credit Systems." SSRN Scholarly Paper ID 3305724. Rochester, NY: Social Science Research Network. https://papers.ssrn.com/abstract=3305724

Mattern, S. 2015. "Mission Control: A History of the Urban Dashboard." *Places Journal*, March. https://doi.org/10.22269/150309

Mitanis, M. 2020. "Sidewalk Labs pulls out of Quayside project." *Urban Toronto*, 7 May. Online at https://urbantoronto.ca/news/2020/05/sidewalk-labs-pulls-out-quayside-project

November, V., and L. Creton-Cazanave. 2016. "Inquiry in Control Rooms – An Analysis through the Lenses of Space, Time and Practice." *HAL Open Science*. Online at https://halshs.archives-ouvertes.fr/halshs-01394719/document

Roth, S.H. 1977. "History of Automatic Vehicle Monitoring (AVM)." *IEEE Transactions on Vehicular Technology* 26 (1): 2–6. https://doi.org/10.1109/T-VT.1977.23648

Rubinstein, D., and J. Anuta. 2020. "City Hall calls Google-backed LinkNYC consortium 'delinquent.'" *Politico PRO*, 3 March. Online at https://www.politico.com/states/new-york/albany/story/2020/03/03/city-hall-calls-google-backed-linknyc-consortium-delinquent-1264966

Sadowski, J., and R. Bendor. 2019. "Selling Smartness: Corporate Narratives and the Smart City as a Sociotechnical Imaginary." *Science, Technology, & Human Values* 44 (3): 540–63. https://doi.org/10.1177/0162243918806061

Sarikaya, A., M. Correll, L. Bartram, M. Tory, and D. Fisher. 2019. "What Do We Talk about When We Talk about Dashboards?" *IEEE Transactions on Visualization and Computer Graphics* 25 (1): 682–92. https://doi.org/10.1109/TVCG.2018.2864903

Shapiro, A. 2018. "The Urban Stack: A Topology for Urban Data Infrastructures." *TECNOSCIENZA: Italian Journal of Science & Technology Studies* 8 (2): 61–80.

Shelton, T. 2017. "The Urban Geographical Imagination in the Age of Big Data." *Big Data & Society* 4 (1): 1–14. https://doi.org/10.1177/2053951716665129

Star, S.L., and K. Ruhleder. 2001. "Steps toward an Ecology of Infrastructure: Design and Access for Large Information Spaces." In *Information Technology and Organizational Transformation: History, Rhetoric, and Practice*, ed. J. Yates and J. van Maanen, 305–46. Thousand Oaks, CA: SAGE.

Stehle, S., and R. Kitchin. 2020. "Real-Time and Archival Data Visualisation Techniques in City Dashboards." *International Journal of Geographical Information Science* 34 (2): 344–66. https://doi.org/10.1080/13658816.2019.1594823

Straube, T. 2016. "Stacked Spaces: Mapping Digital Infrastructures." *Big Data & Society* 3 (2). https://doi.org/10.1177/2053951716642456

Symes, D.J. 1980. "Automatic Vehicle Monitoring: A Tool for Vehicle Fleet Operations." *IEEE Transactions on Vehicular Technology* 29 (2): 235–7. https://doi.org/10.1109/T-VT.1980.23846

Vanolo, A. 2014. "Smartmentality: The Smart City as Disciplinary Strategy." *Urban Studies* 51 (5): 883–98. https://doi.org/10.1177/0042098013494427

Williams, B. 2008. *Intelligent Transport Systems Standards*. Boston: Artech House.

World Bank. 2011. "Intelligent Transport Systems – Dublin, Ireland." Toolkit on Intelligent Transport Systems for Urban Transport. 2011. Washington, DC: World Bank. Online at https://www.ssatp.org/sites/ssatp/files/publications/Toolkits/ITS%20Toolkit%20content/case-studies/dublin-ireland.html

Wyatt, S., and B. Balmer. 2007. "Home on the Range: What and Where Is the Middle in Science and Technology Studies?" *Science, Technology, & Human Values* 32 (6): 619–26. https://doi.org/10.1177/0162243907306085

Ylipulli, J., and A. Luusua. 2020. "Smart Cities with a Nordic Twist? Public Sector Digitalization in Finnish Data-Rich Cities." *Telematics and Informatics*, July. https://doi.org/10.1016/j.tele.2020.101457

13 The Parking Problem and the Limits of Urban Digitalization

KAROL KURNICKI

Urban mobility is one area where the promises of smart cities have come to fruition. New solutions and innovations based on digitalization are being developed to make mobility more accessible, efficient, user friendly, and flexible. Electric scooters are found, rented, and paid for through a smartphone application. Shared cars and taxis can be traced as dots on an online map and booked with one tap. Bikes are fitted with navigation devices and phone holders, thus joining the population of the Internet of Things (Behrend 2019). Planning a journey does not require a bus schedule anymore, just a few swipes on the screen. If Mobility as a Service platforms come to full development as expected, soon every trip will be planned, recorded, and completed with the help of some sort of digital companion (Jittrapirom et al. 2020).

In this chapter I take a closer look at the digitalization of parking. This case shows how processes of digitalization of a simple immobility practice (Kurnicki 2020) interact with existing sociospatial relations in cities in a way that brings problems related to mobility justice to the surface (Sheller 2018a). Although the issues of digital rights and hazards of datafication (Mejias and Couldry 2019) and platformization (Barns 2019) increasingly are recognized in the literature, analyses of systems that transform mobility rights through digital means are still required to shed light on the transformative potential of digitalization. As the growing use of digital data reshapes how people move (de Souza e Silva and Sheller 2015), it also affects people's rights to movement, to freedoms that depend on being anonymous in cities, and to accessibility to spaces and means of transport. There is an inherent danger in the privatization of public spaces and growing inequalities that comes with the implementation of networked technologies in mobility (Sparrow and Howard 2020).

Car parking is no exception when it comes to the adoption of data-based systems, algorithms, and networked devices, both for the management of cars in urban spaces and for the experience of drivers. These systems aim to solve "the parking problem" and make leaving a vehicle predictable, effortless, and economically acceptable for providers and users. The digitalization of parking, not unlike other processes of this kind in the sphere of mobility, mediates an actual everyday experience, and gives it a representational, conceived character (Lefebvre 2010). As with other areas in which Big Data becomes dominant, questions of privacy, objectivity, and accuracy, as well as social and political issues, come into play (boyd and Crawford 2012). But the fact that parking is literally grounded in material, spatial relations makes some obstacles of digitalization more prominent. Vehicles and people are not tweets or memes, but have a very material presence that extends between the digital and the material. Immobile cars as objects cannot be digitized away.

Parking is the area of urban transportation system that, although regulated to a certain extent by local and national laws, has always been organized largely according to free market rules (Shoup 2005). Even where provision of transportation is the responsibility of public administration, parking is often treated as a private matter – only to be regulated, but not necessarily provided by local councils. And where parking is actually provided by public institutions – for instance, in the form of large-scale car parks or on-street spaces – there still exists a large pool of private places in the undergrounds of buildings or on people's properties. This variegated nature of parking, in which private ownership and responsibilities play an important role, makes it a case that forecasts the possible future of the organization of urban mobility if it follows the path towards more privatization.

I identify two issues that arise in the context of the digitalization of parking where its strong private characteristics come into play. First, there is the question of the future of digitalized mobility, in which privatization and its consequences are amplified by platforms run by private companies. What can the ongoing digitalization of parking tell us about the shape of data-based *immobility* in cities? The second question has to do with inequalities that are inherent in mobility in general and calls for increased mobility justice (Sheller 2018b) or work towards the commoning of mobility (Nikolaeva et al. 2019), both of which reflect discussions about digital inequalities. The example of parking shows the negative consequences of digital technology where private ownership and a sense of entitlement – a "right to parking" or a "right to space" (see Taylor 2020) – dominate over the public interest. The example of digitalization of immobility in the context of already unequal access to space highlights the intersection of inequities produced in both of those areas.

Cases: Parking in Two European Cities

To begin, I provide an overview of the main areas of digitalization of parking. Then, to problematize the use of digital solutions to urban immobility, I describe the technologies in use in the two examples of London and Warsaw, with a focus on those areas where these technologies interact with socially unjust and spatially uneven urban processes (Stehlin, Hodson, and McMeekin 2020).

The description draws on a project about car parking practices in London, Warsaw, and Utrecht (the latter of which I have not included in this chapter). The investigation focused mainly on on-street and residential parking to uncover the side of parking that is less organized and controlled than in car parks. I was interested in parking as a social practice, and therefore the investigation was guided by questions such as: what meaning does it have for people? how do they perform it? what objects do they engage in the process? how does it connect with other things they have to do in everyday life? I was also looking at the interrelation of parking with infrastructure: how it is used by cars? how does parking create specific systems and networks on a larger scale? how does infrastructuring (Merriman 2016) occur in everyday circumstances?

In the course of the fieldwork, it turned out that digital technologies feature prominently in how parking is experienced by drivers and how it is organized by companies and administrative entities such as local councils or transportation administrations responsible for its provision. For instance, drivers commonly interact with navigation apps and payment systems on their phones or in machines on the side of the road. Parking providers and local administrations increasingly rely on detailed data gathered through sensors, GPS-fitted smartphones, and cameras to manage daily demand, pricing, and law abidance, and tend to support solutions enabled by technology.

In this chapter I use information gathered during interviews, which included talks with representatives of companies and professional associations in the area of parking and urban mobility, as well as administrative officials on a municipal level. In the course of the project, I have also followed closely the area of parking through professional journals and newsletters – in English – such as *Today's ParkNews* and *Parking News*, and conducted an observation at the Parkex trade fair in Birmingham in April 2019, all of which helped to prepare the overview of the digitalization of parking.

I begin with a description of the different ways in which digital technology augments parking. This is meant to show how digitalization affects the personal, individual experience of parking, and transforms the way dormant vehicles (Spurling 2019) and their infrastructure are organized in cities. I also describe how parking becomes reconfigured alongside the implementation of smart city technologies, which incrementally embed digitalization in the everyday and infrastructural (see González-González, Nogués, and Stead 2020). I then turn to specific examples of some of these technologies based on the fieldwork in cities.

How to Digitize Parking?

At first sight, parking seems to require little technological assistance. It is a very ordinary and mundane practice, done by people habitually at the end of a journey. There is a car, a driver, and a parking space into which the vehicle has to fit. Modern cars, however, equipped with power steering, beeping distance sensors, and rear cameras, make it clear that people look for help in the simple activity of stopping and leaving the vehicle. Drivers like to be supported by additional capacities, provided by technological affordances – extra power for turning the wheels or a set of well-positioned electronic eyes – to be able to park in a safer, more convenient way, or to park at all. These affordances now extend far beyond the vehicle itself and are increasingly dependent on automation, data, and algorithms.

First, there is the car without a driver – a nascent technology of future mobility (Pink, Fors, and Glöss 2017). Parking is at the forefront of the change towards autonomization, and many companies already offer vehicles that can park themselves. Vehicles manoeuvring on their own rely on the fact that parking as an activity is relatively safe. Parking spaces tend to have regular shapes and sizes, and are therefore predictable for digital systems. Most of the objects in the vicinity are stationary. The vehicle has low velocity, and even in case of accident the danger to people is limited, which is the reason these systems are legal even in countries that do not allow autonomous vehicles on the roads yet. In the case of parking automation, the involvement of digital technology, however advanced, is limited to manoeuvring the vehicle. Simply put, the aim here is to reconfigure or even *replace* the driver's skills. Not all skills are involved, however, as a human will still be responsible for finding an appropriate space and making sure that the car can be left there (Hind 2019). But automation of the performance of parking itself, based on technology and digital data, is already available.

More advanced kinds of automation is being developed for car parks, whereby cars are dropped off at the entrance and then self-drive to designated spaces. This solution would eliminate the driver's need to find a specific spot and help to make maximum use of available space, because vehicles without people inside can be fitted more tightly. At the same time, they would require a high level of standardization – such as regular parking space sizes and vehicles fitted with appropriate equipment – and meant to operate in closed, controlled environments. This kind of digitalization relies on the *reduction of messiness*, which is normally inherent in everyday environments and influences how drivers park (Chaniotakis and Pel 2015). This process can be expected to seep from closed car parks into streets with the popularization of automated vehicles, resulting in new management of driving (Dodge and Kitchin 2007) and potentially significant reconfiguration of existing soci-technical relations (Marres 2020).

Once one is out of the car, there will be pieces of digital-data-gathering equipment waiting on the street. Not just cameras, but also sensors built into surfaces, will be widespread in cities. In this case, the idea is that they will transmit real-life information about the occupancy of designated parking spaces. As a result, parking management should come much more under control, as in car parks, where cameras and sensors have been in use for many years. The data will be fed directly to urban control centres and then transmitted to users, who will be able to see the availability of spaces close to their destination on the screens of their smartphones or control panels inside vehicles. The full operationality of this system will require the *standardization* of parking spaces, cars, and the practices of drivers, who can be careless and disregard designated lines. It also will demand coverage of urban spaces with data-gathering equipment that is as complete as possible.

The data gathered by sensors could be combined with those sent by GPS devices installed in cars and smartphones to create a comprehensive, real-life representation of the state of parking in a given area. This is a form of *datafication* (Mejias and Couldry 2019) that will connect parking providers – both public and private – and drivers. It will rely on the integration of different types of data: from the availability of spaces, ideally in real time, to coordinates and destinations of cars as well as payment data. These data, in turn, could be integrated with city-wide technologies that generate data through cameras and sensors. These are the elements that will make the world – in this case, the world of driving and parking – machine readable (Dodge and Kitchin 2005) and possibly more manageable with the use of technology.

With enough data, the proprietary algorithms can discover patterns in daily or weekly traffic flows and predict the availability of particular locations. This opens possibilities for flexible pricing, based on current demand, which is already operational in San Francisco (Chatman and Manville 2014). The integrative trend is one of the instances of *platformization* (Poell, Nieborg, and van Dijck 2019), which occurs when two or more elements of parking become connected through digital means in order to match drivers with service providers. Platformization has become important for automobile manufacturers, which increasingly include digital parking solutions – navigation, payments, and various forms of automation – in the hardware of their cars (for instance, in the central panel), along with other surveillance functions (Gekker and Hind 2019). Platformization is also important in differentiating urban spaces between, for instance, those who can afford to occupy certain spaces and those who cannot. It might also support parking as a money-generating feature of urban streets, thus precluding changes towards active-travel or low-traffic areas.

Beyond the street environment, there are technologies that deal with payments, the processing of penalty notices, complaints, and contestations of fines. They often come as automated front-end and back-end solutions that help to manage customers of public and private parking spaces. One important and contested area where digitalization becomes relevant is enforcement. From automatic identification of offenders through licence plate recognition to algorithm-based systems that guide camera-equipped vehicles through city streets and streamlined processes of payment collection, the control side of parking already makes use of very advanced digital technologies.

The increased efficiency and ease of parking will be accompanied by enhanced mechanisms of control that impose different levels of accessibility to space and automatically identify offenders – those who leave their cars in the wrong place or who do not pay for their time in a given spot. The digitalization of parking promises to increase accessibility and ease of parking for individual drivers, while making control much stricter by technological means. Considering that immobility, including parking, is already fundamentally unequal, as it is often based on privilege, the digitalization of enforcement will intensify this differentiation and increase divisions based on access to mobility and privatization of public spaces.

Digitalization of Parking: Two Examples

Two cities, London and Warsaw, can serve as illustrations of how some of the technologies that deal with the "parking problem" are used. Here they are seen through the eyes of people who are directly involved in their use or management – a representative of a company that provides booking solutions, and employees of a city administration who work on parking issues. I want to use this perspective to shed light on problems with digitalization that result from its private nature, inability to resolve sociomaterial issues as well as spatial inequalities that become visible in the digitalization processes.

London: The Platformization of Parking

The first example is a platform that provides parking solutions to providers and an application that allows drivers to look for, reserve, and pay for a parking space. Its two-way, infrastructural nature is noted by the company's representative, who characterized its role as: "we're literally the pipe in the middle to connect one to the other" – that is, providers to customers.

The service is accessible as a smartphone app and available through a web browser. In both cases, its interface is a map and a search engine that allow the user to select a location of a parking space. There are options to look in garages, on streets, and among privately offered spaces, with different booking possibilities – from an hour to long-term use. Normally, there are several options in the selected area. In London, parking spaces are usually in commercially operated car parks, but there is also a selection of on-street parking, including those managed by councils. The interface and logic of the system are very simple, based on the presentation of available offers in a given area. Price and availability play major roles in how parking is represented.

The platform does not promise more than it can do – it is simply a connective layer between customers and providers that commodifies physical spaces in which cars and parking spaces exist. At the same time, as a kind of digital technology, it holds the potential for being one of the main modes in which parking operates and through which it is managed and organized. It is common in London for parking machines to fall into disrepair, and payments can be made virtually only via the smartphone application, thus feeding the digital databases and potentially excluding some users. The logic of expansion is built into the operationality of the system, which works better as it gets more information. Drivers who do not adjust to the informational format – for instance, those who have no smartphone – encounter problems.

This situation can be problematized in reference to Lefebvre (2010), whose ideas have already been applied to the growing importance of information in cities. As Shaw and Graham (2017) argue, we can talk about an "informational right to the city" or, rather, its insufficiency in a world dominated by global corporations that control data. The case of a driver finding a parking space in London is not as serious and consequential, despite the platforms' clear inequality in control over data. Here, however, I want to point out a different issue that the operation of the platform brings to the surface. This is the problem of representation – in the Lefebvrean sense – which loses connection to the complex urban reality while becoming the dominant perspective through which space is conceived. Flattened and formatted representation of space in the form of a digital platform takes over the complex urban reality in which parking has actually to be performed by people and cars.

Second, representation, as such, attempts to be complete and hide its internal gaps, fissures, and holes. This is where the nature of a platform as an innately colonizing technology (Srnicek 2016) aligns with the logic of representational space. The simplification and standardization, which are also crucial for the success of commodified spatial relations, are probably part and parcel of any data processing despite the fact that space is inherently messy and mediated (Leszczynski 2014). These two qualities are also visible in parking, which, when digitalized, has a strong ordering quality – i.e., it reduces the messiness – that directly influences urban spaces. The logic of representation based on "governing through code" (Klauser and Söderström 2016) means not only that spaces become commercialized, but also that drivers will have to adjust their parking to how the system works. The clear sorting mechanism based on pricing and availability of parking for different groups of users contributes to already existing social differentiations in immobility.

Warsaw: Enforcement of Payments

The second example is the enforcement system introduced in Warsaw to solve the "parking problem." At the time of fieldwork (2019), the city had an obsolete parking zone that did not fulfil its function of regulating parking in areas under most pressure from cars, mainly in the central districts. The size and boundaries of the zone, as well as pricing, had not been updated since the zone's introduction in 2008. Combined with a great increase in the number of cars in the city, this exacerbated the problem of vehicles occupying every available space and taking over pavements and green spaces. Moreover, the existing control and enforcement procedures, based on human controllers walking the streets of Warsaw to check payment slips stuck behind windshields, was inefficient and the penalty of 50 złoty – just over £10 – for the lack of payment was not a sufficient deterrent.

Although changes to the parking zone, including prices, had been planned in the course of 2019 and introduced in 2020, one solution preferred by the city administration was based on digital enforcement in the form of a car fitted with a licence plate recognition system. Drivers who pay for parking must type in their licence number, which is included in the database that the controlling car uses to check if the payment was made.

Even though this kind of system already functions in many cities, the transportation administration of Warsaw presented it in a promotional video[1] as a cutting-edge, high-tech digital solution. The city is also developing (with the help of a private company) its own parking platform, which will be available for users as an application directing cars to available spaces in the central areas and including payment and reservation functionalities. But, as with other solutions of this kind, the success of the application depends on the scale of uptake – the more people use the system, the better it will get. Therefore, the administration is set to prefer digital parking technology over other, possibly more effective, albeit not necessarily digital, solutions simply to make it operational.

More important, however, is that the digitalization of parking does not eliminate the need either to allocate adequate space for cars (increase supply) or to limit their number (decrease demand). The example of Warsaw is especially striking in this respect because the city regularly issues more permits than it has parking spaces within the zone. Even including the daily circulation of cars, this great pressure on space means that the "parking problem" will remain. As my interviewees pointed out, the digital solutions seem to ignore the simple fact that there are too many cars on the streets of Warsaw.

Moreover, space in Warsaw – but also in London – is owned and/or managed by different entities. For instance, larger roads fall under the responsibility of a higher-level administration, while the smaller are the purview of local councils. Things get even more complicated when we consider the scale of illegal parking, which is widespread in Warsaw. If a car is left illegally, the driver might be in danger of a penalty from the police, but does not have to pay a parking ticket. The enforcement cameras mounted above cars cannot control this kind of illegal parking, making the enforcement system highly selective. Considering that many more cars are parked in the centre of Warsaw than there are legal places and that those cannot be extended due to lack of space, economic viability, or environmental issues, digital technologies, if not backed by politically controversial decisions, will always fall short as a solution.

My interviewees from the administration recognized the need to integrate parking policy into the wider planning of transportation in the city. While welcoming the new systems, they stressed, that the main issue was the mismatch between the number of cars and parking spaces, which is unlikely to change in the next few years. At the same time, the implementation of effective solutions would require significant reorganization of mobility policies in the city – for instance, putting limits on car use – which is politically risky and might exacerbate existing social inequalities. Digital solutions cannot alleviate these problems, for reasons extending from the material character of car use to sociospatial relations such as land ownership, lack of appropriate planning, and mixed responsibility for the street layout – not to mention the cunning of drivers who try to avoid parking charges.

Conclusion: Mobility (In)Justice in the Digital World

To conclude, I want to highlight the mobility justice concept and point out that the digitalization of immobility has to be approached critically. Mobility justice extends beyond the accessibility of transport for different groups. Rather, it is an intersectional perspective, which sheds light on a variety of tensions and contestations implicit in how people and things move from everyday to systemic levels (Sheller 2018b). In this context, digitalization brings about its own set of issues – for instance, related to surveillance (Sgibnev and Rekhviashvili 2020) or instability in the labour market in the form of the so-called gig economy (Calo and Rosenblat 2017). At the same time, interpretations of data based on political readings of datafication as capital accumulation or extraction (Sadowski 2019) can be matched in mobility by calls for "commoning" that supports just and inclusive sustainable transitions (Nikolaeva et al. 2019), and includes much more than economic relations in urban spaces.

Technologies for parking are only technologies. They do not solve problems rooted in spatial inequalities. But in light of ongoing and future immobility transformations, questions about the role of digitalization in delivering positive changes are necessary. In the case of parking, relying on digitalization and automation is set to maintain the inequalities and injustices that have roots in wider urban issues, such as ownership of land, planning, or the financialization of every possible urban resource, including space itself.

The marketization and commodification of parking are concomitant with digitalization and support the rule of private ownership of vehicles and the principle of payment for taking up small fragments of cities. Urban space is treated simply as a resource that can be put to economic use. Digital technologies deepen the penetration of markets in the area of mobility and help to lock in the approach that makes space a commodity. As the example of "curbside management" (Marsden, Docherty, and Dowling 2020) makes evident, social interests are usually sidelined by market objectives.

Taking another example from London, where it is possible to rent out private parking spaces through an online application, one could say that technology makes it possible to tap unused resources, share goods, and make urban space more useful and efficient. But in a more critical reading, one might ask if the point of this kind of platformization should be to enable profiting from private resources, therefore maintaining existing divisions based on ownership. Moreover, the shared-economy approach, which works between public institutions and private economic actors, makes it difficult to implement policies that are urgently required to tackle the ongoing climate catastrophe. Digitalization, while disruptive in some ways, entrenches existing spatial relations where immobility transformations are critically important for the survival of urban life. In this context, digital justice has to go hand-in-hand with mobility justice.

NOTE

1 See "Pierwszy dzień z e-kontrolą za nami [The first day of e-control is behind us]," online at https://www.youtube.com/watch?v=qg1LSP9cqag&ab_channel=Zarz%C4%85dDr%C3%B3gMiejskich

REFERENCES

Barns, S. 2019. "Negotiating the Platform Pivot: From Participatory Digital Ecosystems to Infrastructures of Everyday Life." *Geography Compass* 13 (9): 1–13. https://doi.org/10.1111/gec3.12464

Behrendt, F. 2019. "Mobility and Data: Cycling the Utopian Internet of Things." *Mobilities* 15 (1): 1–25. https://doi.org/10.1080/17450101.2019.1698763

boyd, d., and K. Crawford. 2012. "Critical Questions for Big Data: Provocations for a Cultural, Technological, and Scholarly Phenomenon." *Information, Communication and Society* 15 (5): 662–79. https://doi.org/10.1080/1369118X.2012.678878

Calo, R., and A. Rosenblat. 2017. "The Taking Economy: Uber, Information and Power." *Columbia Law Review* 117, 9 March. Online at https:// ssrn.com /abstract=2929643, accessed 14 September 2020.

Chaniotakis, E., and A. Pel. 2015. "Drivers' Parking Location Choice under Uncertain Parking Availability and Search Times: A Stated Preference Experiment." *Transportation Research Part A* 82 (December): 228–39. https://doi.org/10.1016/j.tra.2015.10.004

Chatman, D.G., and M. Manville. 2014. Theory versus Implementation in Congestion-Priced Parking: An Evaluation of SFpark, 2011–2012." *Research in Transportation Economics* 44 (June): 52–60. https://doi.org/10.1016/j.retrec.2014.04.005

de Souza e Silva, A., and M. Sheller. 2015. *Mobility and Locative Media: Mobile Communication in Hybrid Spaces*. London: Routledge.

Dodge, M., and R. Kitchin. 2005. "Codes of Life: Identification Codes and the Machine-Readable World." *Environment and Planning D: Society and Space* 23 (6): 851–81. https://doi.org/10.1068/d378t

Dodge, M., and R. Kitchin. 2007. "The Automatic Management of Drivers and Driving Spaces." *Geoforum* 38 (2): 264–75. https://doi.org/10.1016/j.geoforum.2006.08.004

Gekker, A., and S. Hind. 2019. "Infrastructural Surveillance." *New Media & Society* 22 (8): 1–23. https://doi.org/10.1177/1461444819879426

González-González, E., S. Nogués, and D. Stead. 2020. "Parking Futures: Preparing European Cities for the Advent of Automated Vehicles." *Land Use Policy* 91 (February): 1–19. https://doi.org/10.1016/j.landusepol.2019.05.029

Hind, S. 2019. "Digital Navigation and the Driving-Machine: Supervision, Calculation, Optimization, and Recognition." *Mobilities* 14 (4): 1–17. https://doi.org/10.1080/17450101.2019.1569581

Jittrapirom, P., M. Marchau, R. van der Heijden, and H. Meurs. 2018. "Future Implementation of Mobility as a Service (MaaS): Results of an International Delphi Study." *Travel Behaviour and Society* 21 (October): 281–94. https://doi.org/10.1016/j.tbs.2018.12.004

Klauser, F.R., and O. Söderström. 2016. "Smart City Initiatives and the Foucauldian Logics of Governing through Code." In *Smart Urbanism: Utopian Vision or False Dawn?* ed. S. Marvin, A. Luque-Ayala, and C. McFarlane, 108–24. Abingdon, NY: Routledge,

Kurnicki, K. 2020. "How to Park a Car? Immobility and the Temporal Organisation of Parking Practices." *Mobilities* 15 (5): 1–17. https://doi.org/10.1080/17450101.2020.1802132

Lefebvre, H. 2010. *The Production of Space*. Malden, PA: Blackwell.

Leszczynski, A. 2014. "Spatial Media/tion." *Progress in Human Geography* 39 (6): 1–23. https://doi.org/10.1177/0309132514558443

Marres, N. 2020. "Co-Existence or Displacement: Do Street Trials of Intelligent Vehicles Test Society?" *British Journal of Sociology* 71 (3): 1–19. https://doi.org/10.1111/1468-4446.12730

Marsden, G., I. Docherty, and R. Dowling. 2020. "Parking Futures: Curbside Management in the Era of 'New Mobility' Services in British and Australian Cities." *Land Use Policy* 91 (February): 1–10. https://doi.org/10.1016/j.landusepol.2019.05.031

Mejias, U.A., and N. Couldry. 2019. "Datafication." *Internet Policy Review* 8 (4). https://doi.org/10.14763/2019.4.1428

Merriman, P. 2016. "Mobility Infrastructures: Modern Visions, Affective Environments and the Problem of Car Parking." *Mobilities* 11(1): 83–98. https://doi.org/10.1080/17450101.2015.1097036

Nikolaeva, A., P. Adey, T. Cresswell, J.Y. Lee, A. Nóvoa, and C. Temenos. 2019. "Commoning Mobility: Towards a New Politics of Mobility Transitions."

Transactions of the Institute of British Geographers 44 (2): 346–60. https://doi.org/10.1111/tran.12287

Pink, S., V. Fors., and M. Glöss. 2017. "Automated Futures and the Mobile Present: In-Car Video Ethnographies." *Ethnography* 20 (1): 88–107. https://doi.org/10.1177/1466138117735621

Poell, T., D. Nieborg, and J. van Dijck. 2019. "Platformisation." *Internet Policy Review* 8 (4): 1–13. https://doi.org/10.14763/2019.4.1425

Sadowski, J. 2018. "When Data Is Capital: Datafication, Accumulation, and Extraction." *Big Data & Society* 6 (1): 1–19. https://doi.org/10.1177/2053951718820549

Sgibnev, W., and L. Rekhviashvili. 2020. "Marschrutkas: Digitalisation, Sustainability and Mobility Justice in a Low-Tech Mobility Sector." *Transportation Research Part A: Policy and Practice* 138 (August): 342–52. https://doi.org/10.1016/j.tra.2020.05.025

Shaw, J., and S. Graham. 2017. "An Informational Right to the City? Code, Content, Control, and the Urbanization of Information." *Antipode* 49 (4): 1–21. https://doi.org/10.1111/anti.12312

Shoup, D. 2005. *The High Cost of Free Parking*. Oxford: Routledge.

Sheller, M. 2018a. *Mobility Justice: The Politics of Movement in an Age of Extremes*. London: Verso.

Sheller, M. 2018b. "Theorising Mobility Justice." *Tempo Social* 30 (2): 17–34. https://doi.org/10.11606/0103-2070.ts.2018.142763

Sparrow, R., and M. Howard. 2020. "Make Way for the Wealthy? Autonomous Vehicles, Markets in Mobility, and Social Justice." *Mobilities* 15 (4): 1–13. https://doi.org/10.1080/17450101.2020.1739832

Spurling, N. 2019. Parking Futures: The Relationship between Parking Space, Everyday Life and Travel Demand in the UK." *Land Use Policy* 91 (February): 1–8. https://doi.org/10.1016/j.landusepol.2019.02.031

Srnicek, N. 2017. *Platform Capitalism*. Malden, PA: Polity Press.

Stehlin, J., M. Hodson, and A. McMeekin. 2020. "Platform Mobilities and the Production of Urban Space: Toward a Typology of Platformization Trajectories." *Environment and Planning A: Economy and Space* 52 (7): 1250–68. https://doi.org/10.1177/0308518X19896801

Taylor, E. 2020. "Who's Been Parking on *My* Street? The Politics and Uneven Use of Residential Parking Space." *Land Use Policy* 91 (February): 1–11. https://doi.org/10.1016/j.landusepol.2018.11.011

14 On the Contradictions of the (Climate) Smart City in the Context of Socio-environmental Crisis

LORENA MELGAÇO AND LÍGIA MILAGRES

Smart Cities for Climate Change?

Amid concerns over the environmental crisis, the use of digital technology is often encouraged by supranational organizations (for example, the UN, EC, OECD, and ECLAC[1]) and adopted by national and local governments attempting to tackle, among other issues, climate change. In this context, the smart city framework – broadly presented as the use of digital technology for urban management to provide a better quality of life (European Commission 2012) – becomes a favoured strategy and policy framework by local governments in various cities (Haarstad and Wathne 2019). Pushed by tech conglomerates and nation-states, its growing adhesion may be attributed partially to the private sector's seeking of new markets and local governments' need for efficiency in the midst of austerity and neoliberal agendas (Kitchin, Cardullo, and Di Feliciantonio 2019). Thus, the smart city paradigm is staked as the way to improve urban governance, services, and resource management, delineating a path dependence between sustainability and smart cities as a measure of development (Kaika 2017).

Nevertheless, measured by its success in implementing digital technology, the smart city "has become a Trojan horse for advancing undemocratic ideas" that privilege private over public interests and further the privatization of the public space by global corporations, especially under the "guise of environmentalism" (Mosco, in this volume). Labels such as *climate smart city* or *low-carbon smart city* reflect oversimplified views that convey urbanization as a major driver of climate change, without properly discussing the contradictions of the capitalist production of space. In this chapter we adopt the comprehensive understanding of capitalism proposed by Nancy Fraser (2021), as "a way of organizing the production and exchange in relation to their non-economic conditions of possibility" (99) – namely, its intrinsic relation to nature, social reproduction, and political conditions. She states that the ecological contradiction of capitalism is "entwined with several others, equally endemic" ones (97). Following this logic, the contradictions regarding technology could also be added to this entanglement as specific and intrinsic to capitalism as the others.

Concerns regarding the relationship between urbanization and the environmental crisis are long standing – see, for example, Bookchin (1987); Sander (2019); Schindler and Silver (2019); Schlosberg (2013); Souza (2014, 2020); and Swyngedouw (2004). In this chapter we acknowledge that such a relationship is multiscalar, and should be addressed at two levels. The first considers the relations of dependency at a global scale, which define centres and peripheries based on an international division of labour, concentration of resources, expropriation, uneven conditions of production, social reproduction, and consumption within and among countries, expressed in the context of the regional, national, and global climate crisis. The second accounts for socio-spatial inequalities, segregation, and struggles that are embedded in the economic and political organization of our cities and neighbourhoods, including contradictions of environmental policies at the local level that result in "ecological gentrification" (Dooling 2009), "green gentrification" (Sander 2019), "anti-poor environmentalist measures" (Souza 2014), "carbon gentrification" (Rice et al. 2020), and other expressions of the "environmentalism of the rich" (Fraser 2021).

Environmental injustices, social inequalities, and power asymmetries are closely intertwined, since the poorest and marginalized are those most affected by these consequences. Addressing such intricate relationships requires bridging socio-ecological issues at a local level with a broader and global analysis of the environmental crisis. The latter, however, increasingly has been dealt with from a fragmented perspective, using a *single-issue environmentalism* approach without considering its intertwined sociohistorical drivers (Fraser 2021).

This approach privileges climate change in policy discourse, even if it is clear that addressing the crisis requires confronting structural factors that lead to social injustices. The techno-managerial nature of the smart city agenda and its reliance on solutionism do not challenge the social and political dimensions of urban problems they supposedly seek to solve, even when they are acknowledged[2]. Pushed as a new urban paradigm, the smart city offers a *quick fix* exempting development agencies and states from addressing structural factors (Montero 2018), especially as those often involve contested political choices around the redistribution of risks, resources, and power (Shi et al. 2016), deepening the contradictions of the capitalist production of space.

Recognizing their overtly technocratic aspect, the risk that smart policies will reproduce socio-spatial inequalities (Cardullo and Kitchin 2019; Luque-Ayala and Marvin 2015) and attempt to centre the discussion on societal challenges, a growing critical scholarship has discussed alternative approaches to predominant smart city views (for example, McFarlane and Söderström 2017; Melgaço and Willis 2017). Rosol, Blue, and Fast (2019) delve into the smart city critique by dialoguing with Fraser's three dimensions of justice: redistribution, recognition, and representation. They suggest the need for "a broader understanding of justice based on economic, cultural and political parity of participation" (6) to shift the discussion from alternative and critical approaches of smart city towards a discussion of justice within the city. To name a few, smart cities' injustices include the public subsidizing of the private sector, the creation of new forms of exclusion, the reduction of citizens to consumers and sensors, the corporatization of urban governance, and the narrowing of conditions for participation and decision-making (Rosol, Blue, and Fast 2019). With the deepening climate crisis, there also emerges the need to discuss a broader understanding of justice within the smart city debate by considering the more-than-human (Burns, in this volume).

To expand this discussion, we draw on a decolonial critique, placing the drivers of the current environmental crisis as inherent in a "modern/colonial capitalist/patriarchal world-system" (Grosfoguel 2011, 10), characterized by economic, cultural, and epistemological Western/European hegemony (Escobar 2007), henceforth *capitalist modernity* (Grosfoguel 2011). Focusing on the intertwining of different dimensions of control that subordinate racialized and ethnic groups, the decolonial critique situates modernity in its intrinsic relationship with *coloniality* – that is, "colonial situations" which persist despite the eradication of colonial administration (Grosfoguel 2007).

This decolonial critique expands existing critical literature on smart cities by recognizing the intricate relationship between modernity and coloniality, and showing the need to dislocate this discussion, which remains predominantly Eurocentric and Eurocentred (see Tironi and Albornoz, in this volume). Furthermore, such a critique unveils how the technological imaginary is entwined with notions of development, progress, and growth. Here, development is understood in its interrelation with specific forms of knowledge, power, and subjectivity that regards Latin America, Africa, and Asia as pursuing progress and modernization according to European and North American experiences (Escobar 1992).

Therefore, in the following, we first understand the smart city as a techno-managerial approach that capitalizes on ideas of progress entrenched in developmentalism, and discuss its irreconcilable contradictions to addressing environmental issues. We then highlight the multiscalar interaction of socio-spatial inequalities, power asymmetries, and the environmental crisis. Lastly, we point to the possibilities of imagining other forms of socio-spatial organization when crossing a broader understanding of the socio-environmental crisis with a decolonial critique.

Tracing Determinism and Developmentalism as the Basis of the (Climate) Smart City

The growing critical literature on smart cities scrutinizes the overreliance on *technological fixes* to address urban issues and the tendency to conceal specific urban development agendas. Although corporations leading the implementation of smart technology praise its neutrality and detachment from political agendas, the smart city has been discussed as "a strand of technoscientific urbanism" (Brenner and Schmid 2015), "African urban fantasies" (Watson 2014), and "corporate story-telling" (Söderström, Paasche, and Klauser 2014). Scholars have been calling to attention the smart city's one-size-fits-all approach, the entrepreneurial nature undergirding urban governance, the reproduction of unbalanced power relations, and its centralized and regulatory nature (see also Mosco and Datta, in this volume). By doing so, they evidence the normalization of a specific worldview that attributes a causal relationship between technological development and social transformation. Under this purview, development is not only possible, but a desirable and inevitable means to achieve progress, and should, thus, be embraced (Escobar 1995).

A similar sort of determinism is present in how climate change, understood as "*the most urgent* environmental problem of our time," frames the whole environmental agenda (Crist 2007, 32, emphasis added). Climate is detached from interdependent variables that shape human life in its social, ecological, and political dimensions, and addressed, almost solely, in its technical dimension. Using piecemeal approaches, efforts concentrate on finding the *right* technologies to address the immediate problem, without considering its underlying causes (Crist 2007), leading to solutions that follow simple chains of cause and effect, as observed, for example, in the prevalent focus on reducing CO_2 emission levels (Crist 2007; Hulme 2011). Digital technology becomes a magic bullet to tackle urban issues, aggravating and aggravated by climate change in numerous policy documents and reports. However, these do not consider the power relations shaping the technical systems that affect urban growth, access to services, and our everyday – the way we work, consume (Feenberg 1992), and, we may add, make decisions about the city.

The long-standing critique of development and its relation with technology points to the fact that the technological determinism now underpinning the smart city is historical in capitalism. The discursive formation and strategies for development in place since the sixteenth century gave "rise to an efficient apparatus that systematically relates forms of knowledge and techniques of power," leading to the impoverishment, exploitation, and oppression of the so-called Third World (Escobar 1995, 10). The accumulation of technological infrastructure that endowed the industrial revolution and that continuously enable innovation stems from "an asymmetric exchange of embodied labour time and a massive environmental load displacement" – resources, labour, and energy plundered from colonies to the metropolises – initiated in colonial times (Hornborg 2019, 11).

Moreover, the twentieth century marked the intensification of global socio-economic contradictions caused by an international division of labour and justified through the hegemonic presence of modernization and developmentalist theories at all levels of governance, affecting both social and urban policies. To date, technology has aggravated the structural dependency that subjugates global peripheries to the centres: "when we say that a new technology is introduced in our [dependent] economies, we are evidently misusing such a word. What arrives are the products of a technology, whilst it [technology] is a group of activities, of institutionally organised knowledges and in process of development" (Quijano 2014, 130, authors' translation). Therefore, as Tironi and Albornoz argue in this volume, the smart city needs to be examined through a modernity/coloniality perspective, which evidences the "modern and instrumental thought system," validating its implementation, especially in the global South.

The smart city also reflects more recent changes in the mechanisms of the capitalist mode of production to accommodate the increasing role of knowledge as a prime commodity (Dantas 2014), such as through Big Data and the Internet of Things, by spatializing it. Relying on heavy infrastructure, the smart city promotes specific forms of urban development that might deepen the socio-environmental crisis on local and global scales. It requires significant infrastructural interventions, amounting to the depletion of resources, increased power consumption, and production of large amounts of waste (Mosco, in this volume). Such a process is unequally distributed, with poorer countries still bearing the onus generated by the plundering of resources and the outsourcing of electronic waste (Kaika 2017). Connected to intensive extractivism, the digital economy sets a digital labour chain reliant on all forms of labour exploitation in the different stages of its cycle (Fuchs 2016), characterizing how technology is shaped, developed, and unequally distributed on a global scale. As well, the unequal terms under which the extractivist nature of digital platforms capitalizes on social life co-opt least-protected nations into understanding the adoption of platforms as "an inevitable and civilizing project," as Tironi and Albornoz argue in this volume.

Socio-environmental Crisis, Socio-spatial Inequalities, and Power Asymmetries in Cities

We agree with Escobar, Fraser, and Souza that the environmental crisis is part of a generalized crisis. For Escobar (2007, 197), it is a crisis of modernity and instrumental rationality, revealing itself as the "central contradiction and limit to capital today." Similarly, Fraser (2021, 96), proposes a broader understanding of the environmental crisis that considers capitalism as the "socio-historical driver" and the "core institutionalized dynamic" underpinning a general crisis, which also entails the economy, society, politics, and public health. Lastly, Souza points out that the social crisis of our time is a socio-ecological crisis because it encompasses both social relations and the ways the environment is transformed by them. As such, it requires that we avoid adopting either *urbanophilic* and *urbanophobic* approaches (Souza 2020), which reduce the contradictions of capitalist urbanization to so-called urban problems. Such a reduction also reflects an equally problematic understanding of the urban crisis as a result of the emergence of the city, rather than of urbanization itself – which impacts both city and countryside (Bookchin 1987).

A further level of reduction, often adopted in smart city projects, is the tendency of homogenizing cities in ways that ignore specific historical conditions and socio-spatial realities (Watson 2014). The close relationship between socio-spatial inequalities and environmental injustices has different expressions in cities in the global South and North. Although marked by growing inequalities and neoliberal practices, cities located in the centre of capitalism have a relatively better distribution of basic infrastructure, services, and resources. In contrast, cities at the periphery and semi-periphery of the globe are shaped by more profound socio-spatial inequalities and power asymmetries. These inequalities are present both among cities and regions of the same country as well as within cities, in which considerable parts of the population often lack access to basic infrastructure and resources. For example, according to the last National Survey of Basic Sanitation in Brazil, four out of ten municipalities in the country have no sewage services.[3] Another fundamental part of the urbanization logic is the forced displacement of poor and vulnerable social groups to increasingly more peripheral areas – the history of self-built environments erased throughout controversial participatory processes (Kapp and Baltazar 2012). The uneven access to city resources and infrastructure is accompanied by an abyssal decision-making power asymmetry when it comes to defining the socio-spatial organization of the city. As Harvey (2008) has argued forcefully, the right to the city means far more than access to resources; it means the (collective) power to shape the city.

Besides overlooking the limitations of technological fixes for specific types of inequalities, smart city approaches in the global South are immersed, as Datta (2018, 407) points out, in a "rhetorics of urgency" that builds upon the idea of a constant state of crisis, often represented as "disorders" – such as "population explosion, uncontrolled urbanization and climate change." Exploiting such rhetorics, technocratic approaches at the local level associated with market interests have led to contradictory environmental policies that ultimately reinforce structural asymmetries and, therefore, the socio-ecological crisis. They range from anti-poor environmental policies when it comes to *favelas* in Rio de Janeiro – by using the ecological discourse to promote land-use control and social segregation (Souza 2014) – to "green gentrification" in cities such as Berlin (Sander 2019) – reflecting the lack of mechanisms to prevent the exclusion and displacement of low-income residents when the provision of green and free spaces leads to real estate valorization, even when such spaces are achieved through social struggle.

Among the approaches that consider the inequalities at the root of the environmental crisis, pushing the critical analysis beyond the framework of the "hegemonic-Western idea of modernity," Álvarez and Coolsaet (2020, 55) advocate for the decolonization of environmental justice studies. In their view, the decolonial critique reveals how the capitalist mode of production in relation to nature is heavily connected to colonialism, helping to highlight particular forms of oppression of individuals, groups, and ecosystems located at the periphery of the world-system. The decolonial draws attention to epistemological limitations embedded in theories and practices that might reinforce environmental injustices and create new forms of subjugation – what the authors call the "coloniality of justice" – when transposed from Western to non-Western contexts. Moreover, by assuming "the importance of interactions and conflicts between different forms of knowing" (Álvarez and Coolsaet 2020, 63), it allows researchers to recognize how communities of the global South produce knowledge and conceptualize environmental justice, rather than considering them as mere subjects of study.

As the literature already suggests, it is crucial to acknowledge that specific mechanisms deployed through smart city strategies reproduce and normalize "structural and social violence inherent in urban transformations across the world" (Datta and Odendaal 2019). With those mechanisms in mind, recognizing that the smart city framework is entrenched in the contradictions of capitalist modernity becomes essential, since it reproduces power, knowledge, and technological hierarchies, ways of producing space from a developmentalist approach, and, by extension, shaping the way technology is imagined and employed.

Questioning the (Climate) Smart City Framework and Beyond

Given the push for a climate smart agenda at both supra- and national levels and growing interest in deploying such a strategy at the local level, we focus here on the shortcomings of tackling the socio-environmental crisis from a solutionist and techno-managerial perspective that draws on reductionist and Eurocentric views of environmental justice. As Datta reminds us in this volume, more than the technology, the smart city is a framework that shapes urban transformation while repackaging "all the earlier cliched tropes that we have heard around cities: efficiency, innovation, creativity, sustainability." We departed from an understanding that, embedded in a developmentalist framework, the *climate smart city* congregates technocratic assumptions, detaching the social dimension of technological production, implementation, and use (Cherlet 2011; Feenberg 1992). Moreover, it adopts a "perspective of single-issue environmentalism" (Fraser 2021, 106), by isolating ecological issues from the contradictory capitalist social relations and state institutions underlying them. Therefore, we focus on three issues: technological determinism, developmentalism, and the modernity/coloniality perspective, for they are founding elements and drivers of social inequalities and power asymmetries entrenched in capitalist modernity that linger through the ecological modernization promise present in the discourses and the production of the (smart) city. Three main pointers follow for further exploration:

1 *Climate smart city approaches rely on technological fixes, neglecting the overarching elements of the socio-environmental crisis.* Although critical research on smart cities acknowledges the social dimension of so-called urban problems that question their recurrent techno-managerial underpinning, the grey literature still offers technological fixes for climate change. The various dimensions of the socio-environmental crisis and its inherent socio-spatial inequalities and power asymmetries are often disregarded and risk being deepened with the implementation of *climate smart city* approaches. Pertaining to a general crisis of capitalism, both the environmental crisis and the production of technology demand tackling structural and historically reproduced processes of exploitation and accumulation.
2 *Focusing on urbanization as the (primary) cause of the environmental crisis might conceal the fundamental issue: capitalist modernity and its processes of producing, imagining, and shaping cities.* Reflecting a North-based scholarly interpretation of urbanization and urban problems (Souza 2020), the urbanization rate gains centrality in the grey literature, especially in connection to its negative environmental impacts, while the intertwining of socio-spatial inequalities, power asymmetries, and environmental impacts – reproduced throughout local, national, and global scales – remains indirectly approached. Such intertwining dates to the colonial period and still operates through developmentalism and structural dependency. Disregarding the relationship between urbanization and environmental crisis within the contradictions of capitalist modernity poses a risk of normalizing and reproducing the conditions that only deepen the crisis. In this context, understanding how existing decision-making processes in urban planning and management – which exclude or co-opt vulnerable and marginalized groups in the city – are also drivers that deepen the crisis becomes crucial.
3 *Bridging local and global challenges using a decolonial lens is needed to expand the socio-environmental critique.* Even though there is now general consensus on the connection between local and global aspects of the environmental crisis, a decolonial lens gives a critical perspective on

> the relationship between the mode of production and the social relations forged within modernity/coloniality. It reveals that individuals, groups, and ecosystems located at the periphery of capitalist modernity are those that suffer the most, both from capitalism itself and from its socio-environmental impacts. Thus, it is crucial that any approach to so-called urban problems – whether or not deploying digital technologies – acknowledge the different conditions of cities around the globe and consider their positioning within those power relations. Universalized solutionist approaches often disregard the geopolitics of knowledge production and how it affects local and global urban policies. Such approaches also conceal the power asymmetries among urban actors and their different conditions in making decisions about the city.

Even though we have departed from reflecting on the contradictions of the smart city framework to address the socio-environmental crisis, these contradictions, we maintain, do not pertain only to the smart city. Rather, they are integral to a systematic mode of thinking, producing, and deploying technology that does not challenge capitalist modernity's social and spatial relations. Acknowledging the social and political dimension of the socio-environmental crisis requires overcoming this mode of thinking towards imagining other ways of producing knowledge and technology and shaping cities. It also requires, as Rosol, Blue, and Fast (2019, 5) propose, a shift of "the discussion away from the smart city, even an alternative one, towards the just city and a just urbanism in the digital age." Alternatively, as Souza (2020, 83) argues, to "deeply rethink social relations, technology and the spatial organization of society" is to imagine other forms of socio-spatial organization from a long-term ecological perspective. To our minds, such views might converge with Escobar's (2014), whose work highlights that the ecological crisis brings the opportunity for destabilizing the premises of contemporary thinking on development towards new forms of production and articulations of knowledge that might lead to a new environmental rationale, as a way of overcoming the search for development alternatives, towards the discussion of alternatives *to* development (Escobar 1992). We consider it crucial to engage with a radical critique that enables understanding the present and imagining other futures. We believe that the decolonial critique foregrounds the contradictions of capitalist modernity in which frameworks such as the (climate) smart city are inscribed, unravelling the limits of current approaches to the relationship among digital technologies, the socio-environmental crisis, and the production of space in cities. Nevertheless, to reflect on possible futures, the discussion proposed here should be expanded beyond the academic field and join ongoing debates in the everyday of urban planning and management practices, and, above all, of the social groups involved in urban struggles.

NOTES

1 The United Nations (UN), European Commission (EC), Organisation for Economic Co-operation and Development (OECD), and Economic Commission for Latin America and the Caribbean (ECLAC).

2 Examples abound, such as the United Nations approach to *Climate Smart City on Emerging Economies* (online at https://unfccc.int/climate-action/momentum-for-change/activity-database/momentum-for-change-climate-smart-cities-in-emerging-economies) or the Big Environmental push, a developmentalist approach based on large infrastructure construction to support low carbon growth in Latin America proposed in CEPAL (2016) or in Horizon Europe Mission Area: Climate-Neutral and smart cities (European Commission 2020).

3 See Agência IBGE, "Quatro em cada dez municípios não têm serviço de esgoto no país [Four out of ten municipalities do not have sewage service in the country]," 22 July 2020, online at https://agenciadenoticias.ibge.gov.br/agencia-noticias/2012-agencia-de-noticias/noticias/28326-quatro-em-cada-dez-municipios-nao-tem-servico-de-esgoto-no-pais

REFERENCES

Álvarez, L., and B. Coolsaet. 2020. "Decolonizing Environmental Justice Studies: A Latin American Perspective." *Capitalism Nature Socialism* 31 (2): 50–69. https://doi.org/10.1080/10455752.2018.1558272

Bookchin, M. 1987. *The Rise of Urbanization and the Decline of Citizenship*. San Francisco: Sierra Club Books.

Brenner, N., and C. Schmid. 2015. "Towards a New Epistemology of the Urban?" *City* 19 (2–3): 151–82. https://doi.org/10.1080/13604813.2015.1014712

Cardullo, P., and R. Kitchin. 2019. "Smart Urbanism and Smart Citizenship: The Neoliberal Logic of 'Citizen-Focused' Smart Cities in Europe." *Environment and Planning C: Politics and Space* 37 (5): 813–30. https://doi.org/10.1177/0263774X18806508.

CEPAL. 2016. "Horizontes 2030: A igualdade no centro do desenvolvimento sustentável." Síntese (LC/G.2661/Rev.1). Santiago, Chile. https://repositorio.cepal.org/bitstream/handle/11362/40118/4/S1600753_pt.pdf

Cherlet, J. 2011. "A Genealogy of Epistemic and Technological Determinism in Development Aid Discourses." In *DIME Workshop "Technology, Institutions and Development," Max Planck Institute*. Ghent University, Department of Third World Studies.

Crist, E. 2007. "Beyond the Climate Crisis: A Critique of Climate Change Discourse." *Telos* 141 (Winter): 29–55.

Dantas, Marcos. 2014. "Mais-Valia 2.0: Produção e apropriação de valor nas redes do capital [Surplus-Value 2.0: Production and value appropriation in capital networks]."

Eptic online: revista electronica internacional de economia política da informaçao, da comuniçao e da cultura 16 (2): 85–108. Online at marcosdantas.com.br/conteudos/wp-content/uploads/2014/07/Maisvalia-2-0.pdf

Datta, A. 2018. "The Digital Turn in Postcolonial Urbanism: Smart Citizenship in the Making of India's 100 Smart Cities." *Transactions of the Institute of British Geographers* 43: 405–19. https://doi.org/10.1111/tran.12225

Datta, A., and N. Odendaal. 2019. "Smart Cities and the Banality of Power." *Environment and Planning D: Society and Space* 37 (3): 387–92. https://doi.org/10.1177/0263775819841765

Dooling, S. 2009. "Ecological Gentrification: A Research Agenda Exploring Justice in the City." *International Journal of Urban and Regional Research* 33 (3): 621–39. https://doi.org/10.1111/j.1468-2427.2009.00860.x

Escobar, A. 1992. "Reflections on 'Development': Grassroots Approaches and Alternative Politics in the Third World." *Futures* 24 (5): 411–36. https://doi.org/10.1016/0016-3287(92)90014-7

Escobar, A. 1995. *Encountering Development: The Making and Unmaking of the Third World*. Princeton Studies in Culture/Power/History. Princeton, NJ: Princeton University Press.

Escobar, A. 2007. "Worlds and Knowledges Otherwise: The Latin American Modernity/Coloniality Research Program." *Cultural Studies* 21 (2–3): 179–210. https://doi.org/10.1080/09502380601162506

European Commission. 2012. "Smart Cities and Communities – European Innovation Partnership." Online at https://ec.europa.eu/transparency/regdoc/rep/3/2012/EN/3-2012-4701-EN-F1-1.PDF

European Commission. 2020. "Mission Area: Climate-Neutral and Smart Cities." Online at https://ec.europa.eu/info/horizon-europe-next-research-and-innovation-framework-programme/mission-area-climate-neutral-and-smart-cities_en

Feenberg, A. 1992. "Subversive Rationalization: Technology, Power, and Democracy." *Inquiry* 35 (3–4): 301–22. https://doi.org/10.1080/00201749208602296

Fraser, N. 2021. "Climates of Capital: For a Trans-Environmental Eco-Socialism." *New Left Review* 127: 94–127. Online at https://newleftreview.org/issues/ii127/articles/nancy-fraser-climates-of-capital.pdf

Fuchs, C. 2016. "Digital Labor and Imperialism." *Monthly Review* 67 (8): 14. https://doi.org/10.14452/MR-067-08-2016-01_2

Grosfoguel, R. 2007. "The Epistemic Decolonial Turn: Beyond Political-Economy Paradigms." *Cultural Studies* 21 (2–3): 211–23. https://doi.org/10.1080/09502380601162514

Grosfoguel, R. 2011. "Decolonizing Post-Colonial Studies and Paradigms of Political-Economy: Transmodernity, Decolonial Thinking, and Global Coloniality." *Transmodernity: Journal of Peripheral Cultural Production of the Luso-Hispanic World* 1 (1): 1–36. https://doi.org/10.5070/T411000004

Haarstad, H., and M.W. Wathne. 2019. "Are Smart City Projects Catalyzing Urban Energy Sustainability?" *Energy Policy* 129 (June): 918–25. https://doi.org/10.1016/j.enpol.2019.03.001

Harvey, D. 2008. "The Right to the City." *New Left Review* 53: 23–40. Online at https://newleftreview.org/issues/ii53/articles/david-harvey-the-right-to-the-city

Hornborg, A. 2019. "Colonialism in the Anthropocene: The Political Ecology of the Money-Energy-Technology Complex." *Journal of Human Rights and the Environment* 10 (1): 7–21. https://doi.org/10.4337/jhre.2019.01.01

Hulme, M. 2011. "Reducing the Future to Climate: A Story of Climate Determinism and Reductionism." *Osiris* 26 (1): 245–66. https://doi.org/10.1086/661274

Kaika, M. 2017. "'Don't Call Me Resilient Again!': The New Urban Agenda as Immunology … or … What Happens When Communities Refuse to Be Vaccinated with 'Smart Cities' and Indicators." *Environment and Urbanization* 29 (1): 89–102. https://doi.org/10.1177/0956247816684763

Kapp, S., and A.P. Baltazar. 2012. "The Paradox of Participation: A Case Study on Urban Planning in Favelas and a Plea for Autonomy." *Bulletin of Latin American Research* 31 (2): 160–73. https://doi.org/10.1111/j.1470-9856.2011.00660.x

Kitchin, R., P. Cardullo, and C. Di Feliciantonio. 2019. "Citizenship, Justice, and the Right to the Smart City." In *The Right to the Smart City*, ed. P. Cardullo, C. Di Feliciantonio, and R. Kitchin, 1–24. Bingley, UK: Emerald Publishing.

Luque-Ayala, A., and S. Marvin. 2015. "Developing a Critical Understanding of Smart Urbanism?" *Urban Studies* 52 (12): 2105–16. https://doi.org/10.1177/0042098015577319

McFarlane, C., and O. Söderström. 2017. "On Alternative Smart Cities: From a Technology-Intensive to a Knowledge-Intensive Smart Urbanism." *City* 21 (3–4): 312–28. https://doi.org/10.1080/13604813.2017.1327166

Melgaço, L., and K.S. Willis. 2017. "Social Smart Cities: Reflecting on the Implications of ICTs in Urban Space." *PlaNext – Next Generation Planning* 4: 5–7. https://doi.org/10.24306/plnxt.2017.04.001

Montero, S. 2018. "Leveraging Bogotá: Sustainable Development, Global Philanthropy and the Rise of Urban Solutionism." *Urban Studies*, December, 0042098018798555. https://doi.org/10.1177/0042098018798555

Quijano, A. 2014. "'Polo marginal' y 'mano de obra marginal.'" In *Cuestiones y horizontes: de la dependencia histórico-estructural a la colonialidad/descolonialidad del poder*, ed. A. Quijano, 125–69. Buenos Aires: CLACSO.

Rice, J.L., D.A. Cohen, J. Long, and J.R. Jurjevich. 2020. "Contradictions of the Climate-Friendly City: New Perspectives on Eco-Gentrification and Housing Justice." *International Journal of Urban and Regional Research* 44 (1): 145–65. https://doi.org/10.1111/1468-2427.12740

Rosol, M., G. Blue, and V. Fast. 2019. "Social Justice in the Digital Age: Re-Thinking the Smart City with Nancy Fraser." UCCities Working Paper 1. Preprint. SocArXiv. https://doi.org/10.31235/osf.io/wkqy2

Sander, H. 2019. "Städtische umweltgerechtigkeit: Zwischen progressiver Verwaltungspraxis und sozial-ökologischen Transformationskonflikten [Urban environmental justice: Between progressive management praxis and socio-ecological transformation conflicts]." *Analysen* 58 (1). Online at https://ww.rosalux.de/fileadmin/rls_uploads/pdfs/Analysen/Analysen58_Umweltgerechtigkeit.pdf

Schindler, S., and J. Silver. 2019. "Florida in the Global South: How Eurocentrism Obscures Global Urban Challenges – and What We Can Do about It." *International Journal of Urban and Regional Research* 43 (4): 794–805. https://doi.org/10.1111/1468-2427.12747

Schlosberg, D. 2013. "Theorising Environmental Justice: The Expanding Sphere of a Discourse." *Environmental Politics* 22 (1): 37–55. https://doi.org/10.1080/09644016.2013.755387

Shi, L., E. Chu, I. Anguelovski, A. Aylett, J. Debats, K. Goh, T. Schenk, et al. 2016. "Roadmap towards Justice in Urban Climate Adaptation Research." *Nature Climate Change* 6 (2): 131–7. https://doi.org/10.1038/nclimate2841

Söderström, O., T. Paasche, and F. Klauser. 2014. "Smart Cities as Corporate Storytelling." *City* 18 (3): 307–20. https://doi.org/10.1080/13604813.2014.906716

Souza, M.L. 2014. "O lugar das pessoas nas agendas 'verde', 'marrom' e 'azul': Sobre a dimensão geopolítica da política ambiental urbana [People's place in the 'green,' 'brown' and 'blue' agendas: About the geopolitical dimension of urban environmental politics]." *Passa Palavra* (blog), 4 December. Online at https://passapalavra.info/2014/12/101245/

Souza, M.L. 2020. "The City and the Planet: Notes on Utopias, Dystopias and a Complex Relationship." *City* 24 (1–2): 76–84. https://doi.org/10.1080/13604813.2020.1739907

Swyngedouw, E. 2004. "Globalisation or 'Glocalisation'? Networks, Territories and Rescaling." *Cambridge Review of International Affairs* 17 (1): 25–48. https://doi.org/10.1080/0955757042000203632

Watson, V. 2014. "African Urban Fantasies: Dreams or Nightmares?" *Environment and Urbanization* 26 (1): 215–31. https://doi.org/10.1177/0956247813513705

PART FOUR

Complicated and Complicating Digital Divides

A Dialogue with Ayona Datta

Part Four of the volume critically reimagines digital divides in order to update and discuss uneven and unjust distribution, access, and recognition. Authors in this section challenge traditional understandings of the digital divide, and instead use the lens of the digital to think about social, political, and economic inequalities across a spectrum of foci.

To provide insight on the evolving digital divide, we spoke with Dr Ayona Datta, Professor of Geography at University College London. Dr Datta is the author of numerous books and articles concerning feminist urban futures, smart cities, and digital urbanism. Her work combines interdisciplinary approaches to navigate postcolonial urbanism, marginality, and the complex variegations of smartness.

Editors: Your work has brought postcolonial thinking to bear on smart cities, exploring the complicated intersections between city development, technology, and marginality. In this volume, Alan Smart, Lorena Melgaço and Lígia Milagres, and many others call attention to the complex variegations of "smartness" as a political strategy across the globe and the ways it might reinstate global hierarchies of political and economic power. Additionally, as you note, "smartness" has emerged largely from the global North, and deploys logics of accumulation by dispossession.

If smartness is from the global North, what sorts of logics does it deploy? Is it always-already colonial? Or is it possible for smart cities to depart from the logics of colonialism? For instance, Martin Tironi and Camila Albornoz stress the importance of developing a "Latin American and decolonial reading of smart," and they argue that "a decolonial strategy does not imply rejecting everything that comes from the global North, but rather creatively and selectively appropriating the elements that allow us to build more just and inclusive cities."

AD: My interest in smart cities did not start from technology. I began from an interest in cities and urban transformations, particularly in the global South. I had also looked at the global North cities through migrant experiences. I was keen to understand urban transformations, and smart cities happened to emerge at a moment of enormous urban transformations. It tagged itself onto all the earlier cliched tropes that we have heard around cities: efficiency, innovation, creativity, sustainability – all the words that have been appropriated by neoliberal governments and private corporations. For me, getting into smart was not necessarily about a critique of technology or technological solutionism, as it has been for many other smart cities scholars; it was mainly to understand what sorts of new urban transformations are happening in India, particularly now.

It is not a coincidence that, in India, smart cities began their journey with the populist and nationalist governments. It is particularly interesting because it has the potential to encompass a vision of authoritarian, centralized way of seeing and governing. They can be the eyes of the state, or eyes of private corporations, but it is a very centralized vision of a city, and I think that is what is exciting to many governments in many states in the global South. For governments, smart cities were not radically different; they just conferred extra powers that the state always wanted. The state has always wanted to surveil more. The state has always wanted to take democracy away from the citizenry. So, to me, when I saw *smart* in the context of urbanization, it just kind of made sense that this was what was going on. I thought, now they have another extra tool to do what they had already been doing before, but better.

In India, smart cities began their journey in that sense around 2013, when it was clear that the government would change and a new ruling party would come to power. That's when the whole discussion of efficiency, innovation, and entrepreneurial cities – helped and facilitated by technology – began to enter public discourse. It is definitely a *Western import*, if you want to put it like that. But I think it would be too simplistic to reduce it only to a Western import. My reticence stems from the fact that, when it actually came to India, nobody really knew what it was. Even at the state level, nobody knew. We did so many interviews with state stakeholders, municipal commissioners, lower-level bureaucrats, civil servants, and even mayors and deputy mayors. Nobody knew what it was! These interviewees often asked us, "You have come from the West, have you seen smart cities there? Do you understand what it is? What are we supposed to do?" And many of them signed on basically because they thought they would be getting extra cash in order to implement infrastructure projects that they had always wanted to do. In that sense, it first took root in lower levels of the state as an infrastructure project, and because of that, in a peculiar way, it had decolonial implications.

A key approach to seeing this as decolonial was that, in the smart city guidelines, every city is given power to devise its own projects. And because this was the first time that Indian cities had to come up with a vision, or a concept for their future, it was rather radical. So it was a mix between a global model brought in and, to use Dipesh Chakrabarty's phrase, *provincialized* at various scales of the state. Even with this flexibility, though, it was still essentially conceived from the top down as a centralizing force. It was following the Western model of this integrated command-and-control centre, with everything now able to be surveyed. At the same time, how it rolled out in various contexts was not necessarily the way it was intended or envisioned.

It is important that we examine the smart city not just as something that's going to radically alter things, but also as one moment in the long genealogy of cities – and how cities transform. Already in India we are hearing that, because the Smart City Mission has not achieved its goals, they're rolling out a new agenda called "Livable Cities." This new agenda will prioritize *liveable* criteria rather than smart criteria. And it keeps changing. Before the smart cities rhetoric, it was "global cities" – and smart cities do build in important ways on "global" rhetorics, policies, and discourses. But how smart cities actually roll out, and how people reappropriate that term at different scales, has been a really interesting thing to follow.

Editors: You remind us to allow India to enter this discourse that we are trying to foster not only through the lens of coloniality – that independent processes and practices are happening there that are not distillable with respect to the rest of the world. Often cast as "the places that do not matter" (Rodríguez-Pose 2018), *Yonn Dierwechter argues that neglecting them "implies that smartness theory is still developed from a limited empirical gaze." Specifically, as Alan Smart contends, smart cities are part of a postcolonial gaze that espouses public and moral safety and security through the eradication of slums, street markets, and informal transport.*

At the same time, you acknowledge that there is some relation with the rest of the world – and with the West in particular: you call smart cities an "import" and say that their managerial style, for example, has been similar. Not to neglect the rest of your answer, but building on the contributions of Lorena Melgaço and Lígia Milagres in the previous part of this volume, is there a tension between that relation of coloniality and the more state-centric, top-down approach? Or are they complementary? Are they synergistic or antagonistic?

AD: I do not know if I have a definitive answer to whether or not it is coloniality. I think it is many things at many times and at many scales, and people do change their minds about it. When the smart city came in, there was certainly pushback to what appeared to be taking the colonizing moment forward by importing smart cities. In a strange way it has been a way for cities to really think about what they want from the state, and what their relationship should be with other scales of the state. Many cities, and particularly well-to-do cities, have felt that, with smart cities programs, they were being colonized by the central state, insofar as they were being told how they should conduct their projects. In this sense we might think of smart cities as intervening in a long history of internal colonization. This is happening in tandem with the recent momentum gaining in the ultra-nationalistic movement – such as a re-emergence of a linguistic centralization in which the central federal state has been directing regional states to switch to using Hindi as their official language. The southern and eastern regional states do not speak Hindi, so it has been seen as a colonization from the North.

Smart cities occurred in tandem with this internal colonization, and in many cases the regional states felt pressured and they negotiated; some rejected the imposition. West Bengal, (an eastern regional state) rejected it and none of its projects are being funded by the federal state. The Maharashtra state has a few smart cities, and Navi Mumbai was supposed to become a smart city. It was seen to be progressive and successful, and a straightforward nominee. But Navi Mumbai decided they did not want any of it.

We have to understand colonization, not just as historic and neocapitalist, resource- or labour-extractive processes, but also as a set of very complicated power relationships in which regional states and local governments, or marginalized social groups (e.g., caste, religious, or language groups) are facing extreme disbalances of power. In India internal colonization is far more severe than external colonization for marginal groups. Of course, external companies and foreign investment are certainly part of it, but in the past six years a bigger internal power differential has emerged. I would position smart cities in this long genealogy of internal colonization, but it is also a different kind of colonization. It is a colonization of the mind.

From the start, smart cities have been about making people understand "how good this is for them." But how do you convince an ordinary citizen, who stands to gain nothing from this, actually to support it? At ground level, very few people have said, "I don't want the smart city"; they have just said, "I want the smart city on my terms." Nobody says that they do not want to be efficient, creative, innovative, or to have economic growth. Everybody wants growth and a better economy, but they want it on their terms.

Editors: Smart citizenship is explored in greater depth in Part Five, but authors throughout the volume discuss the manifold ways in which infrastructure, people, and spaces are co-constituted through "smart" technologies and rationalities. In this part, Alan Wiig, Hamil Pearsall, and Michele Masucci suggest that because [youth] will both craft and inhabit the so-called smart city into the future, [their] experience ... is indicative of the urban condition in the smart city." From the perspective of established critiques of the digital divide, one of the contradictions here is that youth are seen as at the forefront of technological adaptation, but remain at the margins of social equity.

We were interested in how the "chatur *citizen" emerges from "the margins" in response to political discourses and city development. Can you speak more to the tensions and relationship between smart city development and citizenship?*

AD: The *chatur* citizen was a very interesting moment in our workshops, when we started talking about "what is smart?" and "how do we understand smart?" It was the early days when there was a lot of confusion about smart cities, as the actual projects had not taken shape.

Many of the participants, without having any reference points, just started translating what smart was in their languages. *Chatur* is the same word in Hindi and Marathi. There is no exact translation of smart, other than *chatur*. Many of them would say, "we do not want to be *chatur*" because *chatur* has the cultural connotation of manipulativeness and shrewdness. There is a deluge of signs and symbols around *chatur* in the Indian psyche with song and movies. It is a very commonly used word, and it is not necessarily always used in a very positive way. It has an ambiguous intent, and because of this positive and negative connotation, people did not want it. *Chatur* is not like some other adjectives in the Indian vocabulary, which often refer to caste and class distinctions. *Chatur* is casteless and classless.

So, in a smart city, what does it mean to be a smart or *chatur* citizen? – "are we going to steal?" "are we going to bypass regulations, bypass laws, policies?" There were all these questions that were coming into their mind, and they would say being *chatur* is something the state does very well. There is so much suspicion with being *chatur* because the state is very capable of taking away your rights, enacting extra judicial processes, not following due process, social justice, or judicial justice, and often enacts an informal arrangement between the privileged and politicians and so on. If you think of *chatur* in that way, it is not a quality to emulate.

But on the other hand, if you consider subaltern lives, being *chatur* is essential to survival. If you are not *chatur*, you do not survive. You have to somehow bypass formal processes because you are excluded from these processes and do not have access to basic services. You need to be *chatur* – making do with less, tapping pipes and cables, vending on the street where the police will be coming at any moment to throw you out. Subaltern citizenship has always been about imbibing some quality of being shrewd, clever, and careful. It is closer to what Partha Chatterjee would talk about as working in a political society: you cannot access civil society because it is elitist. Subaltern citizens use political society, patronage, and connections to bypass processes and to access what is denied to them through the state process. The work of ordinary citizenship has always been to keep *chatur* citizenship alive, but the noun form of *chatur* is an important condition of the state, as it bypasses process to further marginalize the subaltern. So there is a lot of suspicion of *chatur* when it is a word given to them by the state.

I wondered why ordinary citizens would not want to be smart citizens. I think it is because of this top-down determination, and it has all these connotations that arouse suspicion. With marginal people living in slums or poorer neighbourhoods, there is also an internalization of bureaucracy. Increasingly people want legal processes to ensure due process and transparency, so for them *chatur* means that the rules of the game can be bypassed or obscured, which is the forte of the state. This is to the detriment of the poor, so why would they stand up for it? Instead, they say, "we do not want smart cities, we want intelligent cities." Intelligence comes with this veneer of neutrality, a promise of fairness, and premised on information, correct knowledge, and evidence. *Chatur* may imply that the state would bypass all of this.

Editors: One surprising aspect of smart cities or of critical smart city research is that they are purported to create more socially just cities. The notion of developing a smart city is to be more just, as an outcome. Smart cities, then, are a means to an end, rather than a process. Teresa Abbruzzese and Brandon Hillier argue that "everyday practices of care make the public library a crucial site of access, literacy, and social infrastructure." However, the smart city's delivery of services through the digital economy challenges libraries' ability to provide care-full justice. In your experience, what are the implications of smart for how we think of and pursue social justice in cities?

AD: Based on much of the empirical data, I would probably turn it around, since social justice is often in tension with smart cities. Theoretically we can make claims to those being connected, but how smart cities are actually rolled out empirically is very different from a radical smart urban future.

In a theoretical sense, smart offers us a lot in its own ambiguity. Compared to the parameters associated with sustainability or ecocities, smart is not that specific. There are subjective ways of understanding it, which is why there is confusion, but also multiple interpretations. In that sense, theoretically there is a possibility to salvage some sort of radical social emancipation through smart. It can empower or disempower people of different social groups, lower castes, religious communities, poor women, LGBTQ communities. But empirically, I have only seen that to a very small extent.

Recently with the COVID-19 crisis, smart and its association with technology has been deployed in the civil society context to reach out to people locked out of the city under COVID-19, and you see this in both the global North and South. In places like Bangalore, the LGBTQ community, migrants, and sex workers were unable to access welfare services, which are contingent on digital IDs and digital platforms. Civil society organizations, not the state, tried manually to find the names and contact details of these people. Making a list of those left outside and locked out of systems using low-end and frugal technologies for an emancipatory end is smart. Another example was IT engineers using open-source software to create a GIS [geographic information system] platform for the city. They then collaborated with the urban government to provide information and reach out to people who did not have food and supplies, or who were locked out and wanted to get home to their villages.

If we can invert smart from being a top-down, centralized system – the whole premise of smart, which was built upon IBM packaging and selling its ubiquitous Big Data Internet of Things – to what Wendy Willems (2019) refers to as "a politics of things," then that becomes an interesting way of reframing smart. When we think about the politics of things, we can consider how simple things like a smartphone can become a political tool for social justice emancipation.

If we can find a way to define smart through the diversity of decentralized, small, mundane, and fragmented nature of various overlapping technologies, then we can theoretically consider smart as made up of an encounter between digital and material space, which opens up different hybrid moments of cultural encounters in which a new politics can change. If we can unhinge it from corporate-speak, from the regulatory nature of how smart cities have been implemented, then perhaps there is a future for that word. Technology can be a tool of authoritarianism; it can also be a tool of social emancipation.

But, so far, the ways smart cities have been implemented in India is with rhetorics, laws, and policies that are top down and authoritarian, so much so that many cities have lost their freedom and autonomy. Empirically, it is very hard to find evidence that smart can develop into or yield social justice.

Editors: In the previous part of the volume Lorena Melgaço and Lígia Milagres contend that "the decolonial critique foregrounds the contradictions of capitalist modernity in which frameworks such as the (climate) smart city are inscribed, unravelling the limits of current approaches to the relationship among digital technologies, the socio-environmental crisis, and the production of space in cities." Similarly, Martin Tironi and Camila Albornoz (in this volume) see much utility in De la Cadena's (2015) concept of excess: "Excess as an epistemological orientation invites one to decolonize the concept of the digital city and its technologies and to pay attention to its decentred and relational nature, considering forms of resistance and recalcitrance, ambiguities, and idiotic misunderstandings ... that emerge in the encounter between digital platforms and the relational ecology in which they perform."

As "smartness" is rolled out on a global scale and conceptions of social justice, which primarily draw on the Western canon, are we potentially reinscribing epistemic injustice, how do we as scholars (noting an uncoded "us" there) make our discussions of social justice in the smart city more globally or grassroots oriented?

AD: There is a kind of epistemic violence in trying to impose a Western notion of social justice. But there is also a potential for using that notion of social justice at the grassroots level. I am thinking of Nancy Fraser's work on subaltern counterpublics, which is evident in the subaltern moments of making do with less. We can see how incomplete and broken technologies can be used at the grassroots level to create new kinds of counterpublics that can provide alternatives in the absence of the state. This absence of the state is incredibly loud. The state is good at creating these larger models of change, which are almost like social engineering through infrastructure engineering. But at the same time, it is not good at understanding the impacts of these at the grassroots level. Those universal models are associated with the social justice that they think is what the West is saying. Architects and planners I interviewed about designing social justice in their smart city project literally would say, "the smart city is inherently socially just because it is neutral, it is not driven by political agendas." The political agenda, to them, is when we are actively engaging with or being sensitive to particular marginalization based on caste, class, religion, gender, and so on. To them, social justice means everyone gets the exact same thing – not that they do not understand the notion of equity. Everyone gets public Wi-Fi and everyone gets a smart metre, but they will not see the historical injustices, which mean that not everyone will benefit from Wi-Fi or a smart metre.

The epistemic violence of a universalized notion of equality is as present in India as it is in the West. I can see the reason and need to push back against Western, Eurocentric ideas of social justice. But at the same time, Indian middle classes grew up in a Westernized system of education, so those ideas of social justice are internalized. The Dalit rights movement, the gender movement, and now the LGBTQ movement have all highlighted that equality is not about giving the same thing to all people.

To imagine social justice in a smart city is continuously to chip away at the cracks. Provincializing the smart city in India is not just about provincializing Europe, it is about provincializing North Indian culture, upper-caste culture, *Hindutva* culture, masculinity, and so on. I do not see this in a binary way. The Black Lives Movement in the West has highlighted how Black people can reimagine other ways of advancing towards equality. In India, we need to think about ways to encompass sensitivity towards the differential effects of technology on social groups and space. We need an intersectional lens for social justice as well, one that considers spatiality and geography.

Words that are handed to us – from justice to sustainability, efficiency to innovation – have a universalizing force. To understand their impact, we need to examine them, unpack them, chip away at the cracks, or find ways to appropriate them. This is what they have been doing at the grassroots level. This is what James Scott calls the "weapons of the weak," or you can call it subaltern counterpublics. But these are counterpublics, which are quite different from what has been conceived of before. In India the nationalists argue that "the West is no good, everything they have told us or done is about exploitation," which is true, I agree. But they then use this to argue for an indigenous nationalism that goes back to a nostalgic past that never existed. In doing so, we wreak more epistemic violence against people who have been historically excluded and have had extreme violence perpetrated upon them, their bodies, their cultures, their languages, and their religions. I have gone to these smart city conclaves to have meetings with the stakeholders. They are all men, and I am not saying just having women there would make a huge difference, because most of the women stakeholders I have spoken to think in the same terms. There is an ethos of masculinity about these projects: they are *big, bold, fast, efficient*. So we need to examine the rhetorics of these initiatives. We need to push back against that multiscalar epistemic violence, whether it is coming from the West, the centre, or the dominant group. We need to highlight the genealogy of these terms: where are they coming from? who stands to gain? what is the directionality of power?

Smart has its origins in the West with ideas of smart growth, but also with global development agencies like the UN and smart economics. They both have the same inherent efficiency logic, and federal states also use that efficiency logic towards urbanization. There is a logic that has driven the word smart that is about efficiency and speed, and the only way to do that is through technology.

We have to find openings within these terms where other imaginaries can be created that are not so big, not so bold, not so fast, and not so seductive. We have to look to the regular and boring and things we have been doing every day but have not really framed within the rhetorics and discourses of smart.

We can think of the rejection of authoritarian imperatives of technology, while also appropriating them at the grassroots level to learn and empower. We can also rethink the use of technology: what is technology for? what radical potential does it have to garner information, bring about knowledge transformation and raise critical consciousness? To me, that would challenge a unitary vision of urban futures. I think that the violence is when we have a unitary vision of urban futures, and we are not really thinking about urban futures as striated, overlapping, as a palimpsest.

Editors: With the Dalit rights movement and women and LGBT communities in mind, you highlight the need for intersectional and relational analyses of smart cities. Authors in this part of the volume – in particular, Wiig, Pearsall, and Mascucci, as well as Abbruzzese and Hillier – highlight the entanglement of the smart city in these divides, as well as their complicated and often contradictory logics and impacts. Returning to discussion of inequality and inequity, could you speak to the utility of the concept of the "digital divide" or more recent updates of the term?

AD: The digital divide has become such a clichéd term. That is not to say you should not be talking about it. I resist using it myself because it does not capture the complexity. It started with digital divide – "the haves" and "have nots." But we know even if you do not have a phone, you still use and borrow a phone, so it has become very complex.

From my research and my work, the digital divide is also a gender divide. But the digital divide is also a paradox, because it depends on what you mean by "digital." If you are referring to mobile phones, people in the slums are one of the highest consumers of mobile phones, albeit cheaper ones. If you are referring to digital infrastructure, then that is where you see a bigger divide in gaining access to digital space. Many slum areas are at peripheries and have the least number of mobile phone towers.

At the infrastructural level, the digital divide is a spatial geographic divide. On a bigger scale, the digital divide is a geopolitical divide. Some countries are better connected than others, and that depends on the data centres and the fibre optic cables going under the sea. Within cities, the digital divide is so complex, with its infrastructure and data and how access to them determines where you can go digitally. Then there is neoliberal capital capture, with one or two providers and regulators controlling all the digital airways, so then they can charge what they want. It also increases the ease of surveillance through them. To call it a divide is reductive.

Then you have another notion of the digital divide. Even if you have a phone, a network connection and data, you might not have the digital capacity, and I see this a lot working with women in the slums and resettlement colonies. So many of these platform companies, such as Google, Facebook, and Amazon, push the apps first and the browser second, so they discourage you from using the browser. You get trapped in these closed app spaces and move from one app to the other. So when you get misinformation, when you get a forwarded message, how do you check its validity? They do not know how to check its validity. It is easier for misinformation campaigns to spread on WhatsApp because people who have mobile phones, but do not necessarily know how to search on the Internet for legitimate information, will get trapped into these closed echo chambers in WhatsApp, where misinformation is circulated, perpetuated, and reinforced. The digital divide, in that sense, has a very tense relationship with social justice and critical consciousness. When you're well informed you're knowledgeable, but if you do not have that, then you cannot understand your own disempowerment or contextualize your own marginalization within digital spaces. That is the bigger digital divide, beyond access.

The digital divide also impacts how we use urban spaces, and how we participate as citizens in urban spaces. In India, last year just before COVID-19 struck, there were large demonstrations around the new Citizenship Amendment Act. The state was monitoring peoples' locations during these street protests. Once you are using these apps or a particular service provider, you also enable them to track where you are going. It becomes difficult to challenge the state. If you are a dissenting citizen, then how do you enact your active citizenship and call the state to account? How do you ask for transparency from the state if you are not even given that option?

The digital divide is also seen in the normalization of terms and conditions, data privacy, data security, and ubiquitous surveillance. Many young women in the slums are tracking their own movements with GPS trackers. They want to be able to witness and document what is happening to them. In particular, in the case of danger or sexual assault, they want to be able to record it, so they keep it switched on and they are always taking photos. So, again, the technology works both ways. One of the biggest paradoxes in the current digital age is that these women have much better access to mobile phones, much better access to digital space (in that sense they are not on the other side of the digital divide), yet they are totally marginalized so that they seem completely out of place in the city. If something happens to them and they go to the police to launch a file, the first thing the police say is, why were you out so late? They can use the phone as a kind of a tool of documentation and witnessing, yet at the same time they know that witnessing is not going to result in justice. The people who perpetrated that violence could say that footage has been doctored. The women know in the end not much will come of it, but they do it nonetheless.

The digital divide is so complicated. We need to find a new word and I don't have it yet. I have recently been reading about urban encounters, a theme we find in a lot of cities in both the West and the global South.

My earlier work on east European migrants in London (Datta 2008, 2009) explored how urban encounters shape the built environment. Looking at digital spaces, they are also a kind of urban encounter, because urban-social contact is also happening in these digital spaces. In a sense, these digital spaces are what Mary Louise Pratt would call contact zones, where huge amounts of transformations are happening. Mobile phones are contact zones between the digital and the physical spaces of a city. These are moments of transculturation. These are moments of struggle, anger, fear, frustration. In these zones there is the potential for critical consciousness to develop. But there is no rulebook or guidelines by which it can be developed.

If we see the technology as a political tool, both *of the state* and *of the people*, then we can begin to see some politics in that digital divide. It is not just about bringing people on this other side. It is about creating social, political, regulatory context in which that tool can actually become emancipatory.

FURTHER READING

Chakrabarty, D. 2002. *Habitations of Modernity: Essays in the Wake of Subaltern Studies*. Chicago: University of Chicago Press.

Chakrabarty, D. 2009. *Provincializing Europe*. Princeton, NJ: Princeton University Press.

Datta, A. 2008. "Building Differences: Material Geographies of Home(s) among Polish Builders in London." *Transactions of the Institute of British Geographers* 33 (4): 518–31. https://doi.org/10.1111/j.1475-5661.2008.00320.x
Datta, A. 2009. "Places of Everyday Cosmopolitanisms: East European Construction Workers in London." *Environment and Planning A: Economy and Space* 41 (2): 353–70. https://doi.org/10.1068/a40211
Datta, A. 2015a. "New Urban Utopias of Postcolonial India: 'Entrepreneurial Urbanization' in Dholera Smart City, Gujarat." *Dialogues in Human Geography* 5 (1): 3–22. https://doi.org/10.1177/2043820614565748
Datta, A. 2015b. "A 100 Smart Cities, a 100 Utopias." *Dialogues in Human Geography* 5 (1): 49–53. https://doi.org/10.1177/2043820614565750
de la Cadena, M. 2015. *Earth Beings: Ecologies of Practice across Andean Worlds.* Durham, NC: Duke University Press.
Fraser, N. 1990. "Rethinking the Public Sphere: A Contribution to the Critique of Actually Existing Democracy." *Social Text* 25/26: 56–80. https://doi.org/10.2307/466240
Howard, E. 1946.*Garden Cities of Tomorrow*. London: Faber.
Mosco, V. 2005.*The Digital Sublime: Myth, Power, and Cyberspace.* Cambridge, MA: MIT Press.
Mosco, V. 2015. *To the Cloud: Big Data in a Turbulent World.* New York: Routledge.
Mosco, V. 2019. *The Smart City in a Digital World.* Bingley, UK: Emerald Publishing.
Pratt, M.L. 2007. *Imperial Eyes: Travel Writing and Transculturation.* New York: Routledge.
Rodríguez-Pose, A. 2018. "The Revenge of the Places That Don't Matter (and What to Do about It)." *Cambridge Journal of Regions, Economy and Society* 11 (1): 189–209. https://doi.org/10.1093/cjres/rsx024
Scott, J.C. 2008. *Weapons of the Weak.* New Haven, CT: Yale University Press.
Willems, W. 2019. "'The Politics of Things': Digital Media, Urban Space, and the Materiality of Publics." *Media, Culture & Society* 41 (8): 1192–209. https://doi.org/10.1177/0163443719831594

15 Decolonizing the Smart City: Excess and Appropriation of Uber Eats in Santiago de Chile

MARTIN TIRONI AND CAMILA ALBORNOZ

The concept of the smart city and its implementation have been introduced in an especially strong way in Latin America – particularly in Santiago de Chile. Alongside media strategies and a strong "cultural circuit" associated with this paradigm (Tironi and Albornoz 2021), this model of development is sustained by an alleged superiority over other forms of urban development based on the technological capacities presented by digital platforms to resolve various problems in the city. The language around smartness is not only reorganizing discussions of the city (Marvin, Luque-Ayala, and McFarlane 2015), but also imposing a way of conceiving urban futures through discourses, values, and practices from the global North. Smart urbanism emerges through the centrality that instrumental, extractive, and technological rationalities acquire: datafication, automation, and digital monitoring that promise better coordinated, safer, and more efficient forms of urban life. The same notion of the data-driven city (Kitchin 2014), according to which the city's services and infrastructure can be governed based on the objectivity and neutrality of digital data, responds to a modern techno-optimistic narrative that appeals to knowledge and management of the city based on the Big Data revolution. However, the use and application of the smart city have been criticized for their tendency towards technological solutionism and the utopian search for a frictionless city (Sadowsky and Bendor 2018). Furthermore, data and the intervention of multinational technology leaders have tended to corporatize cities (Datta 2015; Hollands 2008; Kitchin 2014) and to promote a depoliticization and standardization of the urban space (Hollands 2008).

As Burns, Fast, and Mackinnon state in the Introduction to this volume, understanding smart as an *empty signifier* or as a concept that has no inherent foundation allows the encounter of resistances, inequalities, resignifications, and other local processes that do not have a place within smart corporatist and top-down approaches. We argue that there is still an unproblematized dimension, related to the strong colonialist character of the values and practices behind the smart city paradigm. Currently, there are few concepts and questions that allow the smart project to be anchored to cultures or specific forms of inhabiting and feeling the city (Farrés and Matarán 2014). While the critiques developed to date around the concept of smart urbanism are necessary, we believe that they still move within a modern/colonial epistemological and ontological order. It is not a question of discarding all Eurocentric urban approaches or those that come from the global North. We propose instead that it is necessary to develop a Latin American and decolonial reading of smart, reflecting on the ways in which smart urbanism is installed as a hegemonic paradigm. In other words, a decolonial strategy does not imply rejecting everything that comes from the global North, but rather creatively and selectively appropriating the elements that allow us to build more just and inclusive cities.

In this chapter we propose a decolonized vision of the smart city, showing how this paradigm reproduces a productivist, anthropocentric, and universalist vision. Specifically, and in dialogue with the notion of *ontological excess* (de la Cadena 2015), we explore how the digital paradigm of the smart city has ontological limits for operationalizing local urban life. Based on ethnographic work conducted with Uber Eats delivery staff in Santiago de Chile, we show how *delivery partners* do not always ascribe to the naturalized logics and values of the platform. In other words, delivery staff members' ways of feeling-thinking the city are always in excess of the enunciations, delimitations, and protocols of Uber Eats, which are configured under the notion of user-client and the search for ubiquitous efficiency through algorithmically designed systems. We also discuss some of the operations that we could call decolonial that delivery staff introduce to grapple with and appropriate technology.

Deconstructing the Biases of the Smart City

The rhetoric of smartness has installed an agenda and a series of suppositions about how the intrusion of technologies could facilitate management tasks in daily life and in teh city (Campbell 2012; Harrison and Donnelly 2011). The colonization of the language of optimization, the promise of sterile solutions (Perng and Kitchin 2018), and the idea of installing smooth cities (Forlano 2016) that are productive and efficient have given rise to a prolific production of reports, articles, and books that critique the deployment of the smart city and the negative externalities of the intrusion of technology into the urban world (Graham and Marvin 2011).

According to the literature, these approaches have positioned forms of neoliberal technocracy in the governance of cities. Smart projects that seek the optimization and acceleration of urban processes through datafication, making the routines of the city traceable and computable, "are highly narrow in scope and reductionist and functionalist in approach, based on a limited set of particular kinds of data" (Kitchin 2014, 9). Other authors have called this type of approach *technological solutionism* (Morozov 2013); as a result, there is a depoliticization of the public, which allows for a space in which smart innovations seek to focus on the user (Campbell 2012; Deakin and Al Waer 2013), agents that produce data that will inform the decision-making process.

Some have questioned how the smart city has installed a strong business rhetoric. Some academics argue that this can be understood only within a neoliberal context (Greenfield 2013; Hollands 2008; Vanolo 2014), which implies "[an] ethos that prioritizes market-led and technological solutions to city governance and development" (Kitchin 2014, 2). This model rests on the commodification of the city's political and social problems, integrating stakeholders – businesses – that are not usually part of the solution (Morozov and Bria 2018, 11). Through the formation of public-private alliances, the public sphere could easily permeate public perspectives, allowing the "private sphere to reshape the public sphere in its own image" (Sadowski and Pasquale 2015, 4).

This aspect has been emphasized in various studies that have referred to the smart city as "the corporate smart city," "a marketing device for city branding" (Hollands 2015, 5), or "a utopia built by the neoliberal model" (Linares-García and Vásquez-Santos 2018, 493, authors' translation). Some have even argued that the right to the city would belong to corporations in that they hold the right to accumulate capital in the smart city (Datta 2015, 49). Many case studies describe and analyse the link between multinational corporations such as CISCO, IBM, and Siemens and the technology-designed role to be implemented in cities (Gabrys 2014; Hollands 2015; Kitchin 2015). It seems that the smart city tends to represent the interests of the business elite (Grossi and Pianezzi 2017) or that it imposes "elite-led governance" (Basu 2019, 84). Even though the literature celebrates bottom-up initiatives, which seek to emphasize the opportunity of this type of project to obtain feedback from "normal" citizens (Harrison and Donnelly 2011), experimental approaches and their materialization through living labs have also been criticized. When their logics are analysed, it is evident that smart initiatives tend to preset the type of participation they seek or the type of citizen they configure (Tironi and Valderrama 2018).

Along these lines, some authors have questioned new forms of information asymmetry between technology companies and users/data producers. This gap has generated a majority – citizens – that generates data and a minority – businesses – that accesses ownership of the technologies and tools necessary to process data (Andrejevic 2014). As an evolution, some claim that smart urbanism would move towards so-called *platform urbanism* (Caprotti and Liu 2020), its emergence taking advantage of "more opportunities for mediating social relations and extracting economic value in large, diverse markets" (Sadowski 2020, 3) – in this way, each platform establishes its own game rules (Srnicek

2017). It is even suggested that platform planning would take the form of a monopoly (Barns 2019), where multinationals offering services such as Uber or Airbnb store large amounts of data through everyday practices of people using their smartphones.

Decolonizing Smartness

The smart city has introduced new epistemological and ontological approaches – ways of making, feeling, and inhabiting the city – associated with certain frameworks that we could call colonial. We argue that the implementation of the smart digital regime in the city cannot be separated from a certain form of colonialism that is guided by systems of thought characteristic of the global North. Various authors argue that colonial structures are and have been resilient to the passage of time, and are constitutive of imaginaries and technological infrastructures that are developmentalist and neoliberal in nature (Couldry and Mejías 2019; Escobar 2019; Maldonado-Torres 2011). Cities and the smart project would not be the exception to coloniality, insofar as they correspond to a top-down form of urbanism impregnated with a modern and instrumental thought system. According to decolonial perspectives, contemporary forms of the colonial legacy are reflected in the dominance of structures framed by dualist and rationalist logic, involving power relations that are unequal in terms of ethnicity, gender, and knowledge (Escobar 2019; Lugones 2010; Quijano 2000). In this direction, Escobar (2008) argues that Eurocentric modernity dilutes local histories and narratives, subalternizing the local, and that "much of what we see in cities today finds its long genealogy in this ascendancy of anthropos and the logos and its steady development through Western heteropatriarchal culture" (Escobar 2019, 136). As such, cities face a "territorial coloniality" – a "set of patterns of power that in territorial praxis serve to hegemonically establish a concept of territory" (Farrés and Matarán 2014, 348, authors' translation). Intelligent systems do not consider any type of historical or spatial specificity despite being solutions that are impregnated with user-centred rhetoric. Many of their discourses, practices, and imaginaries are informed by the principle that digital infrastructure should respond to the needs and concerns of the user/consumer, favouring generic approaches or those that are "one-size fits all ... with buildings and cities being treated as mere generic markets or marketable entities" (Figueiredo, Krishnamurthy, and Schroeder 2019, 10).

To the extent that "smartness" has acquired a totalizing character – encompassing different territories and spheres of life – it has not only homogenized the vision of cities or colonized the imaginations of people about how a city needs and benefits from innovation (Datta, in this volume); it has also reduced multiplicity and pluriversity to certain hegemonic futures, making invisible other practices, sensitivities, and forms of autonomy (Escobar 2008). As such, it is a unidirectional and profoundly anthropocentric future concept in which the value of the future is oriented towards generating cities at the service of the interests of certain humans. We could state that the globalization of smart cities is inscribed in what Fry (2020) has called the problem of *defuturing*, a term used to name the current trend towards the elimination of any possibility of an alternative future. As a result of the massification of certain colonial and Eurocentric epistemological and ontological regimes, futures are captured and manufactured by promises of the large technological corporations of the North, which manage to install certain desired imaginaries and concepts over others (Fry 2020). This means that the challenge is not limited to the decolonization of the epistemologies and ontologies that the smart cities program mobilizes, but includes addressing the problem of defuturing that the smart regime poses, exploring alternative and counterhegemonic futures that emerge from the problems and forms of knowledge that are situated in the global South.

Another approach that connects technologies and colonization is proposed by Couldry and Mejías (2019), who argue that smart technologies that expand in the world to extract, accumulate, and process data constitute the installation of a form of colonialism of data. Digital platforms are the technological media that seek to capitalize on social life, making it traceable, computable, and quantifiable. This extractivist logic, the authors argue, is particularly present in the South, where the poorest and least-protected nations in terms of data use regulation are co-opted by sophisticated platforms and digital services as if they were an inevitable and civilizing project.

In this sense, the promise of salvation that accompanies the introduction of increasingly automated algorithms, platforms, and data underlies a negation of the contextual, relational, and situated knowledge that exists locally (Haraway 2016). This negation rests on a sanctification of the new and of ongoing innovation. By relating to the city from an instrumental logic of data extractivism, the smart paradigm operates by reducing the experience of the city to the surface of an optimized present that is focused on the future. As such, we believe that a decolonial approach should try to restore the pre-existing – those forms of distributed intelligence that force us to think about space and its technologies as unstable, dynamic, and heterogeneous entities co-produced in the linking of different agencies. Decolonizing the smart vision of the city implies recognizing that the urban is always being made and assembled in multiple situated places (Jirón et at. 2020), presences, and processes, and resists being computed and optimized using datafication logics.

Another element of the smart city that should be questioned is its marked anthropocentric nature, in which the technological forms of managing urban life are oriented towards meeting users' needs, ignoring that cities comprise an ecology made up of communities, objects, animals, and other organisms whose lives and futures are inevitably intertwined (Puig de la Bellacasa 2017). From a decolonial perspective, it seems necessary to recognize the value of urban coexistence, beyond solely the club of humans (Stengers 2005), by paying attention to the capacities of non-human entities in the configuration of coexistence in and the experience of the city. At the same time, the decolonial approach recognizes that environmental justice and inequalities are closely intertwined, with historically marginalized populations suffering the most (Melgaço and Milagres, in this volume).

To move towards a decolonial pluriversity of digital cities in the South, we turn to the concept of excess proposed by de la Cadena (2015). In an ethnographic work on the Pacchanta community in the Peruvian Andes, de la Cadena suggests that excess appears when the epistemic tools that modern people use to translate *Ausangate* – a word that means not only a sentient being for the community, but also a mountain or land with extractive value – collide with the ontology that Indigenous people designate to *Ausangate*. Taken to the level of digital technologies, we want to show that situated communities ontologically exceed the regulatory frameworks of the smart platforms founded and codified from the global North through their own ways of being and feeling the city and its technologies. Excess as an epistemological orientation invites one to decolonize the concept of the digital city and its technologies and to pay attention to its decentred and relational nature, considering forms of resistance and recalcitrance, ambiguities, and idiotic misunderstandings (Tironi and Valderrama 2018) that emerge in the encounter between digital platforms and the relational ecology in which they perform. The case of Uber Eats analysed here shows ways of making and feeling the city that exceed the categories that are normalized by platforms. In other words, delivery staff create spaces that are difficult or *impossible* (Blaser and de la Cadena 2009) to codify or that are not recognized by the algorithmic intelligence of Uber.

Delivery Staff and Their Excesses

The utopia of autonomous work on platforms involves long workdays, with delivery staff working six days and sixty hours per week, just delivering by bike. They are paid per delivery completed, and have no health insurance, schedule, or union rights. Many delivery staff are migrants waiting to secure their permanent residency visa and find a stable job. Opening an account as a delivery person is a quick and easy alternative. As part of the minimal bureaucracy of Uber Eats, to open a *delivery partner* account, one must have an identity card or other document that lists one's tax identification number.[1] Many immigrants do not yet have this identification number, and without it they cannot open an account or join the formal labour world.

According to Roberto, a delivery worker we interviewed, borrowing an account is common practice among delivery partners: "It is common to meet a

Venezuelan who has a stable job and will give you an account. 'Take it – use it to work.' And they don't charge for it." Renting or borrowing accounts usually occurs between close contacts such as family members and friends and with delivery staff who meet while waiting for an order to arrive. If someone needs a new account, often both the new user and the person renting it out go to an Uber support and activation centre to open the new account. The person with a tax identification number completes these processes and sits for the photo, which is included in the delivery staff member's profile. Then the two meet up to exchange information for the activated account. When accounts are rented, there is usually a weekly charge, which allows the account owner to generate an extra income.

For delivery staff members, working for an app does not go far towards solving their problems; it merely allows them to obtain income, and is also a way to take care of one another. One delivery staff member reported that account rental occurs because some migrants do not yet have work visas, so "they help each other out," allowing them to have a source of subsistence while they wait for their residency visa. "There are always people who have needs, and they get an account and [rent it]. It might not be to make money. They may not earn much because it is not like the effort that the person who works the account puts in. At the end of the day, people who have documents and are part of the community do it to help out others" (Alexis, interview).

Uber categorizes these types of decolonial practices as fraud, "[allowing] information to be intentionally falsified or to take on the identity of another person" (Uber 2020). To eliminate this excess and redirect the actions of delivery staff, Uber has developed a system for verifying delivery partners' identity, whereby clients can indicate if the delivery person was the same individual who appears in the photograph. It has also developed a facial recognition system for delivery staff. To access the app, the delivery person must take a selfie, and the image must align with the facial characteristics of the photo taken the day the account was activated. If the person's identity is not confirmed, the app cannot be accessed. If the face does not match, the account is deactivated – the person is fired – and the account cannot be recovered, reactivating the account rental system.

Refusing to Update

These spaces not only allow delivery staff to work together to resist protocols and forms of surveillance imposed by the platform; they also challenge the way in which the platform has coded how one should move through the city. Delivery staff members usually use bicycles to deliver orders, but other forms of transport are allowed. Different delivery radii are set for each of them. Bicycles can travel up to three kilometres, e-bikes seven kilometres, and motorcycles approximately eleven kilometres. Delivery staff report that these limits are relative and flexible, and they constantly receive orders that go beyond them, forcing them to travel increasingly long distances and to navigate various municipalities and unfamiliar streets.

As such, many decide to buy an e-bike to travel longer distances, tire themselves out less, and earn more money. However, if delivery workers purchase a new mode of transport, they must update their user profile indicating the new transport. In this way, the application calibrates how far the delivery person can travel. Many of them do not update their accounts and continue to work with a bicycle profile. In this way, they can cover the same distance more quickly and be available for deliveries more frequently. Interviewee Mario explained: "Look, I used an e-bike [even though his account was still a bike account]. I bought an e-bike because the radius has increased on all of the apps. I just did a five-plus-kilometre delivery. It recorded six kilometres [as it is a bicycle account, it should not travel more than three kilometres]."

The app monitors the activities of delivery staff through GPS. If it detects that the person is travelling at a speed of over 40 km/h, the delivery staff member will quickly receive a smartphone message that states, "We have determined that you are making deliveries in a vehicle that does not match your account," and will be asked to update the member's information with the correct means of transport. Members who do not do so are blocked from the application.

Multi-apping and GPS Control

Despite Uber's attempts to decrease excess, some delivery staff combine different operations: "I'm going to replace my bike and leave the bike account because the orders are really long ... If someone rents you another account and you have two phones, you can work the two accounts at the same time." Multi-apping (Barratt, Goods, and Veen 2019), or working with multiple accounts, is an easy way to maximize earnings, and many delivery staff carry two cell phones with portable batteries to stay online with both accounts and without any interruptions. Rather than simply opening more than one account, they open accounts on different applications to increase the likelihood that they will receive orders and take advantage of Uber Eats orders as well as competing apps such as Rappi and PedidosYa.

The platform constantly monitors the delivery partners through GPS, however, and delivery staff note this surveillance: "I've figured out that they know everything you're doing through the GPS. If you get there, if you are far away, if you have completed the delivery ..." (Mario, interview). As such, several riders explain that the Uber GPS provides a preset route to use, and if they deviate from the path the app might recognize that they are making a different delivery. "When Uber was down, I would try to connect to Rappi and orders would come in. So, yeah, it is good to have two applications sometimes, but you don't work on them at the same time because they can block you. The staff at the office use the GPS and investigate you" (Roberto, interview). This has led delivery staff to work multiple shifts, using Uber Eats as the primary application (wearing Uber clothes and carrying the company backpack) and Rappi as a secondary option. They close the application when they find an order to ensure that they do not receive two orders at once and to prevent the company from finding out that they are engaged in other activities by tracking them with GPS.

Final Thoughts

Throughout this chapter, we have discussed how the Uber Eats platform in Santiago de Chile mobilizes certain rationalities that cannot necessarily be applied to inhabitants' ways of being and inhabiting. Uber, a multinational corporation, seeks to generate a perfect alignment between technological platform, delivery staff member, client, and urban space, operating as a form of *microphysics of power* (Foucault 1979) through which it tries to define ways of relating to and inhabiting the ecology of the city. We have tried to show a series of conflicts between what is programmed and uses of the platform that reconfigure its performance. Although the platform was designed to be used through personal accounts, with atomized work for an autonomous user-delivery person, delivery staff deploy mundane and disobedient practices that could be considered a form of "platform cooperativism" that "creatively embraces, adapts, or reshapes technologies of the sharing economy, putting them to work with different ownership models" (Scholz 2017, 14). At the same time, the operations we described constitute a form of technical solidarity that slows down (Stengers 2005) the rationality of optimization that the platform seeks to imprint on delivery staff and on the city. In a context of atomized work with no labour rights, the practices of solidarity that delivery staff deploy show not only a form of appropriation, but also a decolonial redirection of the platform, introducing values based on reciprocal care and collaboration. In the face of a platform that imposes an extraordinarily partialized and rationalized work regime, delivery staff manage to materialize a true ethic of care in their situated practices. As Puig de la Bellacasa (2017) puts it, care is recognition of an unavoidable condition of the interdependence and vulnerability of all the spaces that we inhabit. The act of reclaiming care and solidarity that delivery staff perform represents a way of making an alternative city that differs from the one the platform seeks to inscribe.

The supposedly universal digital technological condition that the platform seeks to ensure is placed under stress by a set of actions of collective redesign and reappropriation of technology and the urban space, sustained by the practical and situated knowledge that experience places at the centre of the right to the city. Through practices of disobedience and recalcitrant excesses, delivery staff appropriate (de Certeau 1996) what has been delineated, using the faults and fissures in the platform's logics. Approaching Uber Eats service from the tactics developed by delivery staff allows one to observe how the platform is disobeyed from a decolonial and critical making, opening the opportunity to promote other forms of living the sensitive experience of the city and its technologies.

To the extent that delivery staff continue to find strategies to redirect platform characteristics to their benefit, Uber Eats will continue to make increasingly intelligent and automated adjustments and controls over activities identified as fraud. Although the platform's smart algorithms will seek to compute all possible actions, the delivery staff's experience of the platform and the city appears as a space for experimentation that is constantly contested and undergoing recomposition, generating excesses that the protocols of technology will seek to discipline and silence.

The fact that the programmers seek to codify these excesses implies that they have recognized them and that there is a denial of this excess. For de la Cadena (de la Cadena, Risør, and Feldman 2018), excess is not recognized because "its recognition alters the ontological conditions of what is known" (authors' translation). As such, excess offers a political vitality for moving towards decolonial resurgence. Recognizing the excess that emerges from the operation of the platform implies recognizing a series of precarities that will have labour, legal, and political consequences. In this sense, decolonization as an onto-epistemic orientation "cannot rest on a celebration of difference, creativity, and possibility" (Irani and Philip 2018, 7). Rather, it aims to implement political and economic changes to the technological infrastructure that governs us. It is impossible to decolonize smartness without questioning the instrumental and extractivist paradigm on which the designs, algorithms, and business models of these platforms are based. In this regard, following Escobar (2019), Costanza-Chock (2018) suggests that the development of smart systems should begin to experiment with collaborative and situated forms of design that question those developed by large digital corporations, based in local places and forms of knowledge, recognizing the plurality and interdependence among all human and non-human entities that exist on the planet. In this way, it will be possible to reconfigure anthropocentrism and ecocentrism through which the imaginary of platform urbanism advances, and to restore the dignity of denied and forgotten worlds, generating other ways of understanding the city.

NOTE

1 A Registro Único Tributario (RUT) or Rol Único Nacional (RUN) is an identification number unique to each person and used for all transactions involving Chilean government agencies.

REFERENCES

Andrejevic, M. 2014. “Big Data, Big Questions: The Big Data Divide.” *International Journal of Communication* 8 (17): 1673–89. Online at https://ijoc.org/index.php/ijoc/article/view/2161/1163

Barns, S. 2019. *Platform Urbanism: Negotiating Platform Ecosystems in Connected Cities*. London: Springer Nature.

Barratt, T., C. Goods, and A. Veen. 2020. “‘I’m my own boss ... ’: Active Intermediation and ‘Entrepreneurial’ Worker Agency in the Australian Gig-Economy. *Environment and Planning A: Economy and Space* 52 (8):1643–61. https://doi.org/10.1177/0308518X20914346

Basu, I. 2019. “Elite Discourse Coalitions and the Governance of ‘Smart Spaces’: Politics, Power and Privilege in India’s Smart Cities Mission.” *Political Geography* 68: 77–85. https://doi.org/10.1016/j.polgeo.2018.11.002

Blaser, M., and M. de la Cadena. 2009. “Introducción en World Anthropologies Network (WAN).” *Red de Antropologías del Mundo* 4: 3–9. http://ram-wan.net/old/documents/05_e_Journal/journal-4/jwan4.pdf

Campbell, T. 2012. *Beyond Smart Cities: How Cities Network, Learn and Innovate*. London: Routledge.

Caprotti, F., and D. Liu. 2020. “Platform Urbanism and the Chinese Smart City: The Co-Production and Territorialisation of Hangzhou City Brain.” *GeoJournal*. https://doi.org/10.1007/s10708-020-10320-2

Costanza-Chock, S. 2018. “Design Justice: Towards an Intersectional Feminist Framework for Design Theory and Practice.” *Proceedings of the Design Research Society*. https://ssrn.com/abstract=3189696

Couldry, N., and U. Mejias. 2019. *The Costs of Connection: How Data Is Colonizing Human Life and Appropriating It for Capitalism*. Stanford, CA: Stanford University Press.

Datta, A. 2015. “A 100 Smart Cities, a 100 Utopias.” *Dialogues in Human Geography* 5 (1): 49–53. https://doi.org/10.1177/2043820614565750

Deakin, M., and H. Al Waer. 2013. *From Intelligent to Smart Cities*. Oxford: Routledge.

de Certeau, M. 1996. *La invención de lo cotidiano: artes de hacer*, vol. 1. México: Universidad Iberoamericana.

de la Cadena, M. 2015. *Earth Beings: Ecologies of Practice across Andean Worlds*. Durham, NC: Duke University Press.

de la Cadena, M., H. Risør, and J. Feldman. 2018. “Aperturas onto-epistémicas: conversaciones con Marisol de la Cadena.” *Antípoda. Revista de Antropología y Arqueología* 32: 159–77. https://doi.org/10.7440/antipoda32.2018.08

Escobar, A. 2008. “Beyond the Third World: Imperial Globality, Global Coloniality and Anti-Globalisation Social Movements.” *Third World Quarterly* 25 (1): 207–30. https://doi.org/10.1080/0143659042000185417

Escobar, A. 2019. “Habitability and Design: Radical Interdependence and the Re-Earthing of Cities.” *Geoforum* 101: 132–40. https://doi.org/10.1016/j.geoforum.2019.02.015

Farrés, Y., and A. Matarán. 2014. "Hacia una teoría urbana transmoderna y decolonial: una introducción." *Polis* 13 (37): 339–436. https://doi.org/10.4067/S0718-65682014000100019

Figueiredo, S., S. Krishnamurthy, and T. Schroeder. 2019. "What about Smartness?" *Architecture and Culture* 7 (3): 335–49. https://doi.org/10.1080/20507828.2019.1694232

Forlano, L. 2016. "Decentering the Human in the Design of Collaborative Cities." *Design Issues* 32 (3): 42–54. https://doi.org/10.1162/DESI_a_00398

Foucault, M. 1979. *Microfísica del poder*. Madrid: Ediciones de La Piqueta.

Fry, T. 2020. *Defuturing: A New Design Philosophy*. London: Bloomsbury Publishing.

Gabrys, J. 2014. "Programming Environments: Environmentality and Citizen Sensing in the Smart City." *Environment and Planning D: Society and Space* 32 (1): 30–48. https://doi.org/10.1068/d16812

Graham, S., and S. Marvin. 2001. *Splintering Urbanism: Networked Infrastructures, Technological Mobilities and the Urban Condition*. London: Psychology Press.

Greenfield, A. 2013. *Against the Smart City (The City Is Here for You to Use, Book 1)*. New York: Do Projects.

Grossi, G., and D. Pianezzi. 2017. "Smart Cities: Utopia or Neoliberal Ideology?" *Cities* 69 (September): 79–85. https://doi.org/10.1016/j.cities.2017.07.012

Haraway, D. 2016. *Staying with the Trouble: Making Kin in the Chthulucene*. Durham, NC: Duke University Press.

Harrison, C., and I.A. Donnelly. 2011. "A Theory of Smart Cities." In *Proceedings of the 55th Annual Meeting of the ISSS-2011*, Hull, UK.

Hollands, R.G. 2008. "Will the Real Smart City Please Stand Up? Intelligent, Progressive or Entrepreneurial?" *City* 12 (3): 303–20. https://doi.org/10.1080/13604810802479126

Hollands, R.G. 2015. "Critical Interventions into the Corporate Smart City." *Cambridge Journal of Regions, Economy and Society* 8 (1): 61–77. https://doi.org/10.1093/cjres/rsu011

Irani, L., and K. Philip. 2018. "Negotiating Engines of Difference." *Catalyst: Feminism, Theory, Technoscience* 4 (2): 1–11. https://doi.org/10.28968/cftt.v4i2.29841

Jirón, P., W.A. Imilán, C. Lange, and P. Mansilla. 2020. "Placebo Urban Interventions: Observing Smart City Narratives in Santiago de Chile." *Urban Studies* 58 (3): 601–20. https://doi.org/10.1177/0042098020943426

Kitchin, R. 2014. "The Real-Time City? Big Data and Smart Urbanism." *GeoJournal* 79 (1): 1–14. https://doi.org/10.1007/s10708-013-9516-8

Kitchin, R. 2015. "Making Sense of Smart Cities: Addressing Present Shortcomings." *Cambridge Journal of Regions, Economy and Society* 8 (1): 131–6. https://doi.org/10.1093/cjres/rsu027

Linares-García, J., and K. Vasquez-Santos. 2018. "Ciudades inteligentes: ¿materialización de la sostenibilidad o estrategia económica del modelo neoliberal?" *El Ágora USB* 18 (2): 479–95. https://doi.org/10.21500/16578031.3134

Lugones, M. 2010. "Toward a Decolonial Feminism." *Hypatia* 25 (4): 742–59. http://www.jstor.org/stable/40928654

Maldonado-Torres, N. 2011. "Thinking through the Decolonial Turn: Post-continental Interventions in Theory, Philosophy, and Critique – An Introduction." *Transmodernity: Journal of Peripheral Cultural Production of the Luso-Hispanic World* 1 (2): 1–15. https://doi.org/10.5070/T412011805

Marvin, S., A. Luque-Ayala, and C. McFarlane, eds. 2015. *Smart Urbanism: Utopian Vision or False Dawn?* London: Routledge.

Morozov, E. 2013. *To Save Everything, Click Here: The Folly of Technological Solutionism*. New York: Public Affairs.

Morozov, E., and F. Bria. 2018. *Rethinking the Smart City: Democratizing Urban Technology*. New York: Rosa Luxemburg Foundation.

Perng, S.Y., and R. Kitchin. 2018. "Solutions and Frictions in Civic Hacking: Collaboratively Designing and Building Wait Time Predictions for an Immigration Office." *Social & Cultural Geography* 19 (1): 1–20. https://doi.org/10.1080/14649365.2016.1247193

Puig de la Bellacasa, M. 2017. *Matters of Care: Speculative Ethics in More than Human Worlds*. Minneapolis: University of Minnesota Press.

Quijano, A. 2000. "Coloniality of Power and Eurocentrism in Latin America." *International Sociology* 15 (2): 215–32. https://doi.org/10.1177/0268580900015002005

Sadowski, J. 2020. "Cyberspace and Cityscapes: On the Emergence of Platform Urbanism." *Urban Geography* 41 (3):448–52. https://doi.org/10.1080/02723638.2020.1721055

Sadowski, J., and R. Bendor. 2019. "Selling Smartness: Corporate Narratives and the Smart City as a Sociotechnical Imaginary." *Science, Technology, & Human Values* 44 (3): 540–63. https://doi.org/10.1177/0162243918806061

Sadowski, J., and F. Pasquale. 2015. "The Spectrum of Control: A Social Theory of the Smart City." *First Monday* 20 (7). https://doi.org/10.5210/fm.v20i7.5903

Scholz, T. 2017. *Uberworked and Underpaid: How Workers Are Disrupting the Digital Economy*. Cambridge, MA: Polity Press.

Srnicek, N. 2017. *Platform Capitalism*. Cambridge: John Wiley & Sons.

Stengers, I. 2005. "The Cosmopolitical Proposal." In *Making Things Public: Atmospheres of Democracy*, ed. B. Latour and P. Weibel, 994–1004. Cambridge, MA: MIT Press.

Tironi, M., and C. Albornoz. 2021. "The Circulation of the Smart City Imaginary in the Chilean Context: A Case Study of a Collaborative Platform for Governing Security." In *Smart Cities for Technological and Social Innovation*, ed. H.M. Kim, S. Sabri, and A. Kent, 195–215. London: Academic Press.

Tironi, M., and M. Valderrama. 2018. "Acknowledging the Idiot in the Smart City: Experimentation and Citizenship in the Making of a Low-Carbon District in Santiago de Chile." In *Inside Smart Cities: Place, Politics and Urban Innovation*, ed. A. Karvonen, F. Cugurullo, and F. Caprotti, 183–99. New York: Routledge.

Uber. 2020. "Guías comunitarias de Uber Latinoamérica y el Caribe." Online at https://www.uber.com/legal/es/document/?name=general-community-guidelines&country=chile&lang=es

Vanolo, A. 2014. "Smartmentality: The Smart City as Disciplinary Strategy." *Urban Studies* 51 (5): 883–98. https://doi.org/10.1177/0042098013494427

16 Does Formalization Make a City Smarter? Towards Post-Elitist Smart Cities

ALAN SMART

This chapter brings together two bodies of research on development strategies rarely considered together: smart city strategies and the formalization of informality. Both are common for cities that want to be seen to be at the cutting edge of urban fashion. Both strategies have also been criticized as top-down and corporate friendly initiatives that undermine local vernacular practices (Greenfield 2013). Formalization has been criticized as counterproductive or unnecessary (Varley 2002). Corporate approaches to smart city strategies emphasize benefits from embedding sensors and other technologies in cities. Does converting informality into formal property and other institutions make cities smarter?

In many smart city strategies in the global South, informality is treated as an obstacle to modernization (Datta 2015); in the global North, it is usually ignored (Sassen 2011; Sennett 2012). By informality, I mean practices that do not follow the rules applicable to that domain (statistical reporting, paying taxes and social security, zoning, etc.), but where the good or service is not itself illegal – it involves housing or cooked food, rather than heroin or contract killing. Despite the skirting of regulation, ample studies show that informal practices are better than formal institutions at meeting many needs of citizens, such as childcare. Cities are frequently ineffective at meeting the needs of low-income people, and urban policies often make things worse (see Datta, in this volume).

Proponents of smart cities usually beg the question of what makes a city smart, and neglect forms of intelligence that do not involve sophisticated technology controlled by political, technical, and corporate elites. For example, making traffic flow more smoothly in a sprawling, auto-dependent urban region is a very limited concept of smartness. The Introduction to this volume suggests that "*smart* is usually taken for granted, seen as self-explanatory and as an inherently and unquestionably good value to pursue, and yet undesirable outcomes are understood to be aberrations from an otherwise positive movement." One response would be to reject *smart* as an empty or unjust descriptor and to pursue other directions to address digital (in)justice. Although there is good reason for this, I find it also useful to demand that smartness be treated seriously and in an open manner that includes all forms of urban intelligence, not just those pushed by elites (Green 2019; Mosco, in this volume).

Cities can be *smarter* – if we mean anything other than the quantity of information and communications technology – in a variety of ways, including (1) citizen engagement; (2) low-tech but effective architectural and urban design; and (3) high-tech, such as distributed cognition by studding cities with sensors used for Big Data analytics. Eradicating or formalizing informal practices with higher-tech alternatives might not make a city smarter in a broader sense. I argue that this is often the consequence of smart city strategies, whether intentionally or through neglect and the prejudice of the elite towards the apparent backwardness of everyday informality.

If informality can make it possible for people to achieve desirable goals (such as employment, adequate food, or shelter) that are not met by existing institutions, this seems to have the effect of increasing the responsiveness and intelligence of urban arrangements: informal governance makes some things better (Morris and Polese 2013). For example, in the first decade after the introduction in 1978, of China's Open Door policy, informal practices facilitated Hong Kong investment that, to a large extent, turned Guangdong into the "workshop of the world" and generated hard-currency exports that prevented a debt crisis such as that which undermined eastern European economic reforms (Smart 2010).

If nothing else, neglect of or active antagonism towards informality in smart city strategies is problematic when we consider that more than half of the work done in the world is informal, whether unpaid domestic or philanthropic work or commercial activity. In some regions, this rate reaches at least 80 per cent (Jutting and de Laiglesia 2009, 13). If smart cities are to be more than islands of privilege and connectivity, we need to consider their impact on people involved in informal activities: most of the world's population. More than simply acknowledging the need to mitigate the negative consequences of the displacement of informality, cities that are more intelligent require consideration of how to work effectively with informality – to harness its capacities, rather than struggle to extinguish it. The need is bluntly displayed by the widespread suffering caused by governmental difficulties in extending income support during the COVID-19 pandemic to those employed informally.

Informality in cities of the global South is often seen, if not always acknowledged, as something that needs to be tamed or eradicated – a vestige of the past that takes up valuable space, impedes traffic, and creates many of the problems that strategies are intended to overcome (Willis 2019). In this way, smart city strategies in the postcolonial world are part of a long genealogy of policies that see slums, street markets, and informal transport as eyesores and threats to public safety, security, and morality. Modernist urban planning usually has been directed at replacing such inefficient and destructive "traditional" informality.

Informality and Smart City Strategies

In many smart city strategies in the global South – for example, India's ambitious 100 Smart City program – informality is seen as an obstacle to modernization (Datta, in this volume; Praharaj, Han, and Hawken 2018). Actually existing smart city strategies tend to require extensive clearance and displacement of informal settlements and practices or the establishment of greenfield new cities. Yet, many studies show that informal practices are better than formal institutions at meeting many real needs of citizens (Chalana and Hou 2016). The smartness that this chapter supports is *post-elitist* – inclusive of diverse forms of local knowledge, rather than just sophisticated technology – and *posthumanist* – rejecting humanist assumptions of anthropocentrism and progress through the application of rationality and technology to manage society (Smart and Smart 2017a).

Cities are frequently not effective at meeting the needs of low-income people. Urban policies often make things worse, as Wilson and Keil (2008) and Peck (2005) have demonstrated for creative city strategies. Institutions frequently fail to meet the diverse needs of low-income individuals, families, and communities. Anti-poor consequences even might be intended: a feature rather than a bug. Elitist smart city strategies that increase or fail to decrease inequality are far from wise: there is clear evidence that inequality is a major factor in the social determinants of health and other social problems, creating pressure for government spending to correct problems that might have been prevented with less polarizing policies. Even when governments make genuine efforts at social justice, failures are common.

The question of whether formalization tends to make a city smarter turns out to be a difficult one. The answer is not immediately apparent and requires a new look at assumptions about what is *smart* about a smart city. In the two decades before the financial crisis that began in 2008, investment banks hired many smart people. The regulatory system changed to allow new forms of finance, and highly skilled personnel were paid handsomely to develop new forms of collaborative technology: financial engineering, particularly in the form of collateralized debt, which bundled together loans of variable quality. Did this make the financial system smarter? Would doing the equivalent to our urban governance and regulatory systems make our cities smarter? Vanolo (2014, 892) warns against the tendency to accept the widespread idea that "technological networks and governmental practices will automatically guarantee better cities, regardless, for example of the development trajectories of local societies." This worry particularly applies to the ways in which the developmental influence of informality on the outcomes of smartification is almost completely neglected. It is likely to be destructive again.

If we do not know what a truly smarter city would be, how can we avoid setting up another situation where people are financially or politically induced to take on extra risk, underinvesting in security, and generally creating a technological failure waiting to happen? (Curran and Smart 2021). How can we make smart cities that will not take us in stupid directions? We need not just to bring in citizen participation, but also to address human needs, as well as the needs of our companion species. Trying to eradicate urban informality through smart technologies – in India, for example, by linking together biometric ID, currency digitization, and value-added taxes through information technology – is only a wise thing to do if it increases the capacities of cities to meet human and environmental needs.

Kristian Hoelscher (2016, 29) argues that India's smart cities agenda is a response "to the aspirational desire to 'escape' urban informality." A key way to do this is the use of greenfield sites, as opposed to built-up urban areas that need to be redeveloped or have new smart infrastructure installed within the existing urban fabric. Rebranded as a smart city, Dholera, for example, "turns its back to the challenges of existing Indian cities struggling with pollution, traffic congestion and slums" (Datta 2015, 4). India increasingly "relies on a 'bypass' approach to urbanisation instead of the urban regeneration approach one finds in western cities" (Bhattacharya and Sanyal 2011, 41). Bypass urbanization recognizes the ability of citizens invested in vernacular urban forms to impede top-down urban plans and produce hybrids that might be less efficient but also less malevolent.

Formalization does not necessarily make a city smarter, even when it uses smart city technology to convert or eradicate informality. There might be circumstances when formalization will make cities smarter – demonstrating so will be challenging, however, given the dearth of information about the impact of smart city projects on people participating in informal urban activities. Before we can venture general conclusions, more empirical work needs to be done, although case studies are suggestive. There are different approaches to formalization. There are also radically different kinds of informalities, all of which require different techniques in order to monitor and regulate them. They can be as distinct as solidarity and reciprocity within the family, corporate tax evasion, and officials squeezing market traders for bribes to look the other way. Future work should look at the varying effects on net urban intelligence of different kinds of formalization programs.

Defining the Nature of the Smartness of Smart Cities

It seems plausible that, if a city is densely interconnected with sophisticated technology, if this meshwork of technical connections is used by the most highly educated and highly paid professionals, and if vast amounts of research and innovation are applied to increasing the efficiency and productivity of this system, a city should become smarter. Yet, these criteria applied to the US financial system before the 2007 financial crisis, and particularly to the financial engineering of derivatives that contributed to the problem (Aalbers and Christophers 2014). The lesson from this cautionary tale, insufficiently considered by proponents of smart city strategies, is that smart hardware used by smart people might not lead to wiser approaches to development and regulation. Deregulation and misguided incentives, which rewarded individuals and corporations for taking on greater risk, created a tragedy of the commons, where even those who knew the system was moving closer to disaster had incentives to take profits before the house of cards collapsed.

Alberto Vanolo (2014, 864) suggests that the term smart city is "basically an evocative slogan lacking a well-defined conceptual core," which "is actually assembled, developed, filled with meaning and implemented by policy-makers." In a review of definitions, Albino, Berardi, and Dangelico (2015, 8) reveal a "cacophony" of distinct meanings, overlapping with analogous terms "such as digital, intelligent, virtual, or ubiquitous city" (focusing on technology) and creative or knowledge cities (focusing on people). The fundamental question of what makes a city smarter is left hanging. I prefer definitions that leave open the possibility that the deployment of sophisticated digital technology could produce a city less capable of doing *smart* things, and thus focus more on the outcomes than on the means or goals.

Associated with the problem of defining smartness is that of measurement. Again, it is usually easier to provide numbers for hard domains than for soft ones. Considering the possibility that informality might make a city smarter is particularly challenging because it is the failure to follow governmental regulations, including reporting to statistical agencies, that makes a practice informal.

One of the more succinct and useful definitions of smart city strategies is "the capacity of clever people to generate clever solutions to urban problems," including, but not restricted to, the use of smart technology (Albino, Berardi, and Dangelico 2015, 12). Amin and Thrift (2007) would suggest, however, that this needs to be expanded to include the whole range of intelligences at work in a more-than-human city, including other species (for example, reincorporating wetlands into drainage management, therapy animals, sniffer dogs for COVID-19, and so on). Smart cities are fundamentally about the use of more-than-human capacities to enable doing things better or doing novel things, but if we only consider machines and sensors as possible allies in this endeavour, we will be neglecting some of the most powerful ways in which human capacities have been enhanced throughout our past, such as alliances with other species and adjusting our microbiomes (Smart and Smart 2017b; Tironi and Albornoz, in this volume). Cities have always been smart through the entanglement of humans with the non-human technologies that make them possible.

Neglecting the legacy of good low-tech urban design in places like Masdar and then adding high-tech thermostats to poorly designed buildings that need massive amounts of air-conditioning to make life possible and Big Data–driven traffic management to urban sprawl relying on private vehicles is a travesty of any meaning of urban smartness. The absence of discussion of technological innovations such as redesigning the microbiome and pandemic surveillance in most smart city strategies suggests a regrettable anthropocentric blindness to how we share our cities with agencies and intelligences other than those the elite control and deploy. The truly smart cities, and societies, that we need to survive the next century will have to find ways to harness the full collaborative energies of all the intelligences bundled together in our more-than-human settlements. Houston's ranking by the IESE Business School as the world's thirty-first-smartest city seems belied by the destruction Hurricane Harvey unleashed on a poorly planned and sprawling city that has destroyed most of the wetlands that could have absorbed much of the catastrophic rain the city received.

The Impact of Formalization: Towards Smart Formalization?

My work on the intersection between smart city strategies and informality began with concern for the process and consequences of formalizing informality in Hong Kong. Although the technologies Hong Kong deployed in its largely successful attempt since the 1980s to minimize open informality such as squatting and street vending were mostly analogue and traditional, such as registration, licensing, and mapping, it attuned me to consider the complex interactions that result. These results cannot be read off from the goals of the policies. These conclusions are applicable as well to more technologically sophisticated interventions.

Technologies, both hard technologies such as embedded GPS-enabled chips and soft technologies such as tax systems, have been and are being deployed in myriad ways to try to formalize informality. Formalization is pursued either through pushing people into conformity with unmodified regulatory systems (eradicating informality) or by modifying economic and urban governance to incorporate informal practices within those systems (regularizing informality), such as by titling squatter dwellings or allowing mobile street traders.

Various institutions argue that reduction of the informal economy would increase tax revenues and avoid unproductive use of various assets. Estimates suggest that, if every country reduced its informal economy by 10 per cent, they would collect $1 trillion in extra tax each year. Scholars are divided, however, between those who hope to unleash the energy of informal firms held back by government regulations and those who see informal firms as parasites unfairly competing with law-abiding formal firms (La Porta and Shleifer 2014). From another perspective, informal practices can take four different relations to formal institutions. If formal institutions are reasonably effective, they can be *complementary* (filling gaps) or *accommodating* (modifying the intended effect of formal rules without directly violating them). In contexts of ineffective formal institutions, informality might be *competing* (ignoring or violating formal rules) or *substitutive* (with aims compatible with the goals of formal institutions that have failed to achieve them) (Horak and Restel 2016). The nature of formal institutions and the character of particular informal practices dialectically determine the outcome of the formalization of informality. If informality is primarily complementary or substitutive, formalization might have "stupefying" consequences (Sennett 2012). If, however, formalization is undertaken in ways that improve formal institutions themselves, rather than simply displacing or suppressing informal practices, then formalization might make the city smarter. Such positive outcomes are more likely when informality is competing or substitutive – for example, when corruption leads to the use of substandard materials in infrastructure projects. We need to consider a wide variety of trajectories of formalization programs and their implications for urban intelligence.

In policy circles, De Soto's prescription for curing poverty by legalizing the informal assets of the poor has been influential (De Soto 2000; Varley 2002). Formalizing informality has emerged as a key development strategy promoted by international agencies. The World Bank has been particularly active, along with over forty other global institutions – for example, UN-Habitat established the Global Campaign for Secure Tenure to pursue the Millennium Development Goals' target on slums (Durand-Lasserve 2006). Formalization programs, particularly titling of squatter property, have been widely adopted around the world. Yet, despite widespread acceptance of De Soto's program for formalization, evidence for the consequences of formalization programs is weak and often neglected (Smart 2010, 2020).

Does formalization of informality achieve positive outcomes, or does informal vitality rely on retaining the advantages of operating in ways distinct from the formal sectors of the economy? What kind of differences exist between distinct approaches to formalization? Ascensão (2016) finds that technologically sophisticated programs in Portugal in the twenty-first century actually have had less progressive and inclusive outcomes than projects undertaken to legalize squatter settlements in the 1970s. Should we try to bring the sectors together by liberalizing formal regulations, or by extending existing regulatory practices to fields currently operating under distinct circumstances? Managing the intersection between formality and informality might be of considerable importance in finding ways not only to create jobs for swelling urban populations in the global South, but also to provide affordable housing and services for them. The tendency of much middle-range formality – most non-elite architecture, streetscapes, and retail outlets – to bland uniformity suggests that keeping the strengths of informality might be crucial to optimizing the diversity, vitality, and excitement of ordinary cities and ordinary neighbourhoods (Chalana and Hou 2016).

Hong Kong's eradication of much of its earlier persistent informality offers lessons on what conditions make this possible, and whether or not it is desirable. One result is that the closing of informal employment and housing opportunities might have contributed to Hong Kong's rapid increase in inequality to be one of the highest among rich cities. How much smarter Hong Kong is at present than during the 1970s, when informality was rampant, is a difficult question to answer. It would be impossible to operationalize a measurement of how much a city's intelligence has changed or to identify the relative importance of the causes of these changes without a clear definition of what is smart about smart cities.

"Smart formalization" can be understood in two distinct ways. First, it might concern only the application of smart technologies to curb or transform informal practices. This is consistent with definitions of smart cities restricted to the intensity of the embedding of sensors and the sophistication of data analysis and such like. Second, however, it could be seen more positively as the formalization of informal practices while maintaining the beneficial aspects of informality. Reducing the damaging dimensions – pollution, congestion, exploitative labour conditions – of how street markets and informal settlements have so far been operated could allow for win-win results. This second understanding is more consistent with a definition of smartness as something with consequences for all city residents, fostering a city that meets more genuine needs with less destructive unintended consequences and misuse of scarce resources of capital, social capital, and environmental assets. I believe we should look towards non-elitist and vernacular solutions and resources in our efforts to deploy rapidly changing technology in ways that improve urban futures for all. If government and its corporate partners in smart city strategies cannot extend their vision statements to include vital everyday requirements and desires of their inhabitants, perhaps we can hope that the creativity and inventiveness with which poor people in poor cities have adopted and adapted technologies such as cell phones will appropriate and transform such elitist technological agendas (Odendaal 2014). But that is a thin thread of hope in a context of collaborative agendas between political, economic, and technical elites, where the trade-offs are made mostly at the expense of those who are not at the table.

Conclusions

I have argued that smart city strategies need to pay more attention to informality if outcomes are to improve the situation for the majority of urbanites. Although citizen participation might go some way towards addressing the elitist imbalance, it also might be token and stage managed. As work on formalization has found, the process of incorporating vernacular practices can destroy the vitality and distinctive outcomes intended to be protected (Ascensão 2016). Informality is just one example of the sources of local knowledge and problem solving that are neglected due to the elitism and high modernism/humanism of smart city strategies (McFarlane and Söderström 2017).

A consensus definition is needed for the fuller development of smart city research, despite the danger of wearisome semantic debates. The stakes are too great to leave implicit assumptions in place to direct smart city strategies, especially since these assumptions almost certainly will be those of the economic, political, and technological elites that promote them (see Mosco, in this volume). For my own research question, whether formalization makes a city smarter cannot really be answered without some kind of consensus on the core meaning of *smart*. I suspect that this definitional need also applies to other efforts to compare the effects of distinct configurations of smart city strategies, and it certainly needs consideration if we are to think about incorporating non-human intelligences, organic as well as artificial, into urban management. While at present *smart* appears to be an empty signifier (Burns, Fast, and Mackinnon, in this volume), the dominance of the smart city strategies approach means rejecting the (potential) utility of the signifier, if we take it on with a rigorous, open assessment of how cities can become more intelligent than they are now (Mosco, in this volume).

We cannot measure smartness based solely on technological inputs, because the outcome of high-technology strategies can be counterproductive, if not "stupid." A useful definition should focus on either the process or the outcomes. My enquiry supports a preference for an outcomes-based definition. At a minimum, a smart city should enhance the livelihood of the vast majority of its population, not just that of the elite. A more environmentally sustainable city might do this, but perhaps not if it is accomplished at the expense of the less powerful – for example, through the closing of small workshops that struggle to meet costly compliance procedures. A key need for our research is a genuine effort to explore the outcomes of smart city strategies for *all* those who reside in and are entangled with the cities where the projects are undertaken. As Ezra Ho (2017) shows for efforts to provide smart elder care in Singapore, those who end up actually doing the work tend to be transnational workers and politically marginalized single females, and if their experiences are not incorporated into the designs, failure is much more likely. I agree with McFarlane and Söderström (2017, 15) that we need to work towards an "alternative knowledge-intensive rather than technology-intensive" smart urbanism, and to do this requires building on the initiatives of activists from marginalized communities. An inclusive social justice needs to engage not just with the distribution of negative effects of smart city strategies (Curran and Smart 2021), but also with non-elite capabilities and livelihoods that might be extinguished through ignorance and disdain by planners.

Although informality is frequently seen as a source of problems for urban planners, there is increasing agreement that "formal institutions are seldom completely effective" and that coordination between formal and informal institutions can help to optimize governance approaches (Horak and Restel 2016, 9). The digital technologies at the core of smart city strategies can be used to undermine informality, but there is no reason they could not be used to foster the positive elements of informal practices while discouraging the negative elements, such as congestion or higher pollution intensity. The interaction (if any) between the experts and the people who are supposed to benefit is crucial here.

I end with a few questions to ask if we want to improve cities by building just and effective informational interfaces with informal situations and actors. First, if technocratic smart city strategies increase the prevalence of formality at the expense of those involved with informality, are there better, non-technocratic approaches? Second, what are the responses of informal actors to the implementation of smart city strategies, and do they result in negative- or positive-sum outcomes? Third, could we make a city smarter by working with, rather than against, informal practices? Fourth, which type of smart city strategy would be better suited for such a task? Fifth, how do smart city strategies affect people outside the political boundaries of the city? Sixth, what methods are most effective for answering these questions? We need to go beyond simply reading off, positively or critically, the proposals of smart city strategists. Ethnographic perspectives that study how people interact with such initiatives or how programs are designed in the first place are as yet quite scarce, but greatly needed.

REFERENCES

Aalbers, M.B., and B. Christophers. 2014. "Centring Housing in Political Economy." *Housing, Theory and Society* 31 (4): 373–94. https://doi.org/10.1080/14036096.2014.947082

Albino, V., U. Berardi, and R.M. Dangelico. 2015. "Smart Cities: Definitions, Dimensions, Performance, and Initiatives." *Journal of Urban Technology* 22 (1): 3–21. https://doi.org/10.1080/10630732.2014.942092

Amin, A., and N. Thrift. 2017. *Seeing Like a City*. Cambridge, MA: Polity Press.

Ascensão, E. 2016. "Interfaces of Informality: When Experts Meet Informal Settlers." *City* 20 (4): 563–80. https://doi.org/10.1080/13604813.2016.1193337

Bhattacharya, R., and K. Sanyal. 2011. "Bypassing the Squalor: New Towns, Immaterial Labour and Exclusion in Post-Colonial Urbanisation." *Economic and Political Weekly* 46 (31): 41–8. http://www.jstor.org/stable/23017875

Chalana, M., and J. Hou, eds. 2016. *Messy Urbanism: Understanding the "Other" Cities of Asia*. Hong Kong: Hong Kong University Press.

Curran, D., and A. Smart. 2021. "Data-driven Governance, Smart Urbanism and Risk-Class Inequalities: Security and Social Credit in China." *Urban Studies* 58 (3) 487–506. https://doi.org/10.1177/0042098020927855

Datta, A. 2015. "New Urban Utopias of Postcolonial India: 'Entrepreneurial Urbanization' in Dholera Smart City, Gujarat." *Dialogues in Human Geography* 5 (1): 3–22. https://doi.org/10.1177/2043820614565748

De Soto, H. 2000. *The Mystery of Capital: Why Capitalism Triumphs in the West and Fails Everywhere Else*. New York: Basic Books.

Durand-Lasserve, A. 2006. "Informal Settlements and the Millennium Development Goals: Global Policy Debates on Property Ownership and Security of Tenure." *Global Urban Development* 2 (1): 1–15. Online at https://www.globalurban.org/GUDMag06Vol2Iss1/Durand-Lasserve%20PDF.pdf

Green, B. 2019. *The Smart Enough City: Putting Technology in Its Place to Reclaim Our Urban Future*. Cambridge, MA: MIT Press.

Greenfield, A. 2013. *Against the Smart City*. New York: Do Projects.

Ho, E. 2017. "Smart Subjects for a Smart Nation? Governing (Smart)Mentalities in Singapore." *Urban Studies* 54 (13): 3101–18. https://doi.org/10.1177/0042098016664305

Hoelscher, K. 2016. "The Evolution of the Smart Cities Agenda in India." *International Area Studies Review* 19 (1): 28–44. https://doi.org/10.1177/2233865916632089

Horak, S., and K. Restel. 2016. "A Dynamic Typology of Informal Institutions: Learning from the Case of Guanxi." *Management and Organization Review* 12 (3): 1–22. https://doi.org/10.1017/mor.2015.51

Jutting, J., and J.R. de Laiglesia, eds. 2009. *Is Informal Normal? Towards More and Better Jobs in Developing Countries*. Paris: OECD Publishing.

La Porta, R., and A. Shleifer. 2014. "Informality and Development." *Journal of Economic Perspectives* 28 (3): 109–26. https://www.aeaweb.org/articles?id=10.1257/jep.28.3.109

McFarlane, C., and O. Söderström. 2017. "On Alternative Smart Cities: From a Technology-Intensive to a Knowledge-Intensive Smart Urbanism." *City* 21 (3–4): 1–17. https://doi.org/10.1080/13604813.2017.1327166

Morris, J., and A. Polese. 2013. "Introduction: Informality – Enduring Practices, Entwined Livelihoods." In *The Informal Post-Socialist Economy: Embedded Practices and Livelihoods*, ed. J. Morris and A. Polese, 1–19. New York: Routledge.

Odendaal, N. 2014. "Space Matters: The Relational Power of Mobile Technologies." *urbe: Revista Brasileira de Gestão Urbana* 6 (1): 31–45. https://doi.org/10.7213/urbe.06.001.SE02

Peck, J. 2005. "Struggling with the Creative Class." *International Journal of Urban and Regional Research* 29 (4): 740–70. https://doi.org/10.1111/j.1468-2427.2005.00620.x

Praharaj, S., J.H. Han, and S. Hawken. 2018. "Urban Innovation through Policy Integration: Critical Perspectives from 100 Smart Cities Mission in India." *City, Culture and Society* 12: 35–43. https://doi.org/10.1016/j.ccs.2017.06.004

Sassen, S. 2011. "Open Source Urbanism." *Domus*, 29 June 2011, online at https://www.domusweb.it/en/opinion/2011/06/29/open-source-urbanism.html

Sennett, R. 2012. "The Stupefying Smart City." LSE Cities. Online at https://lsecities.net/media/objects/articles/the-stupefying-smart-city/en-gb/

Smart, A. 2010. "The Strength of Property Rights, Prospects for the Disadvantaged, and Constraints on the Actions of the Politically Powerful in Hong Kong and China." In *Marginalization in Urban China: Comparative Perspectives*, ed. F. Wu and C. Webster, 90–111. New York: Palgrave Macmillan.

Smart, A. 2020. "Squatter Housing." In *Oxford Research Encyclopedia of Anthropology*. Oxford: Oxford University Press. https://doi.org/10.1093/acrefore/9780190854584.013.222

Smart, A., and J. Smart. 2017a. "Formalization as Confinement in Colonial Hong Kong." *International Sociology* 32 (4): 437–53. https://doi.org/10.1177/0268580917701603

Smart, A., and J. Smart. 2017b. *Posthumanism: Anthropological Insights*. Toronto: University of Toronto Press.

Vanolo, A. 2014. "Smartmentality: The Smart City as Disciplinary Strategy." *Urban Studies* 51 (5) 883–98. https://doi.org/10.1177/0042098013494427

Varley, A. 2002. "Private or Public: Debating the Meaning of Tenure Legalization." *International Journal of Urban and Regional Research* 26 (3): 449–62. https://doi.org/10.1111/1468-2427.00392

Willis, K.S. 2019. "Whose Right to the Smart City?" In *The Right to the Smart City*, ed. P. Cardullo, C. di Feliciantonio, and R. Kitchin, 27–41. Bingley, UK: Emerald Publishing.

Wilson, D., and R. Keil. 2008. "The Real Creative Class." *Social & Cultural Geography* 9 (8): 841–7. https://doi.org/10.1080/14649360802441473

17 The Smart City and COVID-19: New Digital Divides amid Hyperconnectivity

ALAN WIIG, HAMIL PEARSALL, AND MICHELE MASUCCI

Digital Natives in the Smart City

As the first generation "born digital" in the smart city (Palfrey and Gasser 2016), youth today have significant technology skills and computer literacy. Throughout the 2010s, this had the potential to translate into educational advancement and work opportunities in the high-tech, information-focused economy. With social distancing and self-quarantine due to the COVID-19 outbreak beginning in spring 2020, the necessity of both digital connectivity and high-tech competencies became apparent. Whether pursuing remote education or telework while sheltering at home, people trying to cope with the pandemic accelerated, in a sense, the uptake of efforts to digitize the city and its economy. The pandemic, however, simultaneously revealed and reinforced persistent technological inequities (Iivari, Sharma, and Ventä-Olkkonen 2020). Across the United States, many elementary and secondary school students were unable to participate in online, remote learning because they did not have a computer at home or a reliable Internet connection (MacGillis 2020). They also faced other, substantial social or economic barriers such as a lack of housing stability or the need for an adult on hand to assist with educational technology. Beyond the virtual classroom, jobs for youth all but disappeared as low-wage, service-oriented opportunities were unavailable due to the requirements of physical distancing (Fry and Barroso 2020). We argue that, because they will both craft and inhabit the so-called smart city into the future, the experience of youth is indicative of the urban condition in the smart city (Masucci, Pearsall, and Wiig 2020).

In this chapter, we suggest that, with COVID-19, in the short term and even into the long term, these educational and economic issues are likely to mutate and multiply: as work becomes home-bound, particularly for white-collar workers (Iivari, Sharma, and Ventä-Olkkonen 2020), access to digital connectivity – and sufficient domestic spaces in which to work and learn online – will only increase in importance. Combined with the shifting challenges of navigating social distancing and the geographic dislocation required to mitigate COVID-19 transmission, the new digital divide manifests as the urban condition in the smart city. It is a divide shaped both by the lack of place-based resources and the fundamental social and racial inequities that moderate digital connectivity and access to the benefits of the smart city more generally. Simultaneously, though, the patchwork of digital connectivity that failed to deliver robust academic continuity during the pandemic provided connections via social media to protest police violence and mobilize youth as part of a nationwide antiracist agenda. We expand on these points throughout this chapter, in alignment with Ayona Datta's dialogue in Part Four of this volume, where she notes that "smart [is] made up of an encounter between digital and material space," enacting "hybrid moments" where new politics can emerge, including demands for racial justice and that Black Lives Matter.

To make this argument, we draw on research observations from a six-week, summer 2020 educational program involving primarily youth of colour from public high schools in Philadelphia, Pennsylvania. The program, titled Building Information Technology Skills (BITS), strives to ground problem-based learning with the critical thinking skills necessary to build civic leadership and affect positive, neighbourhood-based change (see Masucci, Organ, and Wiig 2016 for a full description of this program). BITS pursues these objectives by engaging participants to use spatial information and digital technology skills to evaluate the problems they define. This long-standing program is housed at Temple University, a large, urban, public university just north of downtown Philadelphia. Our prior work found that most youth do not participate in the BITS program as a means of connecting to high-tech fields per se; instead, they see increasing their tech skills as crucial for job security in whatever field interests them (Masucci, Pearsall, and Wiig 2020). The program thus aims to bridge increased self-efficacy with the acquisition of digital skills. Youth typically identify and address community problems they are familiar with, such as crime, homelessness and drug abuse, the quality of their urban environment, the health of their families and communities, and issues of access to public transportation. They work throughout the summer to propose solutions to these civic issues and to develop their understanding of educational opportunities and careers related to their interests.

This chapter proceeds by first situating our work in the scholarly literature. Next, we discuss the summer 2020 BITS program with a focus on how youth navigated the challenges of remote learning and online collaboration. We conclude by applying our observations from that summer to theorize a new digital divide emerging in the smart city.

Background: Youth and Social Justice in the Smart City

One policy trend in urban governance over the past two decades has been to draw upon digital technologies – such as municipal wireless internet service, video surveillance systems, the use of sensors in utilities networks, and software-coded processes – to support governance and decision-making processes. This so-called smart city (Shelton, Zook, and Wiig 2015) reflects the deep and transformative ways in which the goals and technologies of global corporate entities have become central to the theory and implementation of urban governance, with the touted effects of efficiency driving wholesale changes in the public sphere (Townsend 2013). Because the adoption of digital technologies is not conceived of as imbued with political motives, but rather as a neutral, efficient, and cost-effective upgrade to aging systems and outdated processes, the public often does not engage in the underlying implications connected with these technologies (Wiig 2016). This faith among politicians and policy makers in the ability of digital technologies to drive municipal improvement sidesteps the role these systems play in ignoring or even exacerbating underlying social and racial inequalities. We instead posit that there is little chance for us to envision digital technologies as a factor in driving social equity and change unless we simultaneously envision the type of social (and racial) justice we aim to achieve (Kitchin, Cardullo, and di Feliciantonio 2018).

Similarly, the geographic underpinnings of the smart city need to be unpacked. The premise that digital technologies will erase the inequalities of uneven development, distance, and fractured connectivity have yet to be realized (Gilbert and Masucci 2011, 2018). These legacies spill over into the geography of the smart city: the development of digital systems is necessarily situated in an already-existing set of divides that continues to shape the deployment and geographic impacts of new technological systems that aim to mitigate the inequalities of old systems (Shelton, Zook, and Wiig 2015). Civic resources and services such as schools, libraries, and community centres, which have helped to close the digital divide by providing people with computers, Internet access, and technical support, have never received adequate funding to flourish in the smart city (Wiig 2016). In 2020 these existing resources were removed altogether as the spaces to access them closed due to the COVID-19 pandemic, resulting in an increased digital gap for many.

When asked about addressing the problems they face in their everyday lives, youth felt strongly that smart technologies were doomed to reflect the commercial and corporate desires for efficient service delivery, privileging an entrenched, racialized class system that undermines even the most basic technological democratization efforts (Masucci, Pearsall, and Wiig 2020). Technology skills are often promoted as a way for youth to have good jobs and be part of the digital economy; but in cities like Philadelphia these skills evolve within the context of a legacy of racism that still limits equal access to technologies themselves, let alone the job opportunities or professional networks required to enter and succeed in competitive fields. Although beyond the scope of this chapter, we note that, in the wake of the police killing of George Floyd in Minneapolis in May 2020, Philadelphia's protests were violently repressed (Scott 2020), which permeated the experiences of the BITS participants in summer 2020. Similarly, protests in October 2020 in response to the Philadelphia police killing of Walter Wallace Jr after his mother's request for mental health services for her son were shaping their subsequent academic school year (Newell and Palmer 2020). These two events not only drew attention to the city's long-standing racial injustice and to the ways digital technologies might give citizens the tools to monitor conditions and advocate for systematic change (for example, the use of Facebook by Black Lives Matter Philadelphia to organize[1]); they also affected the shape of technology use paradigms among the youth of the city.

Our work has focused on examining the effects on the everyday lives of youth in Philadelphia of the implementation of a set of technological solutions to civic matters. These ostensible solutions emanated from a local government struggling to position the city to garner global attention and multinational capital. We have pointed out in prior work that the smart city is both a platform and a testbed for using technology as an integral part of strategies to rebrand the city around education, medicine, tourism, and neighbourhood revitalization (Wiig 2015; Wiig and Masucci 2020).

We focus on youth for a number of reasons. First, they are fully enmeshed in the smart city because they have no way to opt out of these technologies and, more than previous generations, they are required to be fluent in digital innovations to navigate their everyday needs. Second, as "digital natives" (Hargittai 2010), they are often also digital ambassadors or conduits for the use and deciphering of digital technologies at both the household and neighbourhood or community scale (Barrett et al. 2017; Connolly and McGuinness 2018). Although the presence of digital technologies –in particular, the "always on" Internet – has permeated their entire lives, youth must be especially efficacious with respect to technology use to navigate the complexities of the constantly reshaped services they access, such as education and health care (Masucci, Pearsall, Wiig 2020). Finally, as the newest generation of residents in a city that is still rebounding from fifty-plus-years of deindustrial urban and economic decline (Howell 2018), our work has shown the importance of understanding how youth reconcile newer and older "ground truths" (Pickles 1995). The city they inhabit, the neighbourhoods they live in, learn and work in, and travel through, do not reflect the vision of city and civic leaders of a revitalized Philadelphia, a point reflected in the statistic that it is the poorest large city in the United States (Saffron 2020).

Strategies for Connecting, Opportunities for Learning during COVID-19

In 2020, the six-week summer program ran in an entirely digital format, following a pattern that began in the spring, when Temple University and local public K-12 schools moved into a distance-learning environment. Late into the spring, Philadelphia's public schools delayed restarting remotely because of significant difficulties securing appropriate technology for students to get online – from ensuring the availability of laptops (Graham 2020) to providing sites to access high-speed wireless Internet.[2] Other than at schools, municipalities typically provide free access to the Internet as well as digital literacy training through public libraries, making these spaces and services the foundation of a just smart city, as Teresa Abbruzzese and Brandon Hillier articulate in this volume. Needless to say, because of the pandemic, Philadelphia's libraries closed as well.

For 2020, this pivot of a summer program that already, inherently, had a strong focus on fostering information technology and digital inclusion skills led to a number of methodological and pedagogical changes. The timeline for implementing lessons and project deliverables went from six twenty-hour work weeks to two concentrated weeks. The focus on project-based learning remained, but was integrated with the requirement to use an online platform by the city-wide provider. This requirement eliminated key elements of youths' experience, resulting in the elimination of in-person group work, hands-on lab experiences, and fieldwork components. The digital confines of the program also led to a hyperfocus on group and individual mentoring and skills building that could be achieved remotely. The instructional approach created a work environment in an "out of school" setting in each youth's home, through the smartphone, tablet, or laptop screen on which the youth relied. Out of this shift, the program aimed to create an online collaborative culture where, as students gained proficiency with platforms designed for collaboration, such as Google Docs, they would be encouraged to work more closely with one another as a spatially dispersed, digitally connected team. This new program model aimed to increase youths' self-efficacy in the use of digital skills and to provide them a venue for extending those skills via new projects in newly relevant online platforms.

BITS program participants are placed in teams and trained as interns on specific projects. They are coached to develop and apply technology skills to their depiction and analysis of problems defined by their team mentors and faculty coaches. The skills they gain are meant to help them to stand out in college applications, to support their educational goals, and to enable them to attain future job security and general upward economic mobility. Our prior work has shown, however, that youth considered these same skills as *unable* to help them gain the assumed benefits of living in, and working in, a smart city in the present day (Masucci, Pearsall, and Wiig 2020). In 2020 the smart city was reduced to the confines of their home, even their bedroom when that was the only quiet and private space where they could focus on work. Their ability to apply these skills was limited by their access to digital technology and infrastructure well as other economic and social factors, such as the need to care for siblings, caregivers who were also essential workers and unavailable to help them with their schoolwork, and even educators who were not able to teach because they had not received adequate training on teaching in a virtual environment or had to care for their own families. The disconnect between the future promise of the smart city and the challenges of gaining the skills to engage in that city became even more apparent during the COVID-19 pandemic.

During the pandemic summer, youth interns connected to the BITS virtual program using home computers, personal laptops, smartphones, and even their home phone lines. These multiple modes of connection highlighted how digital technology, including access to connected devices and the Internet, mediates access to education in a virtual learning environment. Although some interns were able to connect to video calls relatively seamlessly, others had low bandwidth and had to participate asynchronously. Those who participated on their smartphones were required to complete all of the activities on their phones, from video calls to typing essays. Because some activities were designed to be completed on a computer with high-speed Internet access, not all youth were able to engage fully with the activities. Conversations with interns about how their Internet access had changed during the pandemic revealed that most previously had accessed it at school, community centres, libraries, and private businesses, and how the need for social distancing had limited Internet access to inside their homes. Those with only limited Internet access at home now lacked the community resources they might have used before the pandemic, and those without devices suitable for virtual work and/or strong Internet access at home became limited in their ability to participate.

We observed, somewhat paradoxically, that the mobile technologies that are supported through ubiquitous connectivity were increasingly anchoring our interns to their homes and, within the home, to most interns' bedrooms. Creating a learning environment at home became part of the summer curriculum; yet, simultaneously, these views into domestic learning environments raised questions about the privacy of interns, as well as the role of technology in revealing stark economic differences among households. In a brick-and-mortar, traditional learning environment, interns arrive with only a backpack, and their home circumstances are not immediately evident or visible. In a virtual learning environment, video calls allow for repeat views of the interns in their bedrooms, which acted as makeshift workspaces for remote school and their summer internship, with their personal effects and decorations on view. Because interns could see themselves in the video, they were also editing themselves and their environments to show a particular image of themselves and their settings. Their environments, however, could not be arranged fully. Their interactions with family members were captured in the calls, as well as the slipping view of interns' faces as their hands became tired of holding their devices for the duration of meetings. Although the interns were coached to be professional and to design their workspaces to be conducive to learning, not all interns were able to create such spaces or to maintain these spaces for the full day.

Despite the challenges that participants faced, their work and final projects tell a story of adaptation and resilience. Each group made the most of the compressed two-week program to create a final project that engaged deeply with their concerns today, such as how to make life better during the COVID-19 pandemic. Interns drew on different skill sets that they had developed prior to the program, and were self-efficacious in knowing the right approach for their task at hand. They also figured out how to work together in a virtual environment in a short period, and quickly advanced their skills using collaborative technologies, which allowed them to troubleshoot technical issues for different team members. For instance, if one team member was unable to display the presentation, others were ready to jump in and show it on their device. Although resolving technical issues became an everyday occurrence, it also taught interns how to be flexible, creative, and patient. Regardless of the successes and opportunities the interns might have through their ongoing education, the experience of those with whom we worked indicated that, in the everyday reality of the pandemic smart city, digital connection and access to appropriate computers mattered, but so did place. Without adequate space in which people can work and learn, the potential of the smart city is constrained severely, even to the point of irrelevance, a critique we expand on below.

Discussion and Conclusion: The Paradox of the Smart City

Scholarship on the post-COVID-19 smart city undoubtedly will continue, but here we offer early observations of the challenges youth are already facing, stemming from a new digital divide. A first glimpse of this divide comes in a recent study of "teleworkability" in the European context, showing that work-from-home requirements imposed during the pandemic affected an already-unequal remote work structure, with large variations both between and within different employment sectors. The study also found that white-collar workers had much more flexibility to telework (Iivari, Sharma, and Ventä-Olkkonen 2020). Our findings in Philadelphia reflect a similar situation.

As traditionally understood, efforts to reduce the digital divide and foster digital inclusion have focused on increasing access to computers and the Internet, as well as teaching the skills necessary to use these technologies (Servon 2002). Indeed, the BITS Program originated in the early 2000s to work towards these sorts of goals. Marginalized groups in Philadelphia's deindustrialized, disenfranchised, working-class neighbourhoods of colour struggle with lower educational outcomes and subsequent lack of employment opportunities in the smart city (Wiig 2016). From the perspective of established critiques of the digital divide, one of the contradictions here is that youth are seen as at the forefront of technological adaptation, but remain at the margin of social equity (Grant and Enyon 2017). Youth might be digital natives (Hargittai 2010) by acquiring and adopting the digital skills that are the foundation of the smart city (Jones et al. 2010), yet our prior work shows that youth do not feel their voices or their communities are included in smart city proposals (Masucci, Pearsall, and Wiig 2020). The unevenness of access, of device ownership, and of technological efficacy speaks to the continued need for digital inclusion efforts, but also to the rise of a new digital divide, one that has a spatial dimension, even if that spatiality is most notably – with the closure of public schools, libraries, community centres, and offices – the absence of a place in which to work or learn.

The transformation of cities into smart cities over the past decade has not taken into account the needs of youth. The new digital divide sharply highlights the contradiction between technologically capable youth and cities that fail to plan for them as citizens or digital users (van der Graaf 2020). Applying a social justice perspective to the ongoing digitization of the city reveals the inequities inherent in this pivot online in accessing smartphones, laptops, and tablets, affording high-speed Internet, or even having a safe, quiet space at home in which to get online. The situation has only been exacerbated by the challenges of navigating the significant spatial dislocation required to mitigate COVID-19 transmission. The ability of youth to navigate a host of infrastructures – from the aging and outmoded, material ones that undergird their lives to the most advanced digital ones that smart city development integrates into the built urban fabric – illustrates the constant paradoxes they face.

Writing in late 2020, the past year has illustrated how the grid of telecommunication infrastructure became a digital substitute for life in the smart city, even as the material city fractured into millions of socially distant units with or without the ability to access that proxy, let alone successfully navigate it. A new digital divide might be emerging that will turn the smart city into a place of isolation, especially for the most vulnerable and least resourced communities. School (and urban life more generally) became socially distanced and profoundly individualized in 2020. This isolation will have lasting personal and professional consequences for youth who cannot "scaffold" the face-to-face sociability of school into opportunities to advance their educational and employment ambitions and social desires.

The pandemic revealed the paradoxes of the smart city through absence and closure. In a remote educational setting that assumes youth and their families have excellent Internet access to do the basic things they are expected and even legally required to do (such as go to school), this new divide is evident in the patchwork of "solutions" that youth have had to navigate. Many participants in the 2020 BITS program were able to get Internet access, but they had to make an effort above and beyond to remain online for a workday, let alone having to type papers or conduct research on their smartphones, while, for instance, sitting outside in a school parking lot to connect to Wi-Fi or doing work for the program while caring for siblings. The data requirements and hardware and software expectations for their projects were not being met fully, and we suggest this will affect their formal education experiences more broadly as well. Moreover, the actions needed to get online and troubleshoot technical problems are additional, if minor, burdens compared to arriving at school and starting the day's activities with the help of a teacher. The world of opportunity the smart city was expected to bring to Philadelphia shrank to the size of a smartphone screen and remained there.

The irony or paradox of this "patchwork" smart city is that the youth we worked with continued to harness the tech platforms solely to connect within the home environment (or a component of the home environment). The municipal response to the pandemic was to close places of business, highlighting that access to the Internet is embedded within both civic and commercial establishments, rather than ubiquitously available through municipal broadband, as was once the promise of Philadelphia's early smart city (Jassem 2010). Those closed access points, combined with a lack of resources to secure fast home Internet, point to the ways in which reliance on privatized telecommunications systems is failing some of our youth. There is a public/private element to the new digital divide, where privately owned resources – computers or smartphones and, in particular, high-speed Internet connection – are necessary even to access public services such as education. Beyond this, expecting students to work productively at home is challenging when youth have lost all of the additional services that brick-and-mortar schools provide, such as hot breakfast and lunch programs, nurses and counsellors, and teachers who see when students are struggling. Even with excellent digital connectivity, these services cannot be provided in the same way, and they compromise the ability of students to participate. Place still matters when navigating the technologies required to connect to the smart city. When the corner of a bedroom or a kitchen table becomes the substitute for an office or a classroom, the setting dictates the opportunity. The space constrains how technologies are used to support learning, relationship building, and project development and completion. We found that the merger of spaces of learning with the home setting resulted in students becoming geographically isolated and disconnected both from their communities and from the more substantive technologies and infrastructures, as well as the professional social networks, they otherwise would have relied on to support their experiences.

To arrive at a viable social justice framework for the further development of an equitable and inclusive smart city, scholars must grapple with the new digital divide that extends into and cuts across the public and domestic spaces of the city itself. All of that isolation and use of technology to knit together the threads of access to education was thrown into upheaval by the dramatic events in Philadelphia in the aftermath of the George Floyd killing. Students moved from isolation to reconnecting in the streets as part of a massive, nationwide movement to address violence and advance an anti-racist, Black Lives Matter agenda. Much of the organizing was enabled by the same social media platforms and connectivity they used to mitigate their isolation from educational infrastructure and their social networks. These events so directly affected the students and the mentors who coached them that the thematic focus of substantive portions of their summer experience was transformed, and many of their final projects focused on their relationship to those protests and the ever-present threat of police violence that they perceive and experience in their communities. Their reconnection with place centred on the framing of that pivotal moment of urban protest and contestation. Youths' dramatic pivots between the isolation of the stay-at-home requirements to avoid COVID-19, engagement with the protests in Philadelphia both in person and through tracking the activities via the media, and their return to education from home during the summer are illustrative of how technology can shape such dramatically different circumstances as home-bound education and the monitoring of protests as well as serving as a conduit for the reframing of urban politics. These connected technologies ultimately served as a poor proxy for the city, even as the events youth ultimately would respond to were brought to the forefront through their access to the news and to events in Minneapolis and around the country. This digitized access became a driver of their own actions and reactions that ultimately unfolded on the streets of Philadelphia, witnessed and mediated through the technology-enabled communications they shared. One dramatic effect of this is was to illustrate how peripheral their educational use of technology remained in comparison to its other connective purposes.

ACKNOWLEDGMENTS

The authors want to thank Philadelphia Youth Network, the staff of Temple University's Office of the Vice President for Research, the coordination efforts of Jean Akingeneye and TJ Seningen, and the Temple University faculty, student mentors, and interns who worked with the BITS Program in 2020 and in previous years.

NOTES

1 See Black Lives Matter Philly, online at https://www.facebook.com/blacklivesmatterphilly
2 A local digital justice coalition sought to draw attention to these issues with a social media campaign directed at Philadelphia's mayor to document instances of students forced by the closure of their school to seek out Wi-Fi in parking lots (Philly Tech Justice 2022).

REFERENCES

Barrett, N., R. Villalba, E.L. Andrade, A. Beltran, and W.D. Evans. 2017. "Adelante Ambassadors: Using Digital Media to Facilitate Community Engagement and Risk-Prevention for Latino Youth." *Journal of Youth Development* 12 (4): 81–106. https://doi.org/10.5195/JYD.2017.513

Connolly, N., and C. McGuinness. 2018. "Towards Digital Literacy for the Active Participation and Engagement of Young People in a Digital World." In *Perspectives on Youth: Young People in a Digitalised World*, ed. Council of Europe, 77–92. Strasbourg: European Union.

Fry, R., and A. Barroso. 2020. "Amid Coronavirus Outbreak, Nearly Three-in-Ten Young People Are Neither Working Nor in School." Pew Research Center, 29 July. Online at https://www.pewresearch.org/fact-tank/2020/07/29/amid-coronavirus-outbreak-nearly-three-in-ten-young-people-are-neither-working-nor-in-school/, accessed 30 September 2020.

Gilbert, M., and M. Masucci. 2011. *Information and Communication Technology Geographies: Strategies for Bridging the Digital Divide*. Vancouver: University of British Columbia Praxis (e)Press.

Gilbert, M., and M. Masucci. 2018. "Defining the Geographic and Policy Dynamics of the Digital Divide." In *Handbook of the Changing World Language Map*, ed. S. Brunn and R. Kehrein, 1–19. Cham, Switzerland: Springer.

Graham, K. 2020. "Are Philly schools waiting too long to get students online after the coronavirus closures?" *Philadelphia Inquirer*, 15 April. Online at https://www.inquirer.com. https://www.inquirer.com/health/coronavirus/coronavirus-philly-public-schools-report-lag-remote-learning-20200415.html, accessed 5 October 2020.

Grant, L., and R. Eynon. 2017. "Digital Divides and Social Justice in Technology-Enhanced Learning." In *Technology Enhanced Learning: Research Themes*, ed. E. Duval, M. Sharples, and R. Sutherland, 157–68. Cham, Switzerland: Springer.

Hargittai, E. 2010. "Digital Na(t)ives? Variation in Internet Skills and Uses among Members of the 'Net Generation.'" *Sociological Inquiry* 80 (1): 92–113. https://doi.org/10.1111/j.1475-682X.2009.00317.x

Howell, O. 2018. "Philadelphia's Poor: Experienced from Below the Poverty Line." Philadelphia: PEW Trusts. Online at https://www.pewtrusts.org/-/media /assets/2018/09/phillypovertyreport2018.pdf, accessed 20 November 2018.

Iivari, N., S. Sharma, and L. Ventä-Olkkonen. 2020. "Digital Transformation of Everyday Life – How COVID-19 Pandemic Transformed the Basic Education of the Young Generation and Why Information Management Research Should Care?" *International Journal of Information Management* 55 (December): 102183. https:// doi.org/10.1016/j.ijinfomgt.2020.102183

Jassem, H.C. 2010. "Municipal Wi-Fi: The Coda." *Journal of Urban Technology* 17 (2): 3–20. https://doi.org/10.1080/10630732.2010.515090

Jones, C., R. Ramanau, S. Cross, and G. Healing. 2010. "Net Generation or Digital Natives: Is There a Distinct New Generation Entering University?" *Computers & Education* 54 (3):722–32. https://doi.org/10.1016/j.compedu.2009.09.022

Kitchin, R., P. Cardullo, and C. di Feliciantonio. 2018. "Citizenship, Justice and the Right to the Smart City." https://osf.io/preprints/socarxiv/b8aq5/, accessed 30 September 2020.

MacGillis, A. 2020. "The Students Left Behind by Remote Learning." *Propublica* 28 (September). Online at https://www.propublica.org/article/the-students-left-behind -by-remote-learning, accessed 30 September 2020.

Masucci, M., D. Organ, and A. Wiig. 2016. "Libraries at the Crossroads of the Digital Content Divide: Pathways for Information Continuity in a Youth-Led Geospatial Technology Program." *Journal of Map & Geography Libraries* 12 (3): 295–317. https://doi.org/10.1080/15420353.2016.1224795

Masucci, M., H. Pearsall, and A. Wiig. 2020. "The Smart City Conundrum for Social Justice: Youth Perspectives on Digital Technologies and Urban Transformations." *Annals of the American Association of Geographers* 110 (2): 476–84. https://doi.org /10.1080/24694452.2019.1617101

Newell, C., and M. Palmer. 2020. "Overnight curfew lifted in Philly as tensions over the police killing of Walter Wallace Jr. continue." *Philadelphia Enquirer*, 28 October. Online at https://fusion.inquirer.com/news/philadelphia-police-shooting-walter-wallace-jr-protests-unrest-20201028.html

Palfrey, J., and U. Gasser. 2016. *Born Digital: How Children Grow Up in a Digital Age*, rev. ed. New York: Basic Books.

Pickles, J. ed. 1995. *Ground Truth: The Social Implications of Geographic Information Systems*. New York: Guilford Press.

Philly Tech Justice. 2022. "Parking Lot Wifi." Online at https://phillytechjustice.org /campaigns/parking-lot-wifi/

Saffron, I. 2020. "How Philadelphia became the nation's poorest big city, and how we can fix it." *Philadelphia Inquirer*, 13 October. Online at https://www.inquirer .com/business/philadelphia-poverty-unemployment-racism-education -upskilling-20201013.html, accessed 14 October 2020.

Scott, I. 2020. “A Photographer on the Front Lines of Philadelphia’s Protests.” *New Yorker*, 22 June. Online at https://www.newyorker.com/magazine/2020/06/22/a-photographer-on-the-front-lines-of-philadelphias-protests, accessed 30 September 2020.

Servon, L.J. 2002. *Bridging the Digital Divide: Technology, Community, and Public Policy*. Malden, MA: Wiley-Blackwell.

Shelton, T., M. Zook, and A. Wiig. 2015. “The ‘Actually Existing Smart City.’” *Cambridge Journal of Regions, Economy and Society* 8 (1): 13–25. https://doi.org/10.1093/cjres/rsu026

Townsend, A.M. 2013. *Smart Cities: Big Data, Civic Hackers, and the Quest for a New Utopia*. New York: W.W. Norton.

van der Graaf, S., 2020. “The Right to the City in the Platform Age: Child-Friendly City and Smart City Premises in Contention.” *Information* 11 (6): 1–16. https://doi.org/10.3390/info11060285

Wiig, A. 2015. “IBM’s Smart City as Techno-Utopian Policy Mobility.” *City* 19 (2–3): 258–73. https://doi.org/10.1080/13604813.2015.1016275

Wiig, A. 2016. “The Empty Rhetoric of the Smart City: From Digital Inclusion to Economic Promotion in Philadelphia.” *Urban Geography*” 37 (4): 535–53. https://doi.org/10.1080/02723638.2015.1065686

Wiig, A., and M. Masucci. 2020. “Digital Infrastructures, Services, and Spaces: The Geography of Platform Urbanism.” In *Urban Platforms and the Future City: Transformations in Infrastructure, Governance, Knowledge, and Everyday Life*, ed. M. Hodson, J. Kasmire, A. McMeekin, J. Stehlin, and K. Ward. New York: Routledge https://doi.org/10.4324/9780429319754

18 Beyond the Digital Divide: Libraries Enabling the Just Smart City

TERESA ABBRUZZESE AND BRANDON HILLIER

Already interwoven into our private lives – through our memories of childhood, personal intellectual development, experiences of immigration, and mundane acts of parenting – public libraries have in the past few decades further expanded and diversified their programs, services, events, resources, staffing, and environmental design. This is connected to their association with an intensifying array of concerns, barely encapsulated by the following: community, democracy and civic engagement, physical and mental health, commercial opportunities, technology access, education, and childhood development. Concomitantly, the library has become a public institution sought by increasingly diverse and often vulnerable populations as a site of urban care – bolstering its extant imaginary as somewhere one can belong and be seen.

The development of the smart city, with its unapologetically neoliberal agenda, encourages the entrepreneurialization of all manner of public institutions. Public libraries are particularly affected, but also serve as important sites articulating a potential for resistance. In this chapter, we provide new insights into the role of libraries in the production of the smart city – specifically, in their capacities for enabling digital justice. Drawing on semi-structured interviews conducted between 2017 and 2019 with the Transformation Office of the City of Toronto, librarians at the Toronto Public Library (TPL; see Figure 18.1) and Seattle Public Library (SPL; see Figure 18.2), and members of the Toronto Region Board of Trade (TRBOT) – and on participation notes as an academic member of the Toronto Smart Cities Working Group (SCWG) – our chapter brings public libraries to the forefront of conversations on inclusion and equity in the smart city. We studied Toronto and Seattle since both have robust public library systems that are drawn into significant smart city initiatives. Toronto hosted the infamously now defunct Google-Sidewalk Labs project and was an entrant in the federal government's Smart Cities Challenge, while Seattle's municipal government hosts an active Innovation Advisory Council on urban technology and is a node in national smart initiatives such as the MetroLab

Figure 18.1. Toronto Public Library

Network. In each of these places, libraries are recast in smart city narratives as agents in the economy helping to bridge the digital divide. Although we problematize their techno-utopian vision, we focus on the ways the TPL and SPL facilitate digital inclusion and democratic intervention in the smart city.

Our contribution interweaves feminist urban geography engagements on the ethic of care – particularly Miriam Williams's (2017, 2020) concept of "*care-full*" *justice* – with library information science (LIS) conversations on human rights and social justice (Jaeger, Taylor, and Gorham 2015), the entrepreneurialization and shifting mandates of the library under neoliberalization (Mattern 2014, 2018; Stevenson 2009), and the role of public libraries in smart cities (Leorke, Wyatt, and McQuire 2018). This builds on existing critical urban theoretical conversations on the smart city (Angelidou 2017; Grossi and Pianezzi 2017; Kitchin 2015). We argue that everyday practices of care make the public library a crucial site of access, literacy, and social infrastructure (Klinenberg 2018). These mundane practices are important mediations in the reconfiguring of the public library in the data-driven knowledge economy.

We begin our discussion by looking at the role of public libraries in mobilizing narratives of the digital divide, conceived as a discursive strategy used to maintain funding and legitimize private partnerships with technology giants, as seen in the United States, United Kingdom, Australia, Europe, and Canada (Leorke, Wyatt, and McQuire 2018; Mattern 2014; Stevenson 2009). The narrative of the digital divide is disentangled from actually existing social inequities, revealing how the ideological tenets shaping this discourse operationalize the smart city (see Datta, in this volume, for a discussion of problematizing the concept of the digital divide). Public libraries under neoliberal conditions serve as digital agents in smart city projects – providing an entrepreneurial space

Figure 18.2. Seattle Public Library
Source: https://www.flickr.com/photos/myhsu/15094965442/. Courtesy of Ming-yen Hsu.

for self-improvement, business startups, training, and workshops – but they continue to serve the community as places of care and access for those left behind. A feminist ethic-of-care lens highlights the significance of public libraries in providing essential infrastructure in times of deepening inequality and institutional social deficits (see Listerborn, in this volume, for a feminist urban critique examining technocapitalist urban development).

We then bring these concepts together with insights gained from empirical research: on one hand, to recognize the potential for public libraries to serve as trusted entities producing more accessible and more just smart cities; and, on the other, highlighting how their traditional everyday practices of "care-full" justice are increasingly challenged by the shift in mandate and delivery of services in the digital economy.

Libraries as Digital Agents in the Smart City

Echoes of the paternalistic, top-down rational approach that shaped twentieth-century modernist design reverberate in present trends of the smart city. Urban leaders under today's neoliberal conditions – where corporate freedoms reign – are private technology developers, closely tied to the management and marketing of economic development (Greenfield 2013; Karvonen,

Cugurullo, and Caprotti 2018; Shepard 2014). To this end, the marketization of the smart city as a strategy to improve competitive positioning within the hierarchy of global intelligent cities – and as a technological approach to solving urban challenges – preoccupies local, regional, and national governments around the world. The goal of the smart city is to pursue growth and regeneration in manners that are networked, automated, and surveilled. It represents a technological evolution of the creative city in the new digital age of urban development, where the primary focus is on using information and communications technology (ICT) to regenerate and address urban values and issues. The solutions couched in a broader discourse on urban innovation are spatialized and packed as edgy and ultraconnected visions fusing the digital layer with urban development, and as scientifically rational initiatives aligned with development goals of resilience and sustainability.

The entrenchment of the public library within this vision is crucial to its entrepreneurialization in the knowledge economy. While this could be seen as an inevitable transition for libraries to remain relevant in the digital economy, Leorke, Wyatt, and McQuire (2018, 38) note how developments in Australia point to the specifically neoliberal underpinnings of this entanglement: "in talking about libraries' entanglement with the digital visions for their cities, we mean to signal their ambivalent relationship to these agendas, agendas within which they have a compromised degree of independence and agency." The alignment of the public library with the economic mandate of the smart city imaginary signals broader shifts relating to the neoliberal state, repositioning the library as a key actor in the digital economy and making a case for its continued relevance in the twenty-first century. This repositioning is integrated into a broader discourse on the digital divide consistent with the library's traditional mandates around access, literacy, and information sharing; however, within the smart city context, this discourse is used to attract public and private funding for the library to become a hub for technical skills training, startups, and facilitating digital citizenry (Leorke, Wyatt, and McQuire 2018; Mattern 2007; Stevenson 2009). Libraries' strategically positioning themselves at the forefront of smart city initiatives has elevated their status to more than just repositories for books; they are recast as key social and cultural centres in cities, as reflected in library developments in Toronto and Seattle.

More than Borrowing: Libraries as Places of Care in the Smart City

In her work on public libraries in the United States, Shannon Mattern (2014, 2018) conceptualizes libraries as *opportunity institutions*, where people can find books, vote, access the Internet, get information about public services, and work on their entrepreneurial ideas. By extension, librarians multitask as care, knowledge, and information workers – e.g., reading stories to children, teaching digital literacy and writing workshops, fielding references questions, and assisting homeless patrons. While this work acknowledges the expanding and shifting mandates of the public library as an institution, it also points to the increased responsibilities placed on librarians as key actors.

A feminist ethic of care highlights place-based practices of care in everyday life that represent a collective sense of social responsibility (Lawson 2007; Low and Iveson 2016; Williams 2020). Within this framework, the allocation of care responsibilities carries both political and moral dimensions. Extensions of this concept in feminist urban geography build on the foundational work of Joan Tronto and Bernice Fisher. According to Tronto (1993, 113), a feminist ethic of care is a "species activity that includes everything that we do to maintain, continue, and repair our 'world' so that we can live in it as well as possible." Through a critical urban justice lens, Williams (2017, 2020) advances the concept of "care-full" justice, arguing that the ethic of care should be considered alongside any proposals for the just city: "care-full justice provides a framework that is contextually sensitive to actually existing manifestations in society of injustice/justice and carelessness/care" (Williams 2017, 822). In her work, she examines how particular sites in cities, such as daycare centres and libraries, become spaces of care at particular moments and how practices of justice and care help repair the gaps and wrongs wrought by a neoliberalized world. In the smart city, it can be argued, libraries are becoming more "care-full" places, providing needed social infrastructure along with digital and educational infrastructure.

The public library is imagined and described as one of the remaining true public spaces in cities (Leckie 2004). The LIS literature focuses on different aspects of inclusion relating to access and the library's qualities as a public space, an institution of human rights and social justice (Jaeger, Taylor, and Gorham 2015), a producer of digital inclusion (Thompson et al. 2014), and a role player in social welfare (Leckie and Hopkins 2002). Within geography, a growing literature engages with the politics of access and inclusion within the library, referencing the restructuring of public spaces and the public sphere, the spatiality of the public library and management of public space, and uneven access to two-spirit, lesbian, gay, bisexual, trans, and queer (2SLGBTQ) collections and programs between city and peripheral branches (Bain and Podmore 2019; Freeman and Blomley 2019; Lees 1997).

Leorke and Wyatt (2019) demonstrate how tensions emerging from the reconfigured role of the public library in the smart city vision point to institutional critiques around *public* and the library's repositioning as a key agent of the globalized information economy. This tension signals the institution's increasing entrenchment in practices of urban imagineering that disconnect the public library from its community and situate it in a broader global competitive city agenda. The public library's contradictory positioning between these two mandates of the competitive city and the just city is the focus of our empirical section.

Taking Care of the Smart City: The Toronto and Seattle Public Libraries

The annual Intelligent Cities Summit has been based in Toronto since 2016. Salespeople from private corporations, civil servants, politicians, and academics converge to present and exchange ideas on the development of urban smartness and technology. Among the demonstrations of lighting technologies, panels discussing digital transformation, and pitches from consulting firms for full-suite urban data management services, particular attention has been devoted to the role of libraries in the smart city. Chief librarians from different cities around the world have participated on panels focusing on equity and inclusion, providing a revealing window into the material tension between the library as a site of care and as an entrepreneurialized institution. The involvement of TPL at the summit coincides with Toronto's selection by Sidewalk Labs for its smart city project, as well as TPL's hiring of a social worker to address vulnerable groups that seek the library for refuge, safety, and assistance. Seattle likewise recently had become a member of the MetroLab Network – a consortium of city-university partnerships focused on technological approaches to civic engagement. SPL's Civics and Social Services programming is also supported by social workers.

Toronto Public Library

Under Mayor John Tory, the City of Toronto has developed various initiatives and infrastructure investments to make the city *smarter*. Key priorities to date include the development of a smart city strategy to evaluate effective city-wide governance related to digital city building, digitizing local government (through service delivery focused on customer experience, open data, Internet connectivity, and digital literacy), forging partnerships to bring key divisions and city agencies into a broader City of Toronto smart city ecosystem, and building on insights and ideas emerging from two 2017 initiatives – the Toronto Broadband Study (examining the relationship between Internet connectivity and the digital divide) and the federal government's Smart Cities Challenge.

The building of Toronto's smart city image and narrative was enhanced by Waterfront Toronto's (formerly the Toronto Waterfront Revitalization Corporation) partnership with Google affiliate Sidewalk Labs, which was developing a parcel of land in an underdeveloped brownfield portion of Toronto's waterfront known as the Quayside district. The now-defunct smart city neighbourhood project – Sidewalk Toronto – was the technology conglomerate's opportunity to realize its vision of building a city from the *Internet up* (Doctoroff 2016) by experimenting with different technologies and services. The high-tech, sensor-laden neighbourhood included free Wi-Fi, self-driving cars, heated and illuminated sidewalks controlled by sensors, affordable housing, tall timber structures, and other initiatives supporting environmental sustainability. The Toronto Public Library as a stakeholder in the TRBOT and the City of Toronto's Economic Development Committee's Smart Cities Working Group supported this controversial development. In these discussions, TPL's narrative focused on digital inclusion and digital literacy, and on its existing initiatives with tech giants Cisco and Google in delivering online technology training courses.

The modernization discourse that fuels the City of Toronto's smart city strategy promises that a digital government will be more intelligent, efficient, productive, accessible, participatory, and democratic. However, questions of transparency and meaningful participation were at the forefront of critiques of the Sidewalk Toronto development project, mainly because citizens were not consulted about their vision of the waterfront or how technology could serve their needs. An apt illustration of this is how participation was organized around the solicitation of post-it notes on a feedback wall at the development office's headquarters – the 307 (see Figure 18.3) – where people could comment on the proposal after the fact. There was no public deliberation over these notes; only reassurance from Sidewalk Toronto that these comments were being taken into consideration.

In many smart city conversations, libraries are evoked as places of trust where people feel safe, where they feel comfortable accessing information – in contrast to sites such as local government offices. As a librarian at TPL stated, "I think it is because we're trusted in the community. We have a lot of credibility ... We are not stuck in the past ... We don't have an agenda other than welcoming everyone to use our services. So I think that those are two important features: one, we're trusted in the community and two, we've changed with the times" (personal communication 2019).

During this time, public tensions and concerns of privacy, surveillance, and governance were growing around Sidewalk Toronto. To counter these concerns, TRBOT (2019) issued a report recommending that TPL become a civic data trust for the Sidewalk Toronto project. While this recommendation ultimately failed to gain traction, it revealed the power of concepts such as neutrality and public trust in the smart city. The library was seen as a trusted entity that would protect the public's data – presumably defending those data against monetization pressures from corporations. One City of Toronto official noted the problematic conflation between the library and the agendas of private technology corporations as a kind of overstep: "Libraries were seen as the solution [from the private sector] to take these decisions out of government and put them in a trusted place. TPL was chosen because of branding, primarily neutral, well respected, technologically enabled. However, it is one thing to be trusted, another to be a

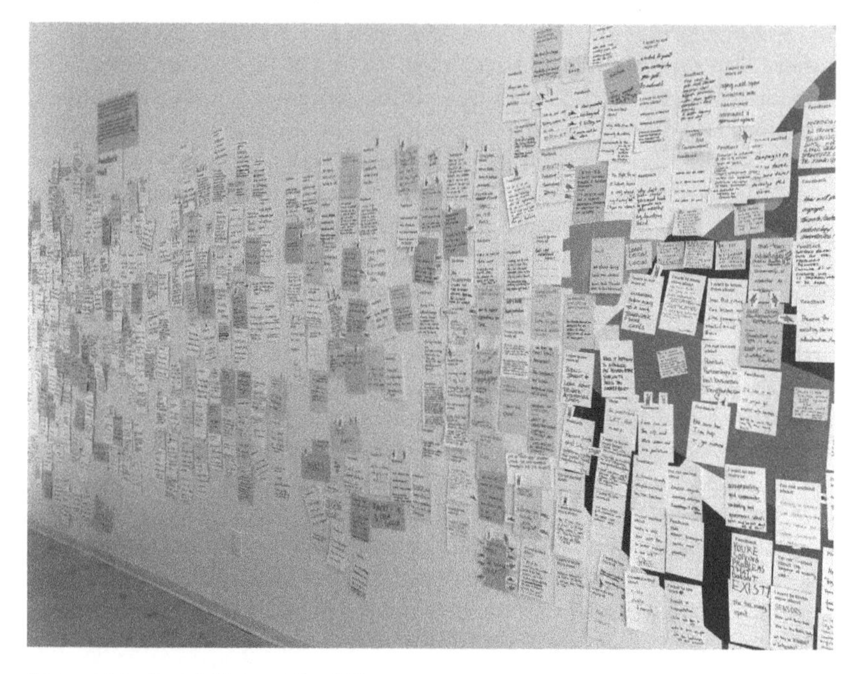

Figure 18.3. Post-It Notes at the 307

tech company and I don't see the library ... as an owner of Big Data. I think cities need to do that ... as a municipal duty" (personal communication 2019).

As Robinson and Ward Mather (2017) highlight, however, public libraries have the potential to partner with municipal staff as civic data infomediaries in the open data ecosystem. Such a partnership would allow data to be accessible to a wider range of users. The TRBOT's recommendation also highlighted the library in the smart city as both an enabler and a channel or site of intervention. This is also reflected in, for example, how some initial public consultations for the federal government's Smart Cities Challenge in 2017 were held at the TPL, acting effectively as an extension of city hall – a place for citizens to ask questions and for project stakeholders to engage with the public.

In the production of smart city narratives, the public library serves as an institutional mediator, bridging the digital divide and equipping citizens with skills and technologies (see Wiig, Pearsall, and Masucci, in this volume, for a discussion on youth and their contradictory positioning in the smart city). The digital divide is a powerful discursive construct that refers not only to uneven access to high-speed broadband Internet, but also to digital literacy (providing the skills required to navigate technology) and digital inclusion (reaching underserved and marginalized populations) (Jaeger et al. 2012). A City of Toronto official illustrated the material realities of the digital divide, and how the library enables digital access and inclusion: "It's not a traditional bookstore. I almost look at [TPL] as social activists. They're part of an inclusive model for society that really looks at underserved populations and, from what I've seen, bringing them access to technology and informing them of opportunities. I think it's an effort to level the playing field ... But I don't think they're primarily about inclusion, but they have taken the place of digital literacy. There are hackathons going on, but they are using technologies to inform citizens on the various issues. They had sessions on how municipal government works, elections, and topical matters like that" (personal communication 2019). This sheds light on the heightened social obligations that libraries are currently experiencing as public institutions under neoliberal conditions, especially with the devolution of social services. In one of our conversations, a city librarian explained that there is a *natural* connection between public sector work, care, and social repair, and argued that no other public institution has the mandate and reach, and thus public libraries are well positioned to allow people to become more connected (digitally and socially). Continued funding, however, relies on the messaging, and we question whether libraries are stretching their social mandate to continue justifying their relevance in the twenty-first century.

Williams (2017) highlights the importance of everyday practices of care and justice in response to injustices in our world. Libraries have different meanings and uses to different publics; they are simultaneously places of refuge, discovery, community hubs, and centres of innovation, technology, and training. Therefore, libraries are mediated sites both enabling *smartness* and digital inclusion, while providing spaces of care for those who are excluded from, or not interested in being part of, networked society.

Seattle Public Library

Commanding a physical presence in the streetscape of downtown Seattle, the eleven-storey central public library, designed by starchitect Rem Koolhaas and his Office for Metropolitan Architecture, communicates the role of libraries in the data-driven economy of the twenty-first century as more than repositories of books, but also as leading centres of innovation, information sharing, and civic engagement. According to a librarian at SPL, the traditional mandate of the library has not changed; instead, as they see it, technology is allowing for new ways of thinking around questions of access and civic engagement – "my job is to give you access to information, so I now have to be creative. If you aren't coming in my doors, I got to figure out a way to get it to you. Our creativity really needs to sit where we deliver services, and how we make you aware of these services" (personal communication 2019). Thus, the social infrastructure embedded within the public space of the library transcends its physical space to reach different groups in the community.

Infrastructure is often used when describing the hard, physical, and organizational structures and facilities needed for the operation of cities and enterprises. However, when discussing the social role of libraries in the community, the public library provides social infrastructures of care, education, repair, and maintenance (Abbruzzese and Riley 2021). The daily patrons of SPL reflect the diversity of needs that coalesce around the library's traditional mandate: as a place of refuge, literary pleasure, training and skills updating, and community. A librarian contextualizes the "daily readers" and patrons of SPL within the city's broader uneven socio-economic geography:

> Seattle is a city of readers ... Amazon, Microsoft, and Eddie Bauer. These are all companies that bring intelligent talent into their communities and jobs. So, we have a very educated workforce and they are all using libraries for either self-pleasure reading, or learning ... We have many colleges and universities, so we have lots of students using our libraries. And then we have lots of children using our libraries.
>
> In addition, we have populations that may be seeking us out ... When they come into our libraries they may be homeless out on the streets, but when they come into our building, they are now "daily readers," and that's how we refer to them ... they may not have any place to go during the day so they are looking for a place to sit. We ask that they engage with the resources that we provide ... sometimes the streets are not as safe as they want so they want to find a space of safety. (Personal communication 2019)

Notwithstanding local critiques of SPL's redevelopment – partially funded by a US$20 million donation from the Bill and Melinda Gates Foundation – as "out of place" from its northwestern context (Mattern 2003, 15), the librarian's perspective did not reflect this crisis of representation between local and global visions. Rather, the library was seen as a trusted entity, with an agenda of social activism, as with TPL, and as a social agency for the community: "[SPL] is a space where we actually see, hear, and understand the challenges that our public is facing. We are a trusted entity to help carry their messages forward, and that is a rightful opportunity that libraries can step into, which is being the voice and ears of their communities ... A patron might say, "Hey, I'm having trouble with my landlord." The library would then step in and say "This is how you address tenants' rights and here are the rights of tenants in Washington State" (personal communication 2019). The redevelopment of SPL signalled the strategic positioning of the library in Seattle's spatial imaginary as a technology city. Recently, Seattle partnered with the University of Washington to join the MetroLab Network, a national network of university-city partnerships focused on applying technology to pressing urban issues. Branded as "tech transfer for the public sector," MetroLab was launched by mayors and university leaders as part of the White House Smart Cities Initiative in 2015. In Seattle, public institutions are pressured to improve their *smartness*, evolving the city's association with innovation, talent, and tech giants such as Microsoft.

Although the librarian spoke to the necessary entrepreneurialization of SPL in order to respond to the growing diversity of the community's needs, this rhetoric was couched within a discourse of emotion that emphasizes an ethic of care enmeshed in the library's sense of place in the community:

> I think that that is the largest thing that libraries contribute to smart cities ... Libraries are very emotional. Every person that I talk to over forty will come in and say "my mother used to bring me to this library" ... They have childhood memories of how great the library was ... And that engenders that trust we built, which is intangible.
>
> The other thing is that we are stepping into spaces where we are an access point for people. For example, the Affordable Care Act: there are lots of people in the community who do not know how to complete the form ... We have the technology so that they can come in and apply for it here. Tax assistance, we do that here too. People have an expectation that, if nobody else will do it, we will ... There is no other profession that centres around answering whatever question you may have. (Personal communication 2019)

This reinforces Mattern's (2014) characterization of public libraries as "opportunity institutions." SPL stands out in smart city conversations as a public library that has successfully partnered smartness with care. Thus, public libraries have a critical role in mediating the neoliberal logic of the smart city, serving as a democratic filter that mitigates the uneven socio-economic and spatial impacts of this technical urban vision. Although the public library has always inherently served this role in the community, its persisting commitment to equity, access, and digital inclusion makes it one of the most crucial public institutions of the twenty-first century.

Conclusion

"Care-full" justice in the smart city goes beyond frameworks of buzzwords and concepts such as the digital divide and civic innovation. In this chapter, we have shed light on both the opportunities and constraints that libraries negotiate in the smart city. Although libraries operationalize their mandate and services increasingly through an entrepreneurial approach to social provision, their everyday practices reflect a persistent urban ethic of care. Libraries are key spaces, providing needed social repair and care, and offsetting the broader retreat of social services from the state under neoliberalization. The public library is used by the smart city discourse to further its own ends, but it also offers a site of prospective resistance, protecting values of access, democracy, public space, education, locality, and social responsibility. The public library upholds these values in the face of competitive, technology-driven, and globally networked visions of urban development and management.

ACKNOWLEDGMENTS

The authors would like to thank Annemarie Gallaugher for her invaluable editing assistance and to acknowledge financial support for this research from a minor research grant awarded by the Faculty of Liberal Arts and Professional Studies at York University.

REFERENCES

Abbruzzese, T.V., and A. Riley. 2021. "The Toronto Public Library as a Site of Urban Care, Social Repair, and Maintenance in the Smart City." In *Care and the City*, ed. A. Gabauer, S. Knierbein, N. Cohen, H. Lebuhn, K. Trogal, and T. Viderman, 183–93. New York: Routledge.

Angelidou, M. 2017. "The Role of Smart City Characteristics in the Plans of Fifteen Cities." *Journal of Urban Technology* 24 (4): 3–28. https://doi.org/10.1080/10630732.2017.1348880

Bain, A.L., and J.A. Podmore. 2019. "Scavenging for LGBTQ2S Public Library Visibility on Vancouver's Periphery." *Tijdschrift voor economische en sociale geografie* 111 (4): 601–15. https://doi.org/10.1111/tesg.12396

Doctoroff, D.L. 2016. "Reimagining cities from the internet up." *Medium* (blog), 30 November. Online at https://medium.com/sidewalk-talk/reimagining-cities-from-the-internet-up-5923d6be63ba

Freeman, L.M., and N. Blomley. 2019. "Enacting Property: Making Space for the Public in the Municipal Library." *Environment and Planning C: Politics and Space* 37 (2): 199–218. https://doi.org/10.1177/2399654418784024

Greenfield, A. 2013. *Against the Smart City*. New York: Do Projects.

Grossi, G., and D. Pianezzi. 2017. "Smart Cities: Utopia or Neoliberal Ideology?" *Cities* 69: 79–85. https://doi.org/10.1016/j.cities.2017.07.012

Jaeger, P.T., J.C. Bertot, K.M. Thompson, S.M. Katz, and E.J. DeCoster. 2012. "The Intersection of Public Policy and Public Access: Digital Divides, Digital Literacy, Digital Inclusion, and Public Libraries." *Public Library Quarterly* 31 (1): 1–20. https://doi.org/10.1080/01616846.2012.654728

Jaeger, P.T., N.G. Taylor, and U. Gorham. 2015. *Libraries, Human Rights, and Social Justice: Enabling Access and Promoting Inclusion*. Lanham, MD: Rowman & Littlefield.

Karvonen, A., F. Cugurullo, and F. Caprotti, eds. 2018. *Inside Smart Cities: Place, Politics and Urban Innovation*. New York: Routledge.

Kitchin, R. 2015. "Making Sense of Smart Cities: Addressing Present Shortcomings." *Cambridge Journal of Regions, Economy and Society* 8 (1): 131–6. https://doi.org/10.1093/cjres/rsu027

Klinenberg, E. 2018. *Palaces for the People: How Social Infrastructure Can Help Fight Inequality, Polarization, and the Decline of Civic Life*. New York: Broadway Books.

Lawson, V. 2007. "Geographies of Care and Responsibility." *Annals of the Association of American Geographers* 97 (1): 1–11. https://doi.org/10.1111/j.1467-8306.2007.00520.x

Leckie, G.J. 2004. "Three Perspectives on Libraries as Public Space." *Feliciter* 50 (6): 233–6.

Leckie, G.J., and J. Hopkins. 2002. "The Public Place of Central Libraries: Findings from Toronto and Vancouver." *Library Quarterly* 72 (3): 326–72. https://doi.org/10.1086/lq.72.3.40039762

Lees, L. 1997. "Ageographia, Heterotopia, and Vancouver's New Public Library." *Environment and Planning D: Society and Space* 15 (3): 321–47. https://doi.org/10.1068/d150321

Leorke, D., D. Wyatt, and S. McQuire. 2018. "More than Just a Library: Public Libraries in the 'Smart City.'" *City, Culture and Society* 15 (December): 37–44. https://doi.org/10.1016/j.ccs.2018.05.002

Leorke, D., and D. Wyatt. 2019. *Public Libraries in the Smart City*. London: Springer.

Low, S., and K. Iveson. 2016. "Propositions for More Just Urban Public Spaces." *City* 20 (1): 10–31. https://doi.org/10.1080/13604813.2015.1128679

Mattern, S. 2003. "Just How Public Is the Seattle Public Library? Publicity, Posturing, and Politics in Public Design." *Journal of Architectural Education* 57 (1): 5–18. https://doi.org/10.1162/104648803322336548

Mattern, S. 2007. *The New Downtown Library: Designing with Communities*. Minneapolis: University of Minnesota Press.

Mattern, S. 2014. "Library as Infrastructure." *Places Journal*, June. Online at https://placesjournal.org/article/library-as-infrastructure/

Mattern, S. 2018. "Maintenance and Care." *Places Journal*, November. Online at https://placesjournal.org/article/maintenance-and-care/

Robinson, P., and L. Ward Mather. 2017. "Open Data Community Maturity: Libraries as Civic Infomediaries." *Journal of the Urban and Regional Information Systems Association* 28: 31–8.

Shepard, M. 2014. "Beyond the Smart City: Everyday Entanglements of Technology and Urban Life." *Harvard Design Magazine* 37: 18–23.

Stevenson, S. 2009. "Digital Divide: A Discursive Move away from the Real Inequities." *Information Society* 25 (1): 1–22. https://doi.org/10.1080/01972240802587539

Thompson, K.M., P.T. Jaeger, N.G. Taylor, M. Subramaniam, and J.C. Bertot. 2014. *Digital Literacy and Digital Inclusion: Information Policy and the Public Library*. Lanham, MD: Rowman & Littlefield.

TRBOT (Toronto Region Board of Trade). 2019. *BiblioTech*. Toronto: Toronto Region Board of Trade. Online at https://www.bot.com/Portals/0/Bibliotech%20-%20Final%20-%20Jan%208.pdf?timestamp=1546987861621

Tronto, J.C. 1993. *Moral Boundaries: A Political Argument for an Ethic of Care*. New York: Routledge.

Williams, M.J. 2017. "Care-full Justice in the City." *Antipode* 49 (3): 821–39. https://doi.org/10.1111/anti.12279

Williams, M.J. 2020. "The Possibility of Care-full Cities." *Cities* 98 (March): 102591. https://doi.org/10.1016/j.cities.2019.102591

19 Struggling Zones, Stagnant Cities, Inner Regions: Just Renewal through Smartness in Saint John, New Brunswick?

YONN DIERWECHTER

Much of what we know about smart cities and wider invocations of smartness, however defined, focuses on reshaping favoured global places or overemphasizes high-profile greenfield experiments. Less is known about the uneven spatialities of smartness in small(er) cities and metro/micropolitan areas, particularly in economic regions of wealthy countries that include what Andrés Rodríguez-Pose (2018) has provocatively called "the places that don't matter." Unlike Vancouver, Calgary, or Toronto, Canada's major conurbations, the old, smaller, regional cities in the Maritime province of New Brunswick – Fredericton, Moncton, and Saint John – rarely feature in discussions of North American urbanism or in global smartness research more generally (although see recent work by Spicer, Goodman, and Olmstead 2021). Such neglect implies that smartness theory is still developed from a limited empirical gaze, a weakness that supports calls in this book and elsewhere to further provincialize engagements with all cities and urban regions, including those in the global North (Sheppard, Leitner, and Maringanti 2013).

This chapter offers an alternative, critically sympathetic reading of recent smart city efforts in smaller regional cities, focusing mostly on Saint John – Canada's oldest incorporated city. Based on fieldwork conducted in 2019, the discussion presents evidence for what I emphasize below as the search for "just renewal." This search challenges scholars to consider not only how smartness discourses shape urban policy dynamics, but also how a city in a province long "pulling against gravity" (Savoie 2001) has territorialized smartness into both the local space-economy and social justice agenda. After an initial discussion of how smartness is increasingly part of the knowledge-based geopolitics of wealthy states (Moisio 2018b), I turn to an analysis of peripheral regions such as New Brunswick – or what the Italian scholar Simonetta Armondi (2022) calls "inner regions." I then explore the core research theme of whether smart city initiatives, which are a response to state-directed funding of smart cities, are more accurately mapped as spaces that attempt to redirect the geographies of growth facilitation associated with the post-Keynesian competition state.

Without reifying the presumed social propinquity of small cities like Saint John, I nonetheless suggest that just renewal is empirically prominent in poorer inner regions within wealthy Canada. This is a provisional claim, of course, which needs further empirical exploration as well as sustained critique from alternative theoretical traditions. This includes Datta's concern (in this volume), for example, with "internal colonization," as well as Abbruzzese and Hillier's interesting concept of "care-full justice" (also in this volume). In short, understanding smart(er) cities in Canada means expanding our atlas to incorporate places frequently left out of conversations and theory building, whatever spaces might be produced.

Framing Smartness and the Variegated Geographies and Politics of Wealthy States

In addition to other qualities debated in this book, smartness is part of what Sami Moisio (2018a) calls the emerging geopolitics of knowledge-based economies. Although focused originally on shifting the material geography of metropolitan growth in city-regions such as Baltimore, Portland, Calgary, Toronto, and Vancouver (i.e., smart *growth*), "smartness" discourses soon conceptually rebranded and technically expanded corresponding discussions of "augmented places," "digital cities," and "electronic cities" (Aurigi and De Cindio 2008). Here, knowledge is increasingly valorized as digitized information through innovations in information and communications technology (ICT).

Seen positively, ICT capabilities in the new millennium, particularly with the diffusion of smartphones, have offered novel ways to harvest data flows and, in theory, better manage major urban systems such as transportation, utilities, and refuse collection. Seen negatively, smart developments in general, and smart city infrastructures in particular, have offered new ways for large corporations and state institutions to monitor populations, undermine privacy, and extract profit for both private and public goals. Inevitably, as with many Big New Concepts, scholarly work on smart cities has reached an oversimplified duality. "The focus, intention and ethos of smart city ideas, approaches and products," Coletta et al. (2019, 2–3) note, "remain quite fragmented and often quite polarized ... Smart city protagonists then are largely divided into those that advocate for the promise or warn of the perils of smart cities." Promise versus peril dichotomies reflect wider tensions within academic and popular urban studies, where the city is often essentialized as Saviour or Satan, rather than both at the same time (Beauregard 2018).

Smart city research, like some work in urban political ecology and other subfields, further suffers from "methodological cityism" (Angelo and Wachsmuth

2014). Smartness is often theorized as a "trait," a geographic characteristic of bounded places with internal agency, rather than "the production of specific and quite heterogeneous spatiotemporal forms embedded within different kinds of social action" (52), or what Harvey (1996) called "urbanization." From this perspective, cities are (re)made from recursive relationships knotted in place that nonetheless often stretch beyond local borders. Globalized commodity chains are one example. Another is bioregional watershed flows managed by public utilities whose operational spaces are usually defined by multiscalar infrastructure and ownership grids. Planetary urbanization, of course, pushes this as far as it can go, elevating epistemologies of variegated process over territorialized place. So conceptualized, smart(er) cities are as much about the "social action" of state-level institutional players in global space as they are about "local" growth coalitions and service providers. Knowledge-based economies in global city-regions, for instance, arguably are better understood as new forms of global geopolitics – what Moisio (2018a) calls emerging processes of "economization." This term captures the political interests of (certain parts of) the central state apparatus in the role that city-regions – e.g., Helsinki, Finland; Toronto, Canada; Johannesburg, South Africa – should play in "national" accumulation strategies.

Armondi (2022) has expanded this argument beyond the global city-regions of Allen Scott's (2001) original archipelago of successful, trendy places that selectively house the digital cognitariat. Instead, she focuses on "inner regions": rural, poor, declining, depopulating, stagnant, and/or simply peripheral zones in otherwise wealthy countries that are often bypassed by the dominant flows of global capital. The two 8.0 earthquakes of 2016 – Brexit and Trump's election – revealed the electoral "revenge" that such places can take on "thriving" regions (Rodríguez-Pose 2018), even as cultural developments in the populist politics of ethnic nationalism and working-class grievance across much of Europe had earlier exposed the growing rifts of uneven development patterns and sociospatial polarization – for example, Germany's AfD, Holland's Party for Freedom, Hungary's Jobbik, France's National Front, Poland's People and Law Party, and see the original "Tea Party" in the United States.

Whereas work on the post-Westphalian rise of city-regions and city regionalism has been largely silent on the question of "inner regions" (although see Herrschel and Dierwechter 2018, 40–1), Armondi (2022, 527) counterargues for seeing such regions as crucial components in what she theorizes as "state-orchestrated processes of territorial distribution around investments in public services, social and physical infrastructures, and, at the same time, envisioning a changing meaning of citizenship in the framing of the national question." Armondi's work allows us to stretch Moisio's (2018b) signature work on the "urbanization of the state" to include smaller cities in inner regions "pulling against gravity," like those in New Brunswick. The re-election of Justin Trudeau as prime minister in 2019 suggests that Canada's populist turn remains more muted than in the United States or Europe. But it is hardly absent. Toronto-based journalist Zach Taylor notes the rise of "two Ontarios," one that matters and one that doesn't, arguing that, "in an economy and society that features hardening 'macro' divisions between regions, as well as growing 'micro' divisions within them, it is only to be expected that political parties [are and will] seek to differentiate themselves from one another by strategically appealing to place-based grievances, lifestyles and problems" (Taylor 2018).

The political and economic geographies of these developments are still unfolding: "micro" divides *within* regions grow, while "macro" divides *between* them expand. Urban electoral districts in Saint John, Moncton, and Fredericton tend to vote either Liberal or Green, and, at the provincial scale, a largely "Liberal north" (much like neighbouring Quebec) challenges more Progressive Conservative support in the south. But the 2018 provincial election nonetheless saw the rise of the People's Alliance Party, which offers a now-familiar right-populist brew of economic conservatism, rural populism, and opposition to bilingualism, with concerning implications for extant patterns of racial inequality and settler colonial relations (Walcom 2018). The populist politics of xenophobia and language purity, should they grow more powerful, threaten not only the pro-immigrant strategies of Saint John, but also settler colonial relations across New Brunswick. Imperative, then, is the question of how the Canadian state is managing new "divides," and how, specifically, smart city investments widen or narrow these divides. As in many countries, uneven developments in Canada have deepened – and thus affect how (geo)political approaches to globalization influence the technological management of state space. Smartness policies and projects are embedded within a territorial cohesion project, projecting intended and unintended discourses of high-tech citizenship that threaten and inform heterogeneities of place.

Recent evidence of these concerns within federal government policy is the launch in 2017 of the Smart Cities Challenge by the Trudeau Liberals. Canada's interest in this type of urban policy approach is distinctive, but comparable to similar "competitions" in other counties, including several annual rounds across India under Prime Minister Narendra Modi (Datta 2015); the explosion of smart city initiatives in China under Premier Xie Jinping (Yu and Xu 2018); and various smart city developments in the United States, largely started during the Obama administration (Ebi 2017). It also reflects the transnational influence of projects such as Triangulum, Remourban, and GrowSmarter in the European Union's Smart Cities and Communities program. Like these projects and initiatives, the Smart Cities Challenge is pitched as a pan-national competition open to all municipalities, local or regional governments, and Indigenous communities (First Nations, Métis, and Inuit). The core idea is to show how smart city approaches can improve the lives of residents through innovation, data, and connected technology that, in theory, encourage all communities "to take risks and think big, identify significant, pressing, and perceived 'un-solvable' problems" (Canada 2020).

Unsurprisingly, the focus here is on lofty promises rather than the emerging perils of smart cities. Administered by Infrastructure Canada – a federal department – the first round attracted over 130 smart city proposals across the provinces and territories, including one from Nunavut that represented all ten municipalities in this inner region. Just over 15 per cent – a total of twenty proposals – were selected in June 2018 for further support with $250,000 capacity-building grants. In April 2019, Infrastructure Canada announced the final winners: the City of Montreal won the $50 million prize; the consolidated proposal by the Nunavut Communities and the joint proposal by the City of Guelph and Wellington County, Ontario, each received $10 million; and the Town of Bridgewater, Nova Scotia, received $5 million. Infrastructure Canada judged winning proposals to be the most "ambitious," yet also "achievable," as well as "innovative, transferable, replicable and scalable" (Canada 2018a, 8). Despite the strong market rationality of a territorial game for public resources that is now so emblematic of post-Keynesian statehood across the world, the parallel insistence on transferability, replicability, and scalability alternatively suggests wider social benefits, not least for communities that ostensibly "failed," yet spent considerable effort reimagining the possibilities of "smartness" in their own local contexts and developmental challenges (Canada 2018b). In fact, the federal government additionally created a Community Support Program, run by a not-for-profit organization, Evergreen, as a strategic complement to the overall Challenge campaign.

With respect to actual policy substance, Nunavut's winning proposal focused on how smart city investments might help facilitate suicide prevention; the City of Guelph and Wellington County jointly aimed to build a circular food economy across the city-region; Montreal's proposal highlighted improved access to mobility services and food; the Town of Bridgewater's proposal sought to reduce energy poverty (Canada 2019). While smart city initiatives worldwide reflect the neoliberal rationalities of the competition state's role in accumulation, these proposals suggest concerns with urban sustainability and social justice or, more simply, just renewal. Yes, Canadian smartness reflects the "perils" of places forced to compete against one another within an otherwise shared polity, but also the "promises" that attend new ways of engaging data and citizens in the work of collective problem solving. Both require attention. Shoehorning variegated empirical concerns with, say, suicide prevention, energy poverty or (as we shall see below) immigrant inclusion into the uniform jackboot of global neoliberalism seems forced and insufficiently attentive to the *variegated geographies* surely produced by the complex empirical intersections between "macro" and "micro" divisions across and within the political economies of places that do (and don't) "matter."

How should one capture these new comparative geographies? Stimulating the economic growth of rural, poor, declining, depopulating, stagnant, and/or simply peripheral zones remains critical to Canada's federal approaches and global economic diplomacy. At the same time, the project creativity of the Smart City Challenge per se, especially in struggling areas of struggling cities in inner regions, sufficiently widens the empirical story of experimentation to require more nuanced representations of the spaces that smart cities make – and unmake. Figure 19.1 provides a visual that captures these themes, incorporating select ideas from, *inter alia*, Coletta et al. (2019), Rodríguez-Pose (2018), Moisio (2018a), and Armondi (2022) within the territorial context of the macro and micro divides emerging nearly everywhere, including across diverse Canada. My argument here is simple: focusing on various smart developments in successful zones of successful cities within vibrant regions (e.g., "the places that matter") *is only one kind of geography*. There are surely many others, including changes in struggling cities within vibrant regions as well as smart city spaces focused on struggling zones in stagnant cities of inner regions that "don't matter," which is my narrower focus in this chapter. Indeed, discourses of success and failure – of mattering, vibrancy, or struggling – reflect powerful assumptions about the preferred possibilities and actual trajectories of territorial development. It may be, however, that depopulating cities, urban zones with manifold social challenges, and/or regions "pulling again gravity" are no less innovative, vibrant, and, ultimately, instructive than is too commonly supposed. We do not yet know; further work is needed on diverse cities (dynamic and stagnant) in diverse regions (vibrant and inner) in many societies to find out what is happening on the ground in a full world, particularly regarding possible concerns with social justice in the context of economic and demographic change.

Understanding different geographies of smart cities in this way does not demand a return to the thick descriptions of ideographic empiricism, although more fieldwork in more places is surely welcome; nor must we embrace nomothetic claims without considerable care, even as efforts to generalize are also important. The role of smartness policies in creating – or trying to create – "a surveillance society," for instance, arguably characterizes community policing in Camden, New Jersey (Wiig 2017), no less than the ill-fated Google experiment in downtown Toronto (Goodman and Powles 2019). Yet comparable theory must try to explain or at least broadly frame diverse practices, and smart city practices are still, I think, understudied in struggling zones of stagnant cities in inner regions, Canada's included. Surveillance space is one possible outcome; justice is quite another. Many questions abound. Do putatively successful zones in "succeeding" cities in the places that *do* matter generate different insights into citizenship, identity, and space than similarly successful zones in struggling cities, however defined? Is the surveillance thesis *too* encompassing,

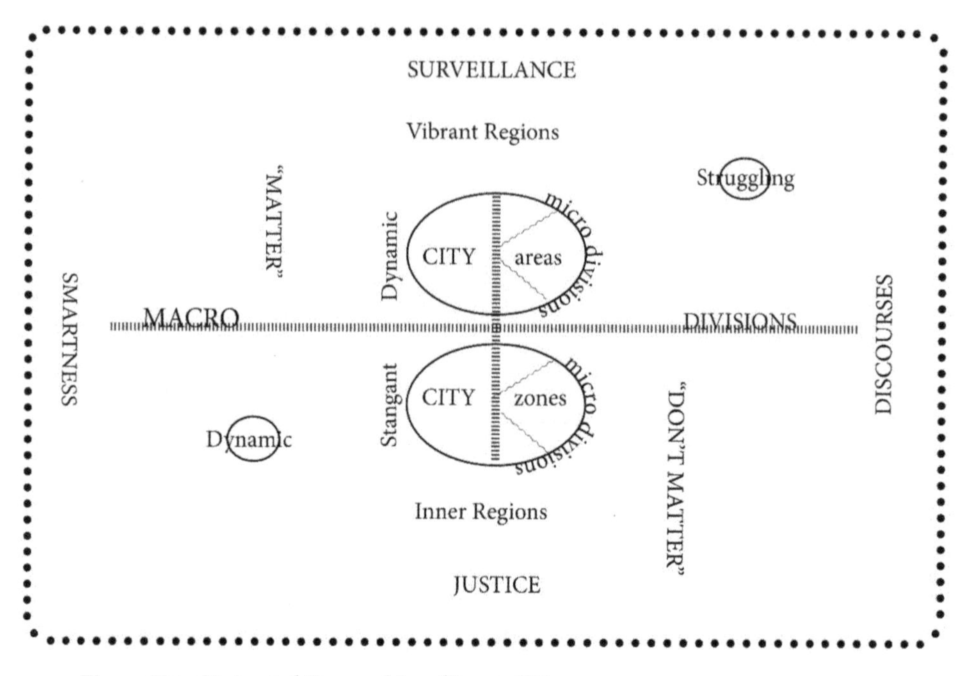

Figure 19.1. Variegated Geographies of Smart Cities

effacing heterogeneous dynamics? Is digital justice, for instance, more likely in some places – or "moments," as Henri Lefebvre might put it – than others and, moreover, in ways that might allow us to move beyond initial empirical work to theory building about different kinds of "smart transitions" in Canadian space and society (Herrschel and Dierwechter 2018)?

Just Renewal in Saint John, New Brunswick?

Once relatively prosperous, maritime- and export-oriented New Brunswick is today among Canada's poorest provinces – the complex outcome of, *inter alia*, local resource endowments, (non)proximity to markets, and the quality and nature of its workforce skills, which nonetheless also reflect the exogenous effects of national decisions on the political economy of Canada's overall spatial development pattern (Desserud 2015). Confederation with much larger Quebec and Ontario led to diminished political clout in Ottawa, and then to tariff, tax, and investment policies over many decades that favoured industrialization, immigration, and accumulation in central Canada. By the late 1950s, New Brunswick – still based on resource-extraction sectors, mainly mining and wood/forestry – had stagnated into a "supplicant economy," reliant on transfer payments from Ottawa to try to stem both the territorial and social effects of uneven economic development (Savoie 2001).

If New Brunswick's provincial politics are framed by the developmental dialectics of economic supplication and the political search for self-sufficiency, the smart turn in Saint John is framed more concretely than perhaps anywhere else in metropolitan Canada by the long-term crisis of municipal decline. Saint John is interested in *smartness* because it has been *smarting* from the painful sting of economic and demographic decay in a prideful place steeped in so much meaningful history. At a 3 April 2018 common council meeting, the mayor of Saint John formally introduced his city's application to the "Smart City Challenge" this way: "I was pleased to be part of the smart city challenge project from its first workshop ... Now we add the final piece to our application: the results of the community's hard work and desire to grow Saint John by submitting [and] reversing population decline using a smart city's approach" (Rogers Communications 2018).

Saint John's application for Canada's Smart City Challenge was not selected for further federal support. The competition state, like sports leagues, always creates more immediate losers than long-term winners. Still, the preparation that went into the application is revealing – and according to several local officials remains a source of discussion (Letson 2018). In its formal application, Saint John sought smart city support to pursue "Growth through Migration" (City of Saint John 2018). The "problem" to which smart city promises might help "to solve" was essentially twofold. On the one hand, the city's population has declined precipitously since its peak of 89,000 in 1971 through emigration, even as death rates have exceeded birth rates. This has reduced unconditional municipal grants from the province, which has further exacerbated local service provision and infrastructure deterioration. This is what is meant by *struggling* in Saint John. Attracting *and then retaining* interprovincial and international migrants to Saint John is, therefore, a favoured way to reignite new rounds of growth, *smart* or otherwise; but, the application specifically argues, this strategy is ironically hamstrung by a locally created condition: "Saint John's much-celebrated connectedness could also be an unintended barrier to making newcomers feel a part of the city – a tightly woven net of personal and professional relationships that can be impenetrable for newcomers" (City of Saint John 2018).

Eminent is the nexus between problem and solution, wherein the "combined force of data, connected technologies and community engagement," as the application puts it, does not aim to destroy the "connectedness" of Saint John's "tightly woven net," but instead the unanticipated "barrier" for would-be migrants to enter and fully enjoy over time an otherwise fecund social space. To save Saint John is to translate (or perhaps better yet, to transduce) a *local social asset* into a *sociotechnical resource* through the improved rollout and strategic integration of data, connected technologies, and community engagement. This is quite interesting in itself. The smartly confected spatiality of this new sociotechnical resource is marked by four key features: openness, integration, collaboration, and transferability. Each feature is rendered a *casual factor* in the attraction, retention, and, most important, urban-developmental benefits of migrants. *Openness* refers, for public authorities, to the development of a common community data platform as well as an open approach to sharing data, both of which presumably facilitate "transparent decision-making," "innovative thinking," and "public trust" between key private, governmental, and non-profit organizations. Largely through the expansion of a "Community Data Repository," *integration* means capturing disparate datasets "in one common place," which hypothetically helps to break down "silos" between organizations. Finally, *collaboration* deepens (and widens) network connections between "traditional and non-traditional partners," while *transferability* means sharing insights, processes, challenges, and successes "with all interested parties," but especially scalability with communities around Canada that face similar growth profiles (City of Saint John 2018).

As a form of local economic development policy, this is a far cry from, for instance, extending ten-year tax abatements to condominium developers or building yet another convention centre and heritage museum. If issue can be taken with identifying the city's "problem" as endogenous rather than exogenous; if the community's internal social divisions and class stratifications are glossed over with the warm colours of connectedness; and if the "force of data, connected technologies and community engagement" seems weak compared with the "gravity" of fickle capital and wider geopolitical economies of uneven territorial development across Canada, Saint John nonetheless presents to the world (and itself) a contingent theory of smart city practice based on an impressive engagement with its social challenges. Saint John wants to use smart city technologies to "delocalize" through "internationalisation," harnessing (retained) growth to the possibilities that a far more cosmopolitan population of actors might just bring, including a recent group of Syrian refugees welcomed in from civil war.

The Community Data Repository – a cloud-based community data analytics platform – constituted a central element in Saint John's "Smart and Connected Community Data Project." Enterprise Saint John is the economic development agency for the Greater Saint John region. It sees this new repository as a foundational element in growing the region economically, and to some extent socially, too (Letson 2018; J. Scott, Director, Community and Economic Development, Enterprise Saint John, author's interview, 20 March 2019). "Anonymized and aggregated datasets" in the repository, Enterprise Saint John suggests, help to drive innovation across the regional economy, even as the wider "Smart and Connected Data Project" is localizing the public wireless network and a new infrastructure of sensors "in the City's urban core to enhance use of digital technologies and deepen understanding of community mobility patterns" (Enterprise Saint John 2017, 6). The much discussed "hardware" of smartness – the "hard" technological interpretation of the smart city – is thus merging with a "softer" view of smartness – of the smart city as a creative space of social innovation across silos, organizations, and scales. Yet the core focus is still how data constantly generated by private and public institutions through their "stuff" and activities can be brought "together" – how buildings, cameras, attendance sheets, lighting, grids, computer monitors, electronic files, solar panels, waste bins, and snow ploughs merge into an eclectic yet common infrastructure and community-building space.

The governance and material challenges are immense, even within otherwise standard municipal services such as snow removal (Scott, interview). Building an effective repository requires "a whole host of factors that have to happen at once, which is really hard. If one variable is missing, it doesn't work" (consultant M. Crevatin, author's interview, 20 March 2019). While Enterprise Saint John had already initiated projects in Big Data, and also explored new regional partnerships with Saint John Energy in developing a smart grid, the smart city challenge project – accelerated and participatory – revealed the importance of "serving newcomers" in new ways: "the newcomer angle" emerged "more clearly than originally expected" through recursive dialogue between diverse actors (Scott, interview). At minimum, the Smart City Challenge expanded local discussions of what a smart city is actually *for*, which had started mainly (and more conventionally) with placing a meshwork of beacons and sensors throughout the Greater Saint John area to collect the disparate sources of data that feed the community's data repository. That done, the *original* discourse also ran, Saint John might better compete with "the Torontos and the Calgaries" for attention (*Wave* 2017). Yet it appears to have evolved over time. A new epistemic community has engaged economic challenges through social responsibilities.

Inward investment and "growth through migration" mean little if the benefits accrue to only a few segments of a general population less battered by demographic decline, if migrants are not retained, and if migrants fail to enter functional economic and social spaces that ultimately provide tangible mobility and opportunity. Within Saint John, the social innovation organization, Living Saint John, focuses mainly on breaking cycles of intergenerational poverty, a mission accelerated by the recent acquisition of a grant from the provincial government (New Brunswick 2017). In the mid-2010s, poverty, education, wellness, and employment measures suggested the overall degree of local concern. Nearly 30 per cent of Saint John's children lived in poverty, nearly three times higher than in Calgary and Edmonton. Only 78 per cent of Grade 2 students met appropriate literacy levels – 12 per cent lower than government targets. Another 30 per cent of adults reported challenges in mental health. Two-thirds of youth surveyed in 2014 did not see themselves living and working in the region in the mid-2020s (Living Saint John 2014). Median household income was more than $104,000 in Calgary, but only $52,000 in Saint John (Townfolio 2017). Micro divisions matter, too.

Living Saint John originally argued for a poverty-reduction and social justice focus, but agreed that growth through more inclusive migration probably made sense in a region long pulling against gravity. Living Saint John's executive director explicitly linked new growth to the long-term success of the city's spatial vision of repopulation: "A poverty reduction strategy is an economic development strategy," she suggested, "... and therefore, when you look at economic development, one of the key drivers of that is population growth. And linking that fact to one of our targets, which is to build mixed income neighbourhoods, we need the population growth. So, I didn't see not having the first focus on poverty as a loss. If I was in that position, I would have made the same decision" (D. Gates, Director, Living Saint John, author's interview, 2 March 2019).

That said, her daily work on poverty reduction involves coordinating dozens of organizations working at multiple scales with diverse perspectives; it is work based on how the social development of disadvantaged populations – particularly children – might be better supported by improved data and by a better overall grasp of "collective impact":

> There's collaboration and then there's collective impact. Collective impact is data driven. We try to make decisions in our community based on what the data are telling us. So, yes, you need the integrated data stream to better do that. We're trying to move away from a traditional, non-profit role that's transactions based to an outcomes-based experience ... We're looking at transitions. It doesn't just mean transition to employment. It could mean out of domestic violence. It could mean [into] after-school programming. Did this after-school program have a positive impact on attendance in school? Are the kids showing up at school more because of that? Are they getting mentoring and homework help? Because kids who live in poverty don't have those kinds of supports. So, we're moving from transactional data collection to transformative data collection. That's how we view it. There's a lot of data. It's out there. But it's not streamed together and that's a goal. But we can't look at everything. So, what are the best indicators of transformation? Non-profits are already strapped to the max. We can't ask them to do that. The question we're really trying to answer is – what is the health of our community? It's poverty reduction. It's social inclusion. But it's [the amount of] loneliness [too], which I just heard someone say at a conference yesterday is "the new tobacco" ... So, people can often be confused by the terminology of smart cities. For me, I would actually say it's a connected city. That's snow ploughs and traffic lights, sure, but are we happy and healthy? And I want to know that neighbourhood by neighbourhood. Do we feel connected to one another? (D. Gates, author's interview, 2 March 2019)

Ultimately, these smart city initiatives seek to produce spaces that at least *attempt* to redirect the digital geographies of growth facilitation associated with the urbanization strategies of the competition state.

Conclusion

Investments that flow into places such as Toronto, Calgary, and Vancouver highlight "the perils" of augmenting physical and social spaces with infrastructures of ICT-enhanced urbanism. New geographies of near-ubiquitous data engagement suggest, as smart city critics have shown for many years, territorialities of surveillance capitalism wherein, for instance, community *deni*zens are reimagined as corporate *data*zens. Yet the variegated geographies of urbanization in wealthy countries such as Canada, shaped by macro and micro divisions, complicate conclusions that underplay what Kian Tajbakhsh (2001) has called "the promise of the city." Smart city research also needs to explore the promise of smartness in struggling zones of stagnant cities located within what Armondi (2022) calls the "inner regions" of wealthy societies – places like neglected neighbourhoods in Saint John, New Brunswick. Such spatialities in places are constructed over many scales of "urban" policy, as local institutional actors like Living Saint John merge their important social justice work with "smart state" projects such as Infrastructure Canada's Smart City Challenge (see Dierwechter 2018).

Although not highlighted in this chapter, new partnerships between municipalities such as Fredericton and neighbouring First Nations further suggest implications for settler colonial relations, not least around both local and national claims of socially inclusive economic transformation (M. McKinley-Nash, Director of Economic Development, Sitansisk Wolastoqiyik – Saint Mary's First Nation, author's interview, 19 March 2019). Here my findings broadly support arguments recently advanced by Spicer, Goodman, and Olmstead (2021), whose own work on small cities in Canada suggests novel pathways of learning and policy exchange that push back against the well-documented powers of corporate storytelling: "Larger municipalities could learn much from small, rural and remote communities. The latter do not try to compete with their neighbours to achieve status in the smart city community. Rather, these municipalities are focused squarely on service provision and improving the living conditions of residents. They see smart city development as a means to an end, not an end in itself" (551).

The danger here, of course, is that settler cities, large *and* small, are simply reimagined as smarter spaces of tolerant liberalism, (e-)democracy, and international migration, concealing their ongoing colonial nature (Porter and Yiftachel 2019). The settler colonial dimension remains hidden, the urbanized place of Indigeneity unclear. Suffering from decades of multifaceted decline, however, Saint John's economic renewal is predicated on demographic recovery. Its practical theory of smartness is to facilitate growth through more inclusive migration, especially of foreigners. This is a welcome public rebuke – by an old Canadian city – of new forms of populist xenophobia now seeping into New Brunswick and Canadian politics. This is a policy effort to pull *against* economic gravity by pulling *in* social others. Smartness in Saint John seeks to break down "barriers" of sedimented civic capital *in loco* by building up new data connections across critical initiatives within and beyond the city. Empirically, this is more than a market-liberal fixation on corporate job creation and public tax generation, however important. It is also about after-school programs, immigrant gateways, domestic violence, mixed-income neighbourhoods, and childhood poverty rates, as well as the accepted banality of integrated traffic lights and efficient snow removal – all public services that edify the public city. It is about *just renewal* that ameliorates local suffering, social exclusion, and populist xenophobia. Widening our empirical lens to include neglected stories is important as we seek to build more compelling critical accounts of smart city spatialities.

ACKNOWLEDGMENTS

The research reported here was conducted as a Canada Fulbright Research Chair while at the University of Calgary in 2019. The project was funded by the Canada-US Fulbright Program's Foundation for Educational Exchange. Award no. 8422-CA | Research Chairs in Social Sciences.

REFERENCES

Angelo, H., and D. Wachsmuth. 2014. "Urbanizing Urban Political Ecology: A Critique of Methodological Cityism." *International Journal of Urban and Regional Research* 39 (1): 16–27. https://doi.org/10.1111/1468-2427.12105

Armondi, S. 2022. "Towards Geopolitical Reading of 'Periphery' in State Spatial Strategies: Concepts and Controversies." *Geopolitics* 27 (2): 526–45. https://doi.org/10.1080/14650045.2020.1792444

Aurigi, A., and F. De Cindio. 2008. *Augmented Urban Spaces: Articulating the Physical and Electronic City*. Burlington, VT: Aldershot.

Beauregard, R. 2018. *Cities in the Urban Age: A Dissent*. Chicago: University of Chicago Press

Canada. 2018a. Infrastructure Canada. "Investing in Canada – Canada's Long-Term Infrastructure Plan." Ottawa.

Canada. 2018b. Infrastructure Canada. "Smart Cities Challenge Helping Shape Canada's Communities of the Future." Ottawa. Online at https://www.canada.ca/en

/office-infrastructure/news/2018/11/smart-cities-challenge-helping-shape-canadas-communities-of-the-future.html

Canada. 2019. Infrastructure Canada. "The Government of Canada Announces Winners of the Smart Cities Challenge." Ottawa, 14 May. Online at https://www.newswire.ca/news-releases/the-government-of-canada-announces-winners-of-the-smart-cities-challenge-859468435.html

Canada. 2020. Infrastructure Canada. "Smart City Challenge." Ottawa, 26 August. Online at https://www.infrastructure.gc.ca/cities-villes/index-eng.html

City of Saint John. 2018. "City of Saint John Smart Cities Challenge Application." Saint John, NB. Online at http://www.saintjohn.ca/site/media/SaintJohn/CSJ_Application.pdf, accessed 19 March 2019.

Coletta, C., L. Evans, L. Heaphy, and R. Kitchin. 2019. *Creating Smart Cities*. Abingdon, UK: Routledge.

Datta, A. 2015. "New Urban Utopias of Postcolonial India: 'Entrepreneurial Urbanization' in Dholera Smart City, Gujarat." *Dialogues in Human Geography* 5 (1): 3–22. https://doi.org/10.1177/2043820614565748

Desserud, D. 2015. "The Political Economy of New Brunswick." In *Transforming Provincial Politics: The Political Economy of Canada's Provinces and Territories in the Neoliberal Era*, ed. B. Evans and C. Smith, 110–34. Toronto: University of Toronto Press.

Dierwechter, Y. 2018. "The Smart State as Utopian Space for Urban Politics." In *The Routledge Handbook on Spaces of Urban Politics*, ed. K. Ward, A. Jonas, B. Miller, and D. Wilson, 47–57. New York: Routledge

Ebi, K. 2017. "Challenge Grant Update: How Orlando's Planning for Smart Growth." Smart Cities Council, 9 June. Online at https://smartcitiescouncil.com/article/challenge-grant-update-how-orlandos-planning-smart-growth

Enterprise Saint John. 2017. "Building Steady Long-Term Economic Growth: 2017 Annual Report." Online at https://edgsj.com/storage/pdfs/resources/IXqsiRbuBPvV7QUt6agiXP4IHrNvQFBHHdA6p45W.pdf, accessed 19 March 2019.

Goodman, E., and J. Powles. 2019. "Urbanism under Google: Lesson from Sidewalk Toronto." *Fordham Law Review* 88 (2): 457–98. https://doi.org/10.2139/ssrn.3390610

Harvey, D. 1996. "Cities or Urbanization?" *City* 1 (1–2): 38–61. https://doi.org/10.1080/13604819608900022

Herrschel, T., and Y. Dierwechter. 2018. *Smart Transitions in City-Regionalism: The Quest for Competitiveness and Sustainability*. London: Routledge.

Letson, C. 2018. "Saint John's Big Data Project a Good Foundation for $10 Million Smart City Challenge." *Huddle*, 6 March. Online at https://huddle.today/saint-johns-big-data-project-good-foundation-10-million-smart-city-challenge/

Living Saint John. 2014. "Social Renewal Strategy." Saint John, NB. Online at https://www.tamarackcommunity.ca/vclibrary/pch-prs-saintjohn

Moisio, S. 2018a. *Geopolitics of the Knowledge-Based Economy*. New York: Routledge.

Moisio, S. 2018b. "Urbanizing the Nation-State? Notes on the Geopolitical Growth of Cities and City-Regions." Urban Geography 39 (9): 1421–4. https://doi.org/10.1080/02723638.2018.1454685

New Brunswick. 2017. Office of the Premier. "Investment of $10 million to address generational poverty in Saint John." Fredericton, 23 May. Online at https://www2.gnb.ca/content/gnb/en/news/news_release.2017.05.0748.html

Porter, L., and O. Yiftachel. 2019. "Urbanizing Settler-Colonial Studies: Introduction to the Special Issue." *Settler Colonial Studies* 9 (2): 177–86. https://doi.org/10.1080/2201473X.2017.1409394

Rodríguez-Pose, A. 2018. "Commentary: The Revenge of the Places That Don't Matter (and What to Do about It)." *Cambridge Journal of Regions, Economy and Society* 11 (1): 189–209. https://doi.org/10.1093/cjres/rsx024

Rogers Communications. 2018. "Saint John Common Council, 23 April." Online at https://www.rogerstv.com/show?lid=12&rid=19&sid=1086&gid=293076

Savoie, D. 2001. *Pulling Against Gravity: Economic Development in New Brunswick during the McKenna Years*. Montreal: Institute for Research on Public Policy.

Scott, A. 2001. "Globalization and the Rise of City-Regions." *European Planning Studies* 9 (7): 813–26. https://doi.org/10.1080/09654310120079788

Sheppard, E., H. Leitner, and A. Maringanti. 2013. "Provincializing Global Urbanism: A Manifesto." *Urban Geography* 34 (7): 893–900. https://doi.org/10.1080/02723638.2013.807977

Spicer, Z., N. Goodman, and N. Olmstead. 2021. "The Frontier of Digital Opportunity: Smart City Implementation in Small, Rural and Remote Communities in Canada." *Urban Studies* 58 (3): 535–58. https://doi.org/10.1177/0042098019863666

Tajbakhsh, K. 2001. *The Promise of the City Space, Identity, and Politics in Contemporary Social Thought*. Berkeley: University of California Press.

Taylor, Z. 2018. "Ontario's 'Places That Don't Matter' Send a Message." *Inroads* 44. online at https://inroadsjournal.ca/ontarios-places-dont-matter-send-message/, accessed 22 March 2019.

Townfolio. 2017. "Demographics." Online at https://townfolio.co/ab/calgary/demographics

Walkom, T. 2018. "Right-wing populism on rise in New Brunswick." *Toronto Star*, 25 September. Online at https://www.thestar.com/opinion/star-columnists/2018/09/25/right-wing-populism-on-rise-in-new-brunswick.html

Wave. 2017. "Enterprise Saint John Launches Community Data Strategy." 25 May. Online at https://www.thewave.ca/2017/05/25/enterprise-saint-john-launches-community-data-strategy/

Wiig, A. 2017. "Secure the City, Revitalize the Zone: Smart Urbanization in Camden, New Jersey." *Environment and Planning C: Politics and Space* 36 (3): 403–22. https://doi.org/10.1177/2399654417743767

Yu, W., and C. Xu. 2018. "Developing Smart Cities in China: An Empirical Analysis." *International Journal of Public Administration in the Digital Age* 5 (3): 76–91. https://doi.org/10.4018/IJPADA.2018070106

PART FIVE

Urban Citizenship and Participation

A Dialogue with Alison Powell

Part Five of the book explores classic themes of participation, citizenship, and community that the smart city might entail and curtail. Contributors in this part examine what constitutes citizenship and participation in the smart city – from tokenistic and paternalistic approaches to smart city participation to citizen-led platforms and community empowerment. Authors also engage with whether the smart city is a worthwhile goal in struggles for justice.

To start this final part of the book, we converse with Dr Alison Powell, Associate Professor of Media and Communication at the London School of Economics and Political Science. Dr Powell's research is a unique examination of the civic use, regulation, and politics of communicate and data technologies. In her recent book, *Undoing Optimization: Civic Action in Smart Cities*, Dr Powell goes beyond binaries of top-down/bottom-up civic action to explore how citizenship and participation shift in response to frameworks and practices of smart cities.

Editors: You conclude your 2014 collaboration with Couldry by noting the murky future of Big Data's impacts on democracy. When you wrote this, it seems as though you held optimistic views regarding the potential of Big Data to express individuals' and collectives' voices. Six years later, with the prevalence of fake news and mass disinformation, have any empirical contexts moved in a direction that you feel captured your optimism? In this part of the book, Orlando Woods analyses Singapore's regulatory repones to disinformation, highlighting the tension between the digital empowerment of citizens and attempts to legislate "smartness." What does smartness mean to you? What do you think is the value of smart or the manifestation of smart in our new digital world?

AP: As Stephen Graham points out earlier in this book, "smartness" has a long history. I'm interested in how smartness is associated with the idea of optimizing some aspect of urban life through an appeal to a technological frame. Smartness can take on different guises associated with the promise of different technologies. So, when you have networking as focus of smartness, then you sell networking as one form of optimization, making access to information (or the Internet) something that you can build an economic identity around. Later, when the technological imaginary is linked to datafication, cities become smart when they are oriented to collecting and profiting from data analysis. There's also an imaginary of a "sensing city," that Nigel Thrift has evoked as the "sentient city" – although, like the smart city, the ideal model here sometimes seems to approach a closed-loop technocratic vision where everything is perfectly calibrated and there is end-to-end control.

Yet there is more to this. There is something interesting about the shared global lived experience of idiocy in the face of smartness. I am alluding to idiocy in the way that Isabelle Stengers uses it. In 2018, I went to this fantastic symposium that was organized at Goldsmiths: *The Idiotic City*. This idea of the idiotic was ever-present in the very concept of the smart city. Based on the idea of not being seduced by the promise that smartness was sufficient, the symposium looked at how idiocy was being continually produced and reproduced through the attempts at smartness. That was the moment when I started thinking about this question around optimization in perhaps a more nuanced way than before – becoming increasingly cynical about whether voice and transparency could really be enhanced through these optimizing dynamics. I had noticed that these frameworks of optimization produced ideal versions of citizenship that configured the means through which people could express themselves, especially to governments. Open data advocacy can be an example of this: citizens do the work of auditing the government, legitimizing an ever-smaller state. This is not mindless extraction, as the critique of "data colonialism" implies; it is an appeal to the logic of datafication and optimization, now directed to the exercise of citizenship.

This can enclose and limit the expression of citizenship. However, these optimizing systems have an idiocy within them; they are incapable of conceiving what their technology might be capable of doing outside the logic in which they are imagined and operated. For example, the smart city of 2015 was essentially parking monitoring –because that is the stuff that is so easy to optimize. How could such banal frameworks capture civic life?

Editors: In Part Three, authors focus on these more mundane elements of the smart city, but push back against the banality foregrounding questions of immobility and access in the city.

AP: It is within these banal examples – parking monitoring – where the questions around justice and participation have the biggest impact. In London there are still parking inspectors – it is an interesting job. There is a great character in John Lanchester's novel *Capital* who is a migrant to London and a parking inspector. In the novel, all the parking inspectors in this area of London are migrants from the same country, so they have a migrant community of women who help one another get parking inspector jobs. There is a wonderful way that Lanchester writes about the affective quality of parking inspection, and how this parking inspector approaches negotiating what she is expected to do (e.g., how many parking tickets she is supposed to give out), who she is, and how she thinks about herself in the novel. She then falls in love with a Polish builder, so there is this other narrative of their connection and the generative relationships that the city produces. John Lanchester's work is so attentive to the bigger changes that are going on; he writes this very complex novel that is all about class and hybridity and the messiness of urban life.

It is these kinds of jobs that are under threat in various ways. The parking inspectors disappear after parking is optimized for the person driving – and all of the affective, complex potential disappears. Similarly, the Polish builder disappears after Brexit. To me this suggests that smartness is intrinsically linked with austerity politics, which I discuss in *Undoing Optimization*.

The smartness of the parking app is therefore totally idiotic because it removes all this intelligence and generativity that is part of the city. It also removes the capacity for people to identify their similarities across differences. It eliminates rather than expands intelligence. By trying to narrow everything down to the things that are easy to pick off and easy for commercial entities to privatize, we miss out on any of the generative capacity to innovate.

Editors: One of the tensions that keeps coming up, is the idiocy versus common sense of the smart city. The word "idiot" has a strong resonance in this field as being a counterweight to smartness because it is seen colloquially as the opposite of smartness. Smartness is the absence of idiocy and vice versa. And so, that gets at one of the key concerns and questions about how words and ideas can be rewritten as a political tactic and rearticulated to express different kinds of goals, logics, and rationalities, thereby leaving open the ways in which they never had a grounding to begin with. They can be rearticulated to mean different things, but they also can be reappropriated for more emancipatory ends. Perhaps a more "ideal city" would embrace the idiotic city fully, recognize idiocy as inherent to urbanism writ large, and pursue that and channel it, cultivate it.

But throughout the volume, we can also see the opposite argument, suggesting that in a lot of ways we need to recognize that smartness is here, that we need to play into smartness, make smartness better for certain kinds of people, and forget about rejecting the term altogether.

As I-Chun Catherine Chang and Ming-Kuang Chung explore in the case of Taipei's "perceptible initiatives," authorities have somewhat "successfully (re-) politicized the smart city away from the top-down, developmental state regime's modernization project and towards commitments to the citizens" – that it is okay to rearticulate smartness, and necessary to play into the structure just because that is where we are at. Could you walk us through how you think through those competing notions?

AP: I love that you are asking this question in this way. Another ongoing dialogue that I have been having around some of these ideas is with Adam Greenfield, who wrote the pamphlet "Against the Smart City" a few years ago. In that pamphlet, he articulates this idea that smartness is this immutable logic that is normalizing and reductive. He takes an architectural metaphor for this, and thinks in terms of the Brownfield Smart City that's coming from nowhere. It is an easy one to argue against. In exactly the way that you have presented you can say, "oh, but actually, that is not what smartness is; smartness is this greater and deeper and more dynamic form of life."

But the other thing he does in that pamphlet that has been more interesting for our conversations is that he identifies how that larger form of smartness is also captured in this back-and-forth dynamic between what you might think of as the civic broadly construed and the optimizing power. The optimizing power is a combination of the promise of commercial intermediation and control in the public sector, and the public sector's obsession with efficiency. It has its roots in both neoliberalism – the idea of the shrinking state – and in the increasing power of technology companies. So, Adam's argument is that the civic notion of smartness has been reappropriated and rearticulated from within the optimizing frame. I take that point because you can see it happening. He points out in work he did after the pamphlet that how the critiques he puts forward in the pamphlet have made their way back into the language and practice of the big architectural consultancies.

My rejoinder to that is that you have to look at what is excessive to the optimizing tendency because, otherwise, you get back into this ancient structure and agency debate, where he is saying, "well, we do not really have agency because the structure is just appropriating it," and my response is yes, and that is going to continue to be the case as long as you have a modernist view of things. If you start thinking about this from a more pluriversal perspective – to go with Arturo Escobar's notion of the world as containing many worlds – then there is something that is excessive. There is excessive smartness to the city that can never be captured, and that is essentially where I go in the end of the book, where I start talking about sensing and where I start trying to invert and reverse the idea of sensing as being smart, not towards idiocy, but towards pluriversal ideas of a multiplication of intelligences.

We are in a real moment of a forced conceptual transformation, partly because of the catastrophic ecological consequences of the COVID-19 pandemic, partly through the knowledge of climate change that we are all beginning to be able to feel and sense, and, I also think, because of real grief in seeing the consequences of neoliberalism and austerity all around us. So, it is a bit like you cannot foreclose it. You cannot make a safe scholarly space where the modern is going to be okay because the rest of the non-modern is coming in at the edges everywhere, like flooding through the floor.

Editors: Shifting explicitly to participation, we are thinking about the ways in which participation specifically can be grounded in infinite numbers of discourses, imperatives, logics, social ontologies, and so on. Participation, like social justice, has been theorized from multiple directions from Paulo Freire to Nancy Fraser to the very famous French students' poster that says "I participate, you participate, they profit." Or Miranda Joseph's engagement with the idea of community in the book Against the Romance of Community, *where she shows how the idea of community is often portrayed as an inclusive, warm sense of belonging, but actually works to exclude as much as it includes.*

As Chang and Chung note, we are seeing a shift, albeit a potentially discursive one, away from a corporate smart city to a citizen-centred one. Marikken Wathne, Andrew Karvonen, and Håvard Haarstad suggest that "smart city activities have turned more towards citizen-centric formulations and narratives. Citizenship in such projects, however, has been shown to be rather limited, as citizens have not been included in the early stages of project development, but rather are expected to function as data points, entrepreneurs, and service users." Nina David, noting these same deficits, calls for smart participation through urban development processes.

We are thinking about chains of equivalents and related terms and ideas that are supposed to give meaning to the term participation. Specifically, what are the limitations and assumptions and uneven implications of grounding participation in terms like community, citizenship, or citizen science, and those kinds of competing values?

AP: In my book, participation and long-standing critiques of "participant washing" come up. I examine the way the dynamics of optimized smart cities structure certain forms of participation as exercises of ideal citizenship. Ideal citizenship has not so much to do with the notion of national belonging, as with an appeal to constant contribution – of time spent online, data generated, sensors activated. That makes it much easier to see the relationship between participation and profit of intermediaries, but it also shows how participation can become so much more than what fits inside this ideal frame.

You can participate in a citizen science project, where your data collection is celebrated but where policy makers reject its conclusions. You can also use your participation in a disruptive or idiotic, unruly way. This is where I like to think about the potential of feral smartness or feral data, which go a little bit wild, go beyond what smartness expects. Like feral foxes or coyotes in cities that flourish because of the excesses of the city, feral smartness suggests the presence and growth of something beyond what the system is supposed to be able to contain. So, instead of participating in the way you ought to, your very presence as a person, as a potential outsider or as a creator of feral data, is challenging the logic of containing participation in the logic of profit, for example.

Community is much more difficult. I've been struggling with the concept of community for a decade. It is an evocative concept but often an exclusive and limiting one. Instead, I work more often with the concept of the commons, building beyond ideas limited to the resource-based commons that Elinor Ostrom talks about in her work. The problem with community is that it is always at risk of enclosing, either from the inside or from the outside. Commons can be enclosed, too, but their very presence acts as a way of resisting this enclosure. Radical commons can model the idea of inclusivity and cooperation. Living in the United Kingdom, I am very aware that community can be a dog-whistle term for "minority" or "other." The use of the word "community" can be violent from the powerful to the less powerful – for example, in explicitly positioning ethnic communities as marginal. So, I am very careful about using "community" here for that reason. I also think the celebration of community so easily becomes a celebration of exclusion, of homogeneity, and of conformity, as opposed to difference.

Editors: In this part of the book, Inka Santala and Pauline McGuirk explore the transformative possibilities of sharing cities. While seeing great potential and capacity –urban agency, proactive citizen subjectivities, and collective capacities – they stress the need for "both practical and ideological supports to foster just and sustainable urban development." Recognizing the tension between citizenship and participation and how it is being co-opted in smart city movements, what implications does this have for the possibility of justice? Do we need to reconceptualize these ideas to integrate justice within the smart narrative?

AP: Again, the concept of citizenship is a difficult one. My favourite conception of how citizenship is evolving is Engin Isin and Evelyn Ruppert's book, *Being Digital Citizens*. The right to claim rights about how you're represented is an enormous part of the groundwork for rethinking citizenship. Especially in this age of re-emerging hard borders, we should be thinking about what, where, and how the right to claim rights emerges in relation to these smart mandates. I think that is a place to start.

In my own work, I adopt Isin and Ruppert's notions of citizenship, and then try to articulate and situate them in a global city context where citizenship is not connected with someone's status as a person in the nation, but instead is connected with someone's stake in the place where they live. I have tried to articulate this capacity to develop the right to claim rights with a real stake in the space where you are living to escape the apparatus of national citizenship, which I think is quite oppressive as well.

Where do we put the idea of citizenship in relation to justice claims? I think if we work with this idea of a deep investment in the place in which you are located, then the claim for justice follows right on the back of that, because that requires you to have solidarity with migrants and with other people who occupy the space along with you, and to have an ecosystem or relational view of all the other beings that are flourishing in that space, which gives you a responsibility to care for the place in which you live.

Editors: This is a great transition to your recent work, which engages with the idea of ethics in practice. Can you talk more about your concept of ethics in relation to some of the other concepts you work with? In the dialogue in Part Two, Rob Kitchin cautions against contemporary forms of ethics washing and the pitfalls of best practices and corporate social responsibility. Why do you advocate for new ethics in relation to smart cities? And what does this mean concretely?

AP: I think about ethics in terms of what we practice, how we navigate our world and our relations. As a result of doing this work on cities and technology, I became interested in what it means to consider the broader flourishing of us as humans and of the places in which we live; this question of flourishing is a deep notion within ethics.

Once I started thinking about flourishing, I began to think of it as a way of situating the different things I become involved in – researching smart cities as well as being involved in activist practices from digital rights activism to work in my neighbourhood. Enacting different relations and examining how they might flourish is also part of the "data walking" research process I've documented and explored. When you're walking in a local area trying to observe different ways that data are produced along with a group of others, and you are each tasked with playing a particular role, this can generate new knowledge as well as break down boundaries. It's not a seamless, optimized process. It is, however, a way to generate new perspectives and enact principles of inclusion, reflection, and agonism.

I was thinking about what connected all these different things together. And it was because, for me, this idea of the generation of potential opportunities for flourishing is something that is practised all the time – which is an ethical stance that goes beyond considerations of consequence. Going back to the smart city, critics will say "well, the smart city is terrible because it's created a surveillance apparatus," and that's a consequence. And then you say, "oh, now we have a surveillance apparatus; what do we do? Oh, we have got to respond to that state. Maybe we can limit the amount of time the cameras are on, to solve the ethical problem." But, for me, the ethical question comes in all the way along – we're always putting things together and we're always making different kinds of possibilities.

My work has always been about multiple possibilities produced at the same time. The ethical questions are to do with how and under what circumstances these possibilities generate practices that tend towards flourishing. I retain an optimism of the spirit about this. So much evidence can be collected about the shortcomings of our practices and the limited capacity of our institutions and forms of social organization, and yet there are still more things being created all the time – the feeling of a city, for example, the spirit that is collectively produced that transcends the efforts at optimization.

Editors: How does this relate, then, to your conceptualizations of justice? Do you see ethics as complementary to or different from ethical considerations?

AP: Justice is a pivot point for me in thinking about ethics in practice. When you are thinking about the multiple possibilities, you're also thinking about possibilities to shift direction. In the work that I did as part of the VIRT-EU research project from 2016 to 2019, we developed a set of tools that oriented practitioners to points of potential transformation, and we talked about them as pivot points. This meant looking at organizations that are making technologies, and asking, "Is there a point at which it could go in a slightly different direction? You could make a different decision and then your product might develop in a different way." We looked at those pivot points as places where another possibility could emerge.

I have just had a pivot point in my own work over the past year, undertaking the *Just AI* project in partnership with the Ada Lovelace Institute. We had been trying to think about different ways to investigate ethics in practice, and then the Black Lives Matter movement exploded into visibility. There was a moment within myself as a thinker, and within the project as a whole, when we realized that this was an invitation to pivot towards justice as an orienting concept for the project and also as something that was intervening with notions of ethics. You could see justice as one consideration within an ethical framework, but you could also see that justice could be oppositional to or fundamentally disruptive of some notions of ethics. Because, if you were taking ethics, for example, as a set of principles of the ideal, those principles might completely exclude perspectives that a justice framework would need to include.

This was what the pivot produced for us. We realized that our thinking had been missing a set of perspectives because they had been systematically marginalized from the conversation. As the project unfolded, it became strongly influenced by perspectives of racial and disability justice. We were able to support a group of fellows working on racial justice issues to develop a space for reflection and work from decolonial, abolitionist, and Black feminist perspectives. Making space for this work shifted the way we understood and practised our own research on data and AI ethics. Making more space for others and using power to protect and lift is a fundamental ethical practice, founded in recognition of potential outside our current frame.

Editors: From historical to future-ing approaches, Güneş Tavmen and Federico Cugurullo highlight the failures of smart cities and AI to learn from earlier iterations. Despite repeated warnings, policy makers and technologies continue to ignore the mounting surveillance, privacy, transparency, and distribution issues that are endemic in the current technological paradigm. Federico Cugurullo notes that "[e]thical issues of intelligent machines that can consciously harm humans and find themselves in the position of having to distribute unavoidable harm are unprecedented and pose new questions that go beyond the smart city." Why would you advocate for studying smart cities and AI?

AP: My answer to that is always that it is much more interesting to study that which people take to have great power over their imagination. We are in a world in technologies are no longer outside our social relations. They are deeply entwined in them, and they are structuring of them and constitutive in certain ways. And I think the mind space that these patterns of thought are occupying is critical. Why do we pay attention to smart cities and AI? Perhaps because we hold out hope that some systems bigger and more complex than humans could improve things for us. That bigger and more complex world is already with us, and we are already part of it. The promise of mastery over time, space, and the natural world, tied up with extractive capitalism, is not the only promise that humanity can fulfil. We should keep thinking about smartness, about intelligence beyond our individual selves, and about the ways that we can connect with one another. But we don't have to be limited by narrow notions of optimization or consequence. There is more.

FURTHER READING

Couldry, N., and A. Powell. 2014. "Big Data from the Bottom Up." *Big Data & Society* 1 (2): 2053951714539277. https://doi.org/10.1177/2053951714539277

Data Walking. n.d. "Data Walking." Online at http://www.datawalking.uk

Escobar, A. 2018. *Designs for the Pluriverse*. Durham, NC: Duke University Press.

Greenfield, A. 2013. *Against the Smart City (The City Is Here for You to Use, Book 1)*. New York: Do Projects.

Isin, E., and E. Ruppert, E. 2020. *Being Digital Citizens*. Lanham, MD: Rowman & Littlefield.

Joseph, M. 2002. *Against the Romance of Community*. Minneapolis: University of Minnesota Press.

Lanchester, J. 2013. *Capital*. New York: W.W. Norton.

Ostrom, E. 1990. *Governing the Commons: The Evolution of Institutions for Collective Action*. Cambridge: Cambridge University Press.

Powell, A.B. 2021. *Undoing Optimization: Civic Action in Smart Cities*. New Haven, CT: Yale University Press.

Stengers, I. 2015. *In Catastrophic Times: Resisting the Coming Barbarism*. Chicago: Open Humanities Press.

VIRT-EU. n.d. "VIRT-EU Project Service Package." Online at https://www.virteuproject.eu/servicepackage/

20 The Challenges of Fostering Citizenship in the Smart City

MARIKKEN WULFF-WATHNE, ANDREW KARVONEN, AND HÅVARD HAARSTAD

Over the past decade, the smart city has emerged as a central driver of urban development in cities around the world. Enhancing the livability, sustainability, and efficiency of cities through the application of information and communications technology (ICT) will have profound, long-term implications for urban governance (Alawadhi et al. 2012; Barresi and Pultrone 2013; Calvillo, Sánchez-Miralles, and Villar 2016; Neirotti et al. 2014). The rise of the smart city has also been the subject of numerous critiques due to an overemphasis on technological innovation at the expense of citizenship and democracy (Hollands 2008, 2015; Marvin, Luque-Ayala, and McFarlane 2016; Vanolo 2016). The focus on citizenship in smart urban development is an attempt to move away from the techno-optimist, solutionist agendas promoted and enacted by large technology companies towards urban futures that are defined by those stakeholders who are directly affected by ICT applications. Responding to the lack of emphasis on citizen engagement, smart city advocates increasingly have emphasized participation, justice, and inclusion as additional criteria for urban digitalization processes (Hill 2013; Joss, Cook, and Dayot 2017). However, it is unclear how residents can participate in smart city projects and the types of citizenships that are produced (Kitchin, Cardullo, and Di Feliciantonio 2019; Shelton and Lodato 2019; Vanolo 2014, 2016). This suggests that there are deeper issues of representative and recognition justice to address in digitally mediated cities.

The aim of this chapter is to explore the challenges of fostering citizenship in smart city projects in Europe. We draw upon empirical findings from three European smart cities projects in Stavanger, Stockholm, and Nottingham to understand how citizenship is framed and enacted through urban digitalization projects. Our findings reveal several intrinsic characteristics in smart city projects that circumscribe citizenship. While planners and other urban elites often have a strong and genuine desire to include citizens in these projects, structural and practical constraints often preclude extensive stakeholder engagement.

Instead, many smart city activities involve superficial forms of inclusion and engagement that do little to enhance citizenship. This suggests that there are significant barriers to ensure that the digitalization of society can support socially just and politically active societies.

The "Citizen Turn" in the Smart City

Over the past decade, there has been a growing body of scholarly work on the smart city phenomenon. Vanolo (2016) argues that the literature on smart cities can be grouped broadly into three strands. The first strand engages with the potential positive and negative outcomes of concrete *smart* engagements (see, for example, Alawadhi et al. 2012; Barresi and Pultrone 2013; Calvillo, Sánchez-Miralles, and Villar 2016; Kramers et al. 2014; Neirotti et al. 2014). The second strand is highly critical of the underpinning values of the smart city that are informed by neoliberal planning ideals, corporate and profit-oriented agendas, and increasing tendencies towards surveillance and control of the general public (Datta 2018; Hollands 2008, 2015; Marvin, Luque-Ayala, and McFarlane 2016; Viitanen and Kingston 2014). And a third strand interprets the smart city as a negotiated concept with highly variegated applications in different contexts (Karvonen, Cugarullo, and Caprotti 2019; March 2018; McFarlane and Söderström 2017). As a whole, these three strands illustrate the promise, peril, and pragmatics of integrating ICT into urban development processes. Moreover, they suggest that smart urbanization is a deeply political endeavour with both positive and negative consequences for a wide range of stakeholders.

In these engagements with the smart city, there has been an increasing emphasis on stakeholder involvement to enhance democratic accountability. Stakeholder involvement concerns *procedural* justice, or the involvement of citizens in formal decision-making processes. It also concerns aspects of *redistributive* justice that question who benefits from smart city developments, as well as *recognition* justice that addresses which cultures and values are acknowledged and affirmed through smart city actions (Fraser 2009). This emphasis on justice requires smart city promoters to extend their gaze beyond technological innovation and the optimization of collective service to engage with issues of social equity and democracy.

In general, European cities tend to have a stronger focus on including citizens in their smart city strategies than do their Asian counterparts (Neirotti et al. 2014). Focusing only on data mining and system optimization is insufficient if the digitalization of cities is intended to benefit a diverse urban population instead of a small segment of technologically skilled *smart citizens* (McFarlane and Söderström 2017; Vanolo 2014). There is a shared understanding that the addition of the human element to the smart city can serve as an antidote to the emphasis on instrumental and functional outcomes of ICT applications (Bakici, Almirall, and Wareham 2013; Barresi and Pultrone 2013; Deakin 2014; Martin, Evans, and Karvonen 2018; Rocco, Carmela, and Adriana 2013). This emphasis on the *softer* elements of citizenship, participation, and inclusion has been particularly pronounced in the past five years, not only with citizen groups and government organizations, but also in the private sector (see also David, in this volume). Kitchin (2014, 133) notes that "smart city vendors such as IBM and Cisco have started to alter the discursive emphasis of some of their initiatives from being top-down managerially focused to stressing inclusivity and citizen empowerment." From this, one could conclude that calls for more inclusive smart city projects have been heard, and that actually existing smart cities are, in fact, beginning to include citizen-centric ideas as key elements (Joss, Cook, and Dayot 2017). In effect, smart urban development practitioners are aligning their approach with the long-standing practices of participatory planning that has been a common approach to urban development over the past five decades (Forester 1999; Healey 1992; Karvonen, Cook, and Haarstad 2020; Parker and Street 2018).

Although the emphasis on citizenship is a welcome addition to smart city debates, many scholars argue, however, that we are still far from realizing a truly citizen-centred smart city. An increasing body of literature has emerged to examine critically how citizenship is enacted through smart city projects (Cardullo and Kitchin 2019; Shelton and Lodato 2019; Zandbergen and Uitermark 2020). For example, Chang and Chung (in this volume) point to the spatial inequalities in inclusionary practices, arguing that the focus on citizen inclusion can lead to favouring certain spatial areas and socio-economic groups over others, creating "new inequalities and social justice concerns." Similarly, Listerborn (in this volume) illustrates how smart technologies are often gender biased despite efforts to use technological innovation to benefit all urban residents. Thus, the notion of the *citizen* is invoked in particular ways in the smart city discourse. As Powell (in this volume) notes, the term citizen "has not so much to do with the notion of national belonging, as with an appeal to constant contribution – of time spent online, data generated, sensors activated." In other words, being a citizen is to follow the standard script in invited spaces of participation. Similarly, Kitchin, Cardullo, and Di Feliciantonio (2019, 9–10) argue that "citizens most often occupy non-participatory, consumer, or tokenistic positions and are framed within political discourses of stewardship, technocracy, paternalism, and the market, rather than being active, engaged participants where smart city initiatives are conceived in terms of rights, citizenship, the public good, and the urban commons." Participation often takes the form of providing and engaging with data through events such as hackathons, where "the belief that urban issues are solvable through technological fixes" encourages citizen participation as "leveraging the innovation capacity of a crowd of talented, technically literate citizens to practice ... 'solutionism'" (Perng, Kitchin, and Donncha 2018, 189).

Cowley, Joss, and Dayot (2018) argue that there are four modalities of citizenship in contemporary smart cities, ranging from the *service user citizen* (benefiting from collective services), the *entrepreneurial citizen* (participating in co-creation and innovation), the *political citizen* (engaging in decision-making and deliberation), and the *civic citizen* (instigating grassroots activities). They argue, however, that citizens in the smart city tend to fall into the first two modalities and the citizen becomes a "data point, both a generator of data and a responsive node in a system of feedback" (Gabrys 2014, 38). Zandbergen and Uitermark (2020) characterize this as *cybernetic*, rather than *republican*, citizenship, where individuals are passive receivers of smart city services, rather than active contributors to their design and development (Cardullo and Kitchin 2019; Kitchin 2019; Shelton and Lodato 2019). Smart city projects are thus formulated to benefit citizens by developing measures to improve their urban lives, but citizens are not engaged directly in the decisions about what is to be done and how. There are no opportunities for contestation, and "space for pluralistic, public deliberation and debate is conspicuous by its absence in the smart city regime, displaced by the overwhelming emphasis on data governance" (Joss, Cook, and Dayot 2017, 42).

In sum, the smart city has been criticized for lacking genuine citizen engagement, and smart city activities recently have turned more towards citizen-centric formulations and narratives. Citizenship in such projects, however, has been shown to be rather limited, as citizens have not been included in the early stages of project development, but rather are expected to function as data points, entrepreneurs, and service users. Cardullo and Kitchin (2019, 813) conclude: "Despite attempts to recast the smart city as 'citizen-focused,' smart urbanism remains rooted in pragmatic, instrumental, and paternalistic discourses and practices rather than those of social rights, political citizenship, and the common good." In other words, the smart city is far from being an emancipatory approach to urban development, but instead tends to reinforce existing inequalities and power structures (Marvin, Luque-Ayala, and McFarlane 2016; Shelton and Lodato 2020).

Circumscribed Citizenship in European Community Smart City Projects

To test the theoretical claims about citizenship in the smart city, it is important to examine how these issues are interpreted and navigated in practice. In the following paragraphs, we briefly summarize empirical findings that reveal the specific ways citizenship is circumscribed in smart cities. The findings are based on in-depth case studies of three smart city projects funded by the European Commission's (EC's) Smart Cities and Communities program from 2014 to 2019. These projects positioned Stavanger, Stockholm, and Nottingham as "Lighthouse" projects and exemplars for other cities around the world to emulate. We conducted interviews with key stakeholders in each city and also participated in site visits and selected public events to understand how the stakeholders attempted to foster citizenship while designing, installing, and monitoring new ICT interventions (Haarstad and Wathne 2019a, 2019b; Wathne and Haarstad 2020). The research activities revealed a combination of structural and practical constraints that restricted the opportunities to support deep and meaningful citizen engagement.

The Tyranny of European Funding Schemes

The EC emphasizes citizen engagement as an essential component of smart city development (Cardullo and Kitchin 2019; European Commission 2020). Paradoxically, multiple structural constraints within the EC's funding schemes prevent such engagement from being realized in practice. European funding applications require a significant amount of work to develop competitive cases for support and research consortia, and the application timeframes tend to preclude opportunities for substantive dialogue with citizens (Valdez, Cook, and Potter 2018). Instead, project teams tend to rely on familiar, reputable individuals and organizations that have a track record of success. In addition, there is an increasing professionalization of EC funding application processes (Büttner and Leopold 2016). It is common for smart city project applications to be led by professional application developers, such as the German research institute Fraunhofer, that have demonstrated competence in the technical aspects of proposal writing, but lack the contextual insights on local issues that are relevant to citizens. This creates an instrumental smart city funding process that prioritizes what is technically achievable at the expense of deep engagement with local stakeholders to define the problems at hand and to deliberate on potential solutions.

In our case studies, respondents noted that they prioritized the recruitment of business and academic partners for their smart city consortia to meet EC funding requirements, rather than involving citizens in defining project objectives and activities. One site manager explained that they had only a few months to build the consortium and write the application: "You have no opportunity whatsoever to have strong citizen input." Another project leader echoed this sentiment, noting that, "In the choice of tasks to do, we did not have much citizen involvement. First of all, we didn't have time for that ... it was more than enough to get a program stitched together with [the partner cities], and also to be sustainable in itself and to build a consortium with so many involved parties from the private sector, academia and the public sector."

Relatedly, the application requirements tended to preclude citizen involvement. As Kitchin, Cardullo, and Di Feliciantonio (2019) note, the focus, objectives, and solutions need to be spelled out clearly at the application stage. In the tough competition for funding, cities vie to present themselves as the most prepared to deliver a feasible project with strong local support. An emphasis on the inherent conflicts of interest and the misgivings of citizens would undermine the case for support. Therefore, including citizens in the application process is impractical and in direct opposition to producing a successful application in a timely manner. As a site manager stated, "I have worked in the past with a lot of bottom-up processes, but we did not have the opportunity with the EC application ... so it has really become more about subsequently bringing the tenants and visitors to the event area and see if we can get more out of a co-creation there." In other words, citizen engagement can be supported, but not until *after* the funding has been secured.

Once funded, successful projects are required to complete specifically defined activities and objectives as promised in their application, allowing little room for input from citizens. Major decisions underpinning smart city interventions had already been made at the application stage by a small group of project stakeholders. During the projects, citizen input was limited to minor, inconsequential details such as the colour of a house, the placement of doors, or the addition of balconies in a retrofitted building. The lack of citizen involvement was particularly pronounced in the Stockholm case study, where the consortium radically *smartified* an apartment building only to find out later that most of the residents were retirees who had little interest in using the newly installed high-tech solutions. As the site manager explained: "When we had chosen the buildings and the area and everything else, it turns out that 60 per cent of those who live there are pensioners (10 per cent are over eighty-five years old). And then there are completely different solutions we should have thought of than having an electric car charging station."

In sum, the EC funding approach for smart cities opens opportunities for urban planners, politicians, and other urban elites to raise funds for key projects and to connect with other cities, but does little to foster citizenship. Funders require well-formulated bids, where the main actors and project activities are clearly delineated and justified. This leaves little time to engage with citizens in the early stages of formulating smart city projects. Consequentially, "the focus, objectives, and solutions are set before any problems and suggestions from citizens can be taken into account, and it is only when the funding is in hand that engagement occurs with local communities" (Kitchin, Cardullo, and Di Feliciantonio 2019, 11). Here, citizens are expected to serve in predefined roles of *smart entrepreneurs* and *civic hackers* (Townsend 2013), rather than influence the aims and objectives of smart interventions.

Paternalistic Approaches to Smart City Interventions

Beyond the constraints imposed by EC funding requirements, several of our respondents described multiple challenges to engaging citizens in their smart city projects. As one project administrator explained, "It's a paradox, because the smart city is about citizen inclusion, but the citizens do not want to be included." In some cases, citizens did not understand how the smart city interventions were personally relevant to them. For example, a respondent described a housing retrofit project where some residents were not convinced they would benefit from the proposed interventions. The project team then spent significant time and energy to engage with the locals and convince them of the benefits of the activities. The respondent noted, "this is a process that takes time because you have to convince people that this is in their interest." In other instances, citizens actively resisted the proposed smart interventions and refused to have their houses refurbished, and, in a few cases, even took the local government to court over the proposed refurbishment plans.

The project managers tended to agree that their projects lacked genuine citizen engagement, but argued that they had the bests interests of the citizens at heart. Rather than seriously engage with the critiques raised by citizens, they saw these disagreements as something to manage and resolve through persuasion. Thus, the project managers engaged in paternalistic practices to advocate for the utility and benefits of predefined smart city objectives. This reflects an underlying embrace of consensus over agonism (Munthe-Kaas 2015; Pløger 2004) in smart city activities and concerted attempts to pacify those who oppose or have diverging opinions about how urban digitalization should be performed (Kitchin, Cardullo, and Di Feliciantonio 2019).

Beyond persuasion, several project managers described how they reframed the very notion of citizen engagement to be about data extraction, rather than substantive democratic engagement. There was thus a substitution of political and civic forms of inclusion with technological inclusion, or an embrace of the cybernetic citizen. For example, one smart city promoter argued that, although citizens were largely unaware of ongoing smart city initiatives, they were still deeply involved because they generated waste heat that was recovered and reused. Other respondents noted that citizens participated by contributing to big datasets that could be used to optimize collective services or benefited from the use of new visualization methods and apps to improve citizen engagement with the local government. All these activities were heralded as important forms of civic engagement while relegating citizens to a *subaltern role* as passive receivers of ICT services (Bina, Inch, and Pereira 2020).

Conclusions

Citizen inclusion is an increasingly prominent goal for smart city proponents. However, including citizens as influential decision makers in smart urban development activities has proven difficult in practice. Two common challenges to citizen involvement are related to structural constraints of funding and the practical difficulties of aligning smart city project aims with citizens' wants and needs. EC smart city projects tend to be delineated in advance of any substantive citizen engagement, which restricts the involvement of citizens to providing their opinions on superficial preferences and serving as users of smart tools. This results in a particularly narrow form of citizenship that lacks deep engagement with the substantive processes of procedural, redistributive, and recognitional justice (Fraser 2009).

To address issues of justice in the smart city, there is a desperate need for genuine citizen engagement, to acknowledge and support divergent perspectives, and to support forms of participatory planning and implementation that include all stakeholders in the design and implementation of smart cities. Such activities would go well beyond empty slogans about "putting citizens at the centre of the smart city" to foster deeper and democratically infused forms of citizenship in the digitally mediated city. Powell (in this volume) speaks of using participation in the smart city as an opportunity to support citizen engagement in *disruptive* and *unruly* ways. Such actions would expand the currently circumscribed space of smart citizenship and open possibilities to steer technological innovation towards applications that can enhance social justice. This requires time and funding that is not currently available in the global competition to be the smartest city in the world, but is sorely needed to ensure that the long-lasting effects of smart urban development are not only technically viable but also socially and politically just. It is only with a concerted effort by all smart city stakeholders to the development of the *actually existing smart citizen* (Shelton and Lodato 2019) that urban digitalization applications can contribute to the fostering of citizenship and the realization of a more just and equitable world.

REFERENCES

Alawadhi, S., A. Aldama-Nalda, H. Chourabi, J.R. Gil-Garcia, S. Leung, S. Mellouli, T. Nam, T.A. Pardo, H.J. Scholl, and S. Walker. 2012. "Building Understanding of Smart City Initiatives." In *Electronic Government: 11th IFIP WG 8.5 International Conference, EGOV 2012, Kristiansand, Norway, September 3–6, 2012. Proceedings*, ed. H.J. Scholl, M. Janssen, M.A. Wimmer, C.E. Moe and L.S. Flak, 40–53. Berlin: Springer.

Bakici, T., E. Almirall, and J. Wareham. 2013. "A Smart City Initiative: The Case of Barcelona." *Journal of the Knowledge Economy* 4 (2): 135–48. https://doi.org/10.1007/s13132-012-0084-9

Barresi, A., and G. Pultrone. 2013. "European Strategies for Smarter Cities." *TeMA: Journal of Land Use* 6 (1): 61–72. https://doi.org/10.6092/1970-9870/1455

Bina, O., A. Inch, and L. Pereira. 2020. "Beyond Techno-Utopia and Its Discontents: On the Role of Utopianism and Speculative Fiction in Shaping Alternatives to the Smart City Imaginary." *Futures* 115: 102475. https://doi.org/10.1016/j.futures.2019.102475

Büttner, S.M., and L.M. Leopold. 2016. "A 'New Spirit' of Public Policy? The Project World of EU Funding." *European Journal of Cultural and Political Sociology* 3 (1): 41–71. https://doi.org/10.1080/23254823.2016.1183503

Calvillo, C.F., Á. Sánchez-Miralles, and J. Villar. 2016. "Energy Management and Planning in Smart Cities." *Renewable and Sustainable Energy Reviews* 55 (March): 273–87. https://doi.org/10.1016/j.rser.2015.10.133

Cardullo, P., and R. Kitchin. 2019. "Smart Urbanism and Smart Citizenship: The Neoliberal Logic of 'Citizen-Focused' Smart Cities in Europe." *Environment and Planning C: Politics and Space* 37 (5): 813–30. https://doi.org/10.1177/0263774X18806508

Cowley, R., S. Joss, and Y. Dayot. 2018. "The Smart City and Its Publics: Insights from across Six UK Cities." *Urban Research & Practice* 11 (1): 53–77. https://doi.org/10.1080/17535069.2017.1293150

Datta, A. 2018. "The Digital Turn in Postcolonial Urbanism: Smart Citizenship in the Making of India's 100 Smart Cities." *Transactions of the Institute of British Geographers* 45 (3): 405–19. https://doi.org/10.1111/tran.12225

Deakin, M. 2014. "Smart Cities: The State-of-the-Art and Governance Challenge." *Journal of University-Industry-Government Innovation and Entrepreneurship* 1 (1): 1–16. https://doi.org/10.1186/s40604-014-0007-9

European Commission. 2020. "Smart Cities: Cities Using Technological Solutions to Improve the Management and Efficiency of the Urban Environment." Online at https://ec.europa.eu/info/eu-regional-and-urban-development/topics/cities-and-urban-development/city-initiatives/smart-cities_en, accessed 24 June 2021.

Forester, J. 1999. *The Deliberative Practitioner: Encouraging Participatory Planning Processes.* Cambridge, MA: MIT Press.

Fraser, N. 2009. *Scales of Justice: Reimagining Political Space in a Globalizing World.* New York: Columbia University Press.

Gabrys, J. 2014. "Programming Environments: Environmentality and Citizen Sensing in the Smart City." *Environment and Planning D: Society and Space* 32 (1): 30–48. https://doi.org/10.1068/d16812

Haarstad, H., and M.W. Wathne. 2019a. "Are Smart City Projects Catalyzing Urban Energy Sustainability?" *Energy Policy* 129 (June): 918–25. https://doi.org/10.1016/j.enpol.2019.03.001

Haarstad, H., and M.W. Wathne. 2019b. "Smart Cities as Strategic Actors." In *Inside Smart Cities: Place, Politics and Urban Innovation*, ed. A. Karvonen, F. Cugurullo, and F. Caprotti, 102–16. London: Routledge.

Healey, P. 1992. "Planning through Debate: The Communicative Turn in Planning Theory." *Town Planning Review* 63 (2): 143–62. https://doi.org/10.3828/tpr.63.2.422x602303814821

Hill, D. 2013. "On the Smart City; or, A 'Manifesto' for Smart Citizens Instead." *City of Sound*, 1 February. Online at https://www.cityofsound.com/blog/2013/02/on-the-smart-city-a-call-for-smart-citizens-instead.html, accessed 24 June 2021.

Hollands, R.G. 2008. "Will the Real Smart City Please Stand Up? Intelligent, Progressive or Entrepreneurial?" *City* 12 (3): 303–20. https://doi.org/10.1080/13604810802479126

Hollands, R.G. 2015. "Critical Interventions into the Corporate Smart City." *Cambridge Journal of Regions, Economy and Society* 8 (1): 61–77. https://doi.org/10.1093/cjres/rsu011

Joss, S., M. Cook, and Y. Dayot. 2017. "Smart Cities: Towards a New Citizenship Regime? A Discourse Analysis of the British Smart City Standard." *Journal of Urban Technology* 24 (4): 29–49. https://doi.org/10.1080/10630732.2017.1336027

Karvonen, A., M. Cook, and H. Haarstad. 2020. "Urban Planning and the Smart City: Projects, Practices and Politics." *Urban Planning* 5 (1): 65–8. https://doi.org/10.17645/up.v5i1.2936

Karvonen, A., F. Cugurullo, and F. Caprotti, eds. 2019. *Inside Smart Cities: Place, Politics and Urban Innovation*. London: Routledge.

Kitchin, R. 2014. "Making Sense of Smart Cities: Addressing Present Shortcomings." *Cambridge Journal of Regions, Economy and Society* 8 (1): 131–6. https://doi.org/10.1093/cjres/rsu027

Kitchin, R. 2019. "Toward a Genuinely Humanizing Smart Urbanism." In *The Right to the Smart City*, ed. P. Cardullo, C. Di Feliciantonio, and R. Kitchin, 193–204. Bingley, UK: Emerald Publishing.

Kitchin, R., P. Cardullo, and C. Di Feliciantonio. 2019. "Citizenship, Justice, and the Right to the Smart City." In *The Right to the Smart City*, ed. P. Cardullo, C. Di Feliciantonio, and R. Kitchin, 1–24. Bingley, UK: Emerald Publishing.

Kramers, A., M. Höjer, N. Lövehagen, and J. Wangel. 2014. "Smart Sustainable Cities – Exploring ICT Solutions for Reduced Energy Use in Cities." *Environmental Modelling & Software* 56 (June): 52–62. https://doi.org/10.1016/j.envsoft.2013.12.019

March, H. 2018. "The Smart City and other ICT-led Techno-Imaginaries: Any Room for Dialogue with Degrowth?" *Journal of Cleaner Production* 197 (October): 1694–703. https://doi.org/10.1016/j.jclepro.2016.09.154

Martin, C.J., J. Evans, and A. Karvonen. 2018. "Smart and Sustainable? Five Tensions in the Visions and Practices of the Smart-Sustainable City in Europe and North America." *Technological Forecasting & Social Change* 133 (August): 269–78. https://doi.org/10.1016/j.techfore.2018.01.005

Marvin, S., A. Luque-Ayala, and C. McFarlane, eds. 2016. *Smart Urbanism: Utopian Vision or False Dawn?* London: Routledge.

McFarlane, C., and O. Söderström. 2017. "On Alternative Smart Cities: From a Technology-Intensive to a Knowledge-Intensive Smart Urbanism." *City* 21 (3–4): 1–17. https://doi.org/10.1080/13604813.2017.1327166

Munthe-Kaas, P. 2015. "Agonism and Co-design of Urban Spaces." *Urban Research & Practice* 8 (2): 218–37. https://doi.org/10.1080/17535069.2015.1050207

Neirotti, P., A. De Marco, A.C. Cagliano, G. Mangano, and F. Scorrano. 2014. "Current Trends in Smart City Initiatives: Some Stylised Facts." *Cities* 38 (June): 25–36. https://doi.org/10.1016/j.cities.2013.12.010

Parker, G., and E. Street. 2018. *Enabling Participatory Planning: Planning Aid and Advocacy in Neoliberal Times.* London: Policy Press.
Perng, S.-Y., R. Kitchin, and D.M. Donncha. 2018. "Hackathons, Entrepreneurial Life and the Making of Smart Cities." *Geoforum* 97 (December): 189–97. https://doi.org/10.1016/j.geoforum.2018.08.024
Pløger, J. 2004. "Strife: Urban Planning and Agonism." *Planning Theory* 3 (1): 71–92. https://doi.org/10.1177/1473095204042318
Rocco, P., G. Carmela, and G. Adriana. 2013. "Towards an Urban Planners' Perspective on Smart City." *TeMA: Journal of Land Use* 6 (1): 5–17. https://doi.org/10.6092/1970-9870/1536
Shelton, T., and T. Lodato. 2019. "Actually Existing Smart Citizens: Expertise and (Non) Participation in the Making of the Smart City." *City* 23 (1): 35–52. https://doi.org/10.1080/13604813.2019.1575115
Townsend, A.M. 2013. *Smart Cities: Big Data, Civic Hackers, and the Quest for a New Utopia.* New York: W.W. Norton.
Valdez, A.-M., M. Cook, and S. Potter. 2018. "Roadmaps to Utopia: Tales of the Smart City." *Urban Studies* 55 (15): 3385–403. https://doi.org/10.1177/0042098017747857
Vanolo, A. 2014. "Smartmentality: The Smart City as Disciplinary Strategy." *Urban Studies* 51 (5): 883–98. https://doi.org/10.1177/0042098013494427
Vanolo, A. 2016. "Is There Anybody Out There? The Place and Role of Citizens in Tomorrow's Smart Cities." *Futures* 82 (September): 26–36. https://doi.org/10.1016/j.futures.2016.05.010
Viitanen, J., and R. Kingston. 2014. "Smart Cities and Green Growth: Outsourcing Democratic and Environmental Resilience to the Global Technology Sector." *Environment and Planning A: Economy and Space* 46 (4): 803–19. https://doi.org/10.1068/a46242
Wathne, M.W., and H. Haarstad. 2020. "The Smart City as Mobile Policy: Insights on Contemporary Urbanism." *Geoforum* 108 (January): 130–8. https://doi.org/10.1016/j.geoforum.2019.12.003
Zandbergen, D., and J. Uitermark. 2020. "In Search of the Smart Citizen: Republican and Cybernetic Citizenship in the Smart City." *Urban Studies* 57 (8): 1733–48. https://doi.org/10.1177/0042098019847410

21 Structuring More, Inclusive, and Smart Participation in Planning: Lessons from the Field

NINA DAVID

Smart governance, or participatory governance, has been identified as a key component in most definitions of smart cities. Caragliu, Del Bo, and Nijkamp (2011) believe a city to be smart when investments in human and social capital, and traditional and modern information and communications technology (ICT) infrastructure support sustainable development, through *participatory governance.* Lombardi et al. (2012) propose that smart cities include six key dimensions: a smart economy, smart mobility, smart environment, smart people, smart living, and smart governance. Vanolo (2014) emphasizes these dimensions as well, and expands further on the role of citizens in smart cities and smart governance. The notion that smart governance and participation are essential components of smart cities has become so ubiquitous that Castelnovo, Misuraca, and Savoldelli (2016, 729) conclude: "Participatory governance and citizen involvement ... are key concepts in many smart city frameworks and even researchers that do not give governance such a central role in smart cities at least include it as one of the dimensions that should be targeted by smart city initiatives."

This is not surprising given the shift in thinking across academics and practitioners alike, normatively, on the "what, how, and who" of governmental policy making. This shift is evident in terms of how we define "what" constitutes the public interest today (unitary to plural); "how" we determine the public interest (rational and scientific versus communicative and collaborative); and "who" plays a role in determining the public interest (value-neutral technocrats versus the public). Many smart cities, in turn, tout citizen-centrism as part of their brand, particularly highlighting citizen-centric initiatives to solve pressing problems relating to sustainability, housing, infrastructure, and services (Joss, Cook, and Dayot 2017). Although the literature is generally scant, scholars working in this area conclude to date that the vast majority of claims of citizen-centrism and inclusion in smart cities are skin deep, participation is tokenistic, the process is technocratic, and the agenda is mostly neoliberal

(Cardullo and Kitchin 2019; Joss, Cook, and Dayot 2017; Lee, Woods, and Kong 2020; Levenda et al. 2020; Tomor et al. 2019). More work is needed to evaluate the extent to which cities are citizen-centric and promote inclusion in their policy making (see David and Benson 2021).

In this chapter, I provide a model for how to structure and combine digital and non-digital participation to achieve inclusivity in smart city governance and policy making. Particularly, I focus on urban land-use planning, a significant area of substantive policy interest for cities (smart cities included) and the administrative area through which most sustainability goals are advanced. In the dialgoue in this part of the book, Alison Powell articulates the importance of *place* as the grounding for how we think of citizenship, community, and justice. Place is the object of the urban planners' focus, so with land-use planning as an example, I develop a framework for smart participation in the "pre-," "during," and "post-" stages of the comprehensive land-use planning process to build capacity, expand participation opportunities, and secure a public role in plan implementation. I then identify examples of cities from the United States that have provided the listed types of participation opportunities.

Inclusion and justice

In his seminal article, Davidoff (1965, 332) wrote, "If the planning process is to encourage democratic urban government then it must operate so as to include rather than exclude citizens from participating in the process." There is a caveat here, however: neither inclusion nor democracy is guaranteed to produce justice or the just city as an outcome – that is, "there is no necessary link between greater inclusiveness and a commitment to a more just society" (Fainstein 2014, 13). This is because pluralism in reality might mean irresolvable conflicts of interest (Young 2000), conflicts in values (Fainstein 2014), polarization (Innes and Booher 2004), and contentious processes (Davidoff 1965). Those participating in planning processes are not all equal in terms of access to information and resources. Further, information and communications technology (ICT) can reinforce and exacerbate inequities rather than reduce them. For example, access to computers, the Internet, and digital skills might vary across populations and geographies (Johnson and Halegoua 2014) and data can be biased (Green 2019; Kontokosta and Hong 2021). What, then, is the value of inclusiveness and what role can ICT play in promoting inclusion?

Young (2000) argues that process matters, and that planning decisions are "normatively legitimate only if all those affected by it are included in the process of discussion and decision-making" (22). Inclusiveness can lead to greater representation of stakeholders at the table. This diversity of stakeholders, at the table, in deliberative settings, can allow stakeholders to learn about one another's interests and perceptions to the extent where this knowledge can result in a "revising [of] what each participant thinks about each other's and their own interests" (Healey 1992, 157). This can lead to a better understanding of positions and interests that are different from one's own (Jordan and Maloney 1996). Inclusiveness therefore could expand the array of policy alternatives available to address planning problems (Davidoff 1965).

Rob Kitchin, in his dialogue in this book, talks about the importance of justice in terms of not only outcomes (instrumental), but also policymaking and political processes – that is, procedural justice and representational justice. In other words, when used well, ICT can give the public a seat at the table, access to and representation at key decision points, and a role in how problems are conceived and, consequently, which solutions "are entertained and which are dismissed" (Dewey 1938, cited in Green 2019, 4). Green (2019) warns that what we need are not smart cities that rely on ICT for engagement and inclusion, but rather "smart enough" cities that use ICT to enhance innovations and solutions. That is, the goal would be to ensure that smart cities tap into both active users of ICT who might still be underrepresented in policy processes (e.g., young people and working parents) *and* those who traditionally might be unplugged (e.g., lower-income, rural, and older residents) (David and Buchanan 2015; Tomor et al. 2019). Consistent with this, in the next few sections, I provide a framework for inclusion through the use of both digital and non-digital participation techniques.

Participation across the Planning Cycle

Before the Plan

Forester (1982, 67) argues that information is a source of power in the planning process for several reasons. First, the distribution of information can rationalise pre-existing structures of power and control, therefore reinforcing structural inequalities – that is, those who do not have access to information will be unable to effect structural change. Second, information might be used strategically by those in power (e.g., misinformation), resulting in insurmountable political and organizational barriers for those who are underrepresented to weed through the noise. Third, information can help find solutions to technical problems, and enable its possessors to ask the right questions during the planning process (see Davidoff 1965, 332). Finally, knowing where to get information and whom to contact makes it easier to acquire the intellectual capital needed to participate in policy making.

ICT can be used to disseminate large quantities of information to the public, effectively liberating information from the hands of the few into the homes and personal devices of the many at relatively low cost (Bonsón, Royo, and Ratkai

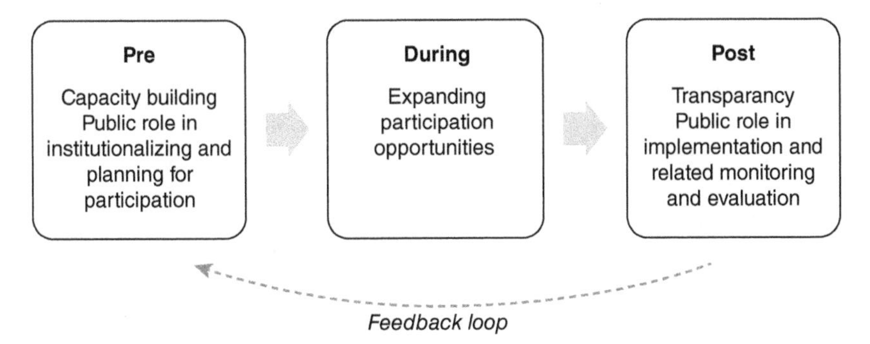

Figure 21.1. A Schema of Inclusion in the Planning Process

2015). ICT thus can increase transparency (Kim and Lee 2019) and inform the public of its rights, responsibilities, and options (Arnstein 1969, 219), thereby building the public's capacity to participate. In a review of the current literature on ICT-enabled participation, Tomor et al. (2019) conclude that, although evidence of effective one-way communication exists, meaningful and deliberative engagement of the public in governance is harder to find. One way to work towards more deliberative public engagement is actually to plan for it and institutionalize it so that two-way engagement in governance is systematic. Such plans for participation should also be co-produced and co-designed with the public (Castelnovo, Misuraca, and Savoldelli 2016), so that the public has a role in determining what sorts of participation opportunities should be provided during the planning process (Utz and Shipley 2012). I suggest, therefore, that engagement in the planning process should occur even before the planning process has begun (see Figure 21.1). This pre-participation stage would consist of cities' efforts to train and build the capacity of the public to participate, and the adoption of formal participation ordinances and plans co-produced with the public that outline what participation in the planning process would look like.

During the Plan

A city's choice of how to engage the public in the planning process will depend on several variables – for example, financial resources and staff capacity, the geographical scope of the issues for which participation is sought, and the time frame of the planning process (see David 2017). The selection of participation techniques will also be affected by what a city hopes to accomplish through participation – whether it seeks public input or feedback, seeks to understand public perceptions, create dialogue, build consensus, build partnerships, or foster deliberation (see Innes and Booher 2004). Finally, the deployment of techniques can also depend on what stage in the planning process participation is sought – whether during problem definition or visioning at the beginning, alternative selection and policy formulation in the middle, or during implementation and evaluation at the end (see Utz and Shipley 2012).

The code of ethics of the American Institute of Certified Planners emphasizes that the public interest should be "formulated through continuous and open debate." Further, the code advocates giving the public "the opportunity to have a meaningful impact on the development of plans and programs that may affect them." It further underlines that "participation should be broad enough to include those who lack formal organization or influence." I propose, therefore, that, in the second stage, when the plan is actually written, the goal should be to provide opportunities for the public to be involved widely and continuously. Here, one-way and two-way interactive techniques might be used to provide a continuum of opportunity for the public to engage throughout the planning process. Today, when used *in addition* to traditional techniques, digital technologies can allow planners to expand participation opportunities to traditionally underrepresented groups (Tomor et al. 2019). Digital techniques also allow the public to participate without a commitment to participate physically, therefore expanding participation of those for whom in-person participation might be a limiting factor (see David 2017; Wilson, Tewdwr-Jones, and Comber 2017).

After the Plan

In the public policy process, policy legitimation, policy implementation, and policy evaluation all occur after policy formulation. Similarly, in the planning process, a plan's policies are implemented by a range of development management regulations (codes and ordinances) and everyday development decisions (e.g., rezoning, development review and approval), meaning that the development of the plan does not bookend the planning process. Most interactive public participation opportunities, however, end when the plan is adopted. Should there be a role for the public in implementation? For example, Painter (1992, cited in Lane 2005) posits that, "we wouldn't know if public participation has had an impact on policymaking until a decision has been rendered that conforms with or contradicts the public opinion on the issue." Aladalah, Cheung, and Lee (2015) argue that, in order to be empowered to participate, the public should feel a sense of impact (i.e., the belief that it is able to influence action) and feel its engagement has been meaningful. ICT can be used here to promote accountability and both inward and outward transparency (David, Justice, and McNutt 2015)

Methods

Stage 1

I first operationalized the inclusion schema from Figure 21.1 as follows:

CAPACITY BUILDING (PRE-)

In the pre-planning phase, capacity building can be achieved by providing the public information and training opportunities. Examples of information provision might include digital platforms that convey information to the public (one-way interaction) on planning related meetings and events, and soliciting responses from the public on the information that is being provided (two-way interaction). Training opportunities might be provided in digital and non-digital forms: webinars and learning videos (one-way), and citizen training academies (two-way).

INSTITUTIONALIZING AND PLANNING FOR PARTICIPATION (PRE-)

The public can also be engaged in the creation of planning documents that will guide participation in the planning process. Examples include the creation of a public participation ordinance and/or a public participation plan.

EXPANDED PARTICIPATION OPPORTUNITIES (DURING)

In the plan-writing phase, expanded participation opportunities might be provided in digital and non-digital formats. Broad participation might be achieved using techniques that allow for both one-way and two-way communication. Examples of traditional participation opportunities include surveys and public hearings (one-way), and focus groups, charrettes, forums, workshops, steering committees, and advisory groups (two-way). Examples of digital participation opportunities might include online and text surveys (one-way), and other web 2.0 and civic technology platforms such as Facebook, Twitter, blogs, discussion boards, and Neighborland (two-way).

TRANSPARENCY (POST-)

In the post-plan phase, transparency might be achieved by continuing to maintain an active web presence dedicated to the comprehensive plan post-adoption. Plan implementation can be tracked online by using indicators and other evaluative measures (one-way).

THE PUBLIC'S ROLE IN IMPLEMENTATION (POST-)

Everyday planning decisions and development projects arising from the plan can be connected explicitly to the plan through information (one-way) presented on a website (e.g., e-trackit, links to upcoming planning commission meetings, digitally available staff reports on development proposals, notices). Feedback can be solicited on these everyday decisions through traditional opportunities such as public hearings (one-way), and focus groups, charrettes, workshops (two-way), and digital tools such as discussion boards and Neighborland (two-way).

Stage 2

I then identified examples of cities for each of the key elements (see stage 1) of the inclusion framework. To do this, I reviewed the academic and professional literatures and focused on awards given, between 2015 and 2020, by entities such as the American Planning Association, Planetizen, the Center for Digital Government, and the National League of Cities for planning-related digital leadership, open government, and citizen engagement. Once, I then accessed each of the examples' websites to gather more information about the public engagement opportunities provided. I also read the community engagement chapters of comprehensive plans and any other supporting material that was provided (e.g., appendices and other material links) to document how engagement occurred. For governments that used online platforms, I also accessed these platforms to decipher what information was being provided in open data portals, dashboards, discussion boards, feedback forms, and civic technology sites, and how these were being used by the public. For example, did the public submit comments on discussion boards and did planners respond? Did the websites contain supporting material for upcoming public meetings? I cite these examples, not to evaluate them per se, but to present them as examples of instances where digital and non-digital participation can be, and has been, used by US cities to institute a more inclusive planning process.

Examples of Inclusive Participation Techniques

Capacity Building

I highlight three cities as examples in capacity building: Lakewood, Washington,[1] for transparency and information provision; and Philadelphia, Pennsylvania, and Indianapolis, Indiana[2] for their citizen training programs.

Lakewood employs a web-based platform, LakewoodSpeaks.org, that enables the public to participate in the city's planning commission hearings two weeks before the live meeting. Staff reports, presentations, applicant files, and all other relevant documents are uploaded to the platform weeks in advance. In addition to reading the documents, the public may ask questions of staff and project applicants, and comment on the proposed actions prior to the meeting. Analysis from 2020 (Lakewood 2020) show that 69 per cent of those who participated in the planning commission meetings digitally were in the 25–54 age group, while only 13 per cent of those who participated in-person were in this age group – meaning that 87 per cent of those who participated in person were above age 55.

Philadelphia and Indianapolis both created planning academies to educate and train the public on the comprehensive planning process. Philadelphia's Citizen Planning Institute[3] attempts to "demystify" planning for the public through an eight-week semi-annual education program. The focus is on core topics of land use, zoning, development, and elective topics that change from year to year. To date, 570 participants representing over 125 different neighbourhoods have completed the program. Program alumni are also invited back to talk to new participants (Philadelphia 2020; see also the website at phila2035.org).

Indianapolis's Peoples Planning Academy was geared specifically towards training the public, through six separate workshops on planning, zoning, development regulations, and transportation to support the city's plan-writing process in 2017. Those who participated in the workshops were invited to join the Land Use Plan Stakeholder Committee to advise the city on the plan. Archival material (e.g., learning videos) are hosted on the city's website. Because of its success, this program was implemented again in 2019 (see the website at plan2020.com).

The Public's Role in Institutionalizing and Planning for Participation

Here, I highlight two examples, Portland, Oregon,[4] and Minneapolis, Minnesota, both for their community engagement plans.

Portland has institutionalized community engagement in at least two ways: first, by dedicating an entire chapter in the comprehensive plan to community engagement; and second, by establishing a formal body called the Community Involvement Committee. Consisting of volunteer members from Portland, the committee's role is institutionalized in the City Code, and the committee is tasked with maintaining a community engagement manual and working with and advising city staff on community engagement in land use and transportation (see website at Portland.gov).

Minneapolis has developed an engagement strategy called "Blueprint for Equitable Engagement." The most notable aspect of the strategy is a dashboard that provides the goals and strategies for community engagement, indicators, and measures of whether the goals and strategies have been achieved. The dashboard also allows planners to indicate where the city has fallen short in terms of engagement (e.g., the diversity of representation in the planning commission) and to develop plans to address the gaps. The dashboard's greatest benefit is its transparency, which allows the public to track whether publicly developed community engagement goals are being met (see website at minneapolismn.org).

Expanded Participation Opportunities

In this case, I highlight two examples: Oklahoma City, Oklahoma,[5] for the range of opportunities provided, and San Francisco, California,[6] for its integrated use of digital and non-digital platforms.

Oklahoma City's comprehensive planning effort is called "planokc." To maximize citizen engagement, the city has prioritized two actions: to reach underrepresented populations (minorities, low-income families, and families with small children) and to go to places where people gather. The city provided a range of one- and two-way, digital and non-digital, opportunities for the public to participate in the planning process. These included workshops, stakeholder groups, meetings-in-a-box – where planners provide all the materials needed for members of the public and neighbourhoods or communities to organize their own planning meetings and report back to the planners – as well as surveys, focus groups, and the use of civic technology platforms such as CrowdGauge. Before the start of the planning process, the planners also embarked on a presentation circuit to different neighbourhoods (50 groups and over 1,300 people) explaining what to look forward to and how to participate. It is estimated that a total of 20,000 people participated. The city continues to maintain a website for all plan-related information, including tying each goal and policy in the plan to implementation, indicating whether the policy is being implemented and how. The plan is an interactive digital plan that can also be downloaded. The website also shows a progress report. For regulatory changes that are being developed out of the comprehensive plan, the website directs the public to the planning department's webpage, where people can leave feedback via a survey on other plans under development (American Planning Association 2018; Oklahoma City 2015; see also website at planokc.org).

The engagement of participants in San Francisco's "Central Waterfront/Dogpatch Public Realm Plan" started with planners' holding preliminary stakeholder meetings with area businesses, institutions, and neighbourhood associations. Then, planners held public workshops where cognitive mapping, dot voting, and other interactive techniques allowed the public to envision changes to the built environment. Ideas generated from public workshops were uploaded to the civic technology platform Neighborland, where the public could provide feedback, ask questions, and offer support for ideas. Planners also sought feedback through e-mail surveys. They then used both the digital and non-digital feedback to identify priority projects, and held more focus groups and workshops to develop design scenarios, with feasibility analyses for the projects. Neighborland was used, again, to solicit online feedback on design scenarios. All handouts and material distributed and collected as part of this community engagement are provided on the city's website. In all, about 1,500 people participated online (Neighborland 2017; San Francisco Planning 2018; see also the website at sfplanning.org).

Transparency and a Public Role in Implementation

Here, I combine the two post-plan categories and highlight three examples: Boulder, Colorado,[7] and Austin, Texas,[8] for their open data and online engagement platforms, and Philadelphia for its provision of public comment opportunities at the scale of neighbourhood parcels of land.

Boulder's comprehensive planning effort, in 2015, included both digital and non-digital, and one- and two-way techniques. Digital opportunities typically consisted of polls, online surveys, social media responses, and e-mail comments, while non-digital opportunities included surveys, focus groups, listening sessions, forums, public hearings and targeted outreach through community organizations. The city continues to maintain a comprehensive plan website with easily accessible and detailed supporting documents that feed into the planning process. The highlights are twofold: easy access to the open data portals and an online engagement platform called Be Heard Boulder, which links the public to all major projects on the docket. It also allows the public to learn about the projects, submit questions to staff about them, and submit feedback directly through the website. All relevant information about the live projects is easy to find, and a "who's listening" insert provides the contact information of the relevant planner who serves as the point of contact for the project (see websites at bouldercolorado.gov; beheardboulder.org).

Austin's comprehensive planning effort, Imagine Austin, resulted in the participation of about 25,000 people through digital and non-digital, one- and two-way techniques. These included surveys (online and paper), community forums (online and in-person), meetings-in-a-box, and travelling planners' circuits. The engagement process was overseen monthly by a Citizens Advisory Task Force and a committee of the planning commission. The city continues to maintain a website dedicated to the plan, and links the plan to an implementation dashboard that shows progress towards achieving plan goals and provides indicators for tracking goals. The indicators further link to the city's open data portal, where data related to each of the indicators are provided and can be downloaded. There is also easy access to live planning commission meetings and archival material (including videos). The highlight, however, is an online engagement portal, SpeakUp Austin, that links the plan to active opportunities (on current projects) for the public to interact with planners online via discussion boards. SpeakUp Austin also has links to surveys, supporting material on the topics under discussion and so on. A review of the discussion boards shows planners responding to all comments posted on the platform (see websites at austintexas.gov; speakupaustin.org).

Philadelphia's comprehensive planning website has a section called "Updates" that lists current projects and provides background information on the projects and public meeting–related information. Online comments are solicited through surveys, e-mail, and ArcGIS-enabled platforms – for example, it is possible to provide comments on a proposed neighbourhood-level rezoning at the scale of each parcel by clicking on a rezoning map and simply entering feedback (see website at phila2035.org).

Participation and Justice in Practice

None of the examples examined in this chapter addressed inclusivity comprehensively from the beginning to the end of the planning process. That said, taken altogether, they provide a model for how public engagement can be structured to be continuous, ongoing, expansive, and inclusive.

Lakewood's online presence increases transparency by making available all information relevant to all petitions considered by the planning commission. It also allows the public to ask questions of staff and petitioners before public hearings. This transparency allows the public to experience the plan in action and to participate actively in the public planning commission meetings without leaving the comfort of home. Unlike most other websites that provide live streams of public meetings (passive participation), the innovation here is in the public's ability to gain information, educate themselves, and participate virtually. It also allows the public to learn through participation, by doing, from experience. That is, while Philadelphia and Indianapolis build the public's planning capacity through education and training, Lakewood builds capacity through experiential learning: the public accumulates knowledge by experiencing and building a "repertoire of cases" (Watson 2002, 182). The Philadelphia and Indianapolis models not only build the intellectual capacity of the public through the training program; they also have the potential to enhance social connectivity and networks (see Innes 1996; Putnam, Leonardi, and Nanetti 1993) in two ways: first, they introduce participants to key institutional players and, therefore, formal connections in government (Forester 1982); and second, they introduce participants to others in the community who, like them, are interested in planning.

Portland and Minneapolis have institutionalized engagement by adopting plans specially dedicated to public participation. The highlight here is that both cities have committed to a public role in the implementation of engagement, albeit in different ways. In Portland, a public body comprised of citizens advises on the implementation of the engagement plan; in Minneapolis, a website allows the public to track the city's progress towards inclusion. The Portland and Minneapolis models are noteworthy because institutionalizing engagement offers several benefits. It codifies engagement, formally designates a governmental role in and a commitment to engagement, leads to greater probability that engagement will be sustained through political changes, and allows for greater accountability and resource dedication to engagement (see David and Buchanan 2015).

Oklahoma's planokc was organized around inclusion, particularly expanding the participation of underrepresented groups in the planning process. To this end, the city used a wide range of participation techniques (digital and non-digital) to "include" the public. San Francisco's subarea plan offers a good example of how digital and non-digital participation can be integrated seamlessly in phases to leverage the advantages and minimize the disadvantages of each of these two types of participation platforms. The San Francisco effort allowed for multiple feedback loops from the non-digital participation opportunities to the digital (civic technology) platforms.

The Boulder, Austin, and Philadelphia examples highlight how governments can facilitate transparency and post-plan implementation-related engagement. The use of open data portals and online engagement platforms, and the provision of opportunities to comment on built environment changes at the level of neighbourhood blocks and parcels expands participation opportunities for those who are unable to attend in-person meetings. Both Boulder's and Austin's online platforms make it abundantly clear as to who is listening to public feedback; Austin's discussion boards facilitate two-way exchange between planners and the public.

In all, these examples have embraced "digital technology as a means by which citizens can more readily understand planning and raise awareness of the opportunities for involvement," therefore removing "barriers to citizen engagement in the planning process," and allowing "a more accessible method for the public to potentially shape their neighbourhood's future" (Wilson, Tewdwr-Jones, and Comber 2017, 288). The examples also provided for technology-mediated participation in all four categories identified by Bugs et al. (2010) – namely, information provision, transparency, participation, and consensus building.

However, although each of these cases is exemplary at particular stages on the inclusion schema (i.e., the planning stages), they have not all been comprehensively inclusive from the beginning of the planning process to the end. For example, Lakewood's expanded participation opportunities are facilitated "after" the planning commission agenda has been decided upon, but there is no interactive online platform that facilitates discussion "before" the agenda is developed. And although Oklahoma City expanded public participation in the writing of its plan, the plan website does not allow two-way interaction with planners (i.e., it not linked to an online engagement platform). The website for planokc does link to current projects on the docket and provides general municipal social media information and e-mail accounts, but, again, these mechanisms do not allow open interaction between planners and the public. Both Boulder and

Austin facilitate two-way interaction online between planners and the public, but neither provides Lakewood's level of online engagement when projects reach a decision point at the planning commission. Philadelphia's semi-annual Citizen Planning Institute builds civic capacity to participate, and participation is facilitated online, but not as interactively as in Boulder or Austin or as extensively as in Lakewood in the lead-up to routine planning commission meetings.

Conclusion

Using the examples examined in this chapter, taken together, it is possible to imagine what the inclusion schema that I provided at the outset might look like in practice. The cases also demonstrate the value of using digital technologies in conjunction with other participatory mechanisms based on local contexts, and leveraging technology to expand the tent of inclusion (see Green 2019). Smart cities can promote digital justice by enabling, engaging, and empowering the public so that members of the public are not just consumers of services, clients, or data points, but active participants and partners in governance (Bonsón, Royo, and Ratkai 2015). Although I have provided a bird's-eye view based on the compilation of examples, future research could focus on providing a more detailed and nuanced understanding of the impact of ICT in planning decisions, and whether the combined use of digital and non-digital platforms affects the quality of participation, plans, and planning decisions.

NOTES

1 Lakewood Speaks won an American Planning Association (APA) National Achievement Award for Best Practice in 2019.
2 The People's Planning Academy in Indianapolis is an APA National Award winner for Public Outreach in 2019.
3 The Philadelphia City Planning Commission won the 2016 APA National Excellence Award for a Planning Agency, as well as a Special Achievement Award for Philadelphia2035, the city's comprehensive plan, from the Pennsylvania Chapter of the APA.
4 The City of Portland's Comprehensive Plan won the Oregon APA Award.
5 Oklahoma City was the recipient of the 2018 APA Daniel Burnham Award for a Comprehensive Plan.
6 San Francisco is a Center for Digital Government and the National League of Cities digital leadership award winner for 2016.
7 The City of Boulder and Boulder County received a 2018 APA Colorado Chapter Honor Award for the updated Boulder Valley Comprehensive Plan, adopted in 2017.

8 Austin is a National League of Cities Bright Spot community for community engagement in planning.

REFERENCES

Aladalah, M., Y. Cheung, and V. Lee. 2015. "Enabling Citizen Participation in Gov. 2.0: An Empowerment Perspective." *Electronic Journal of e-Government* 13 (2): 77–93.

American Planning Association. 2018. "National Planning Awards: Award Recipients." Chicago. Online at https://www.planning.org/awards/2018/planokc/

Arnstein, S.R. 1969. "A Ladder of Citizen Participation." *Journal of the American Institute of Planners* 35 (4): 216–24. https://doi.org/10.1080/01944366908977225

Bonsón, E., S. Royo, and M. Ratkai. 2015. "Citizens' Engagement on Local Governments' Facebook Sites, an Empirical Analysis: The Impact of Different Media and Content Types in Western Europe." *Government Information Quarterly* 32 (1): 52–62. https://doi.org/10.1016/j.giq.2014.11.001

Bugs, G., C. Granell, O. Fonts, J. Huerta, and M. Painho. 2010. "An Assessment of Public Participation GIS and Web 2.0 Technologies in Urban Planning Practice in Canela, Brazil." *Cities* 27 (3): 172–81. https://doi.org/10.1016/j.cities.2009.11.008

Caragliu, A., C. Del Bo, and P. Nijkamp. 2011. "Smart Cities in Europe." *Journal of Urban Technology* 18 (2): 65–82. https://doi.org/10.1080/10630732.2011.601117

Cardullo, P., and R. Kitchin. 2019. "Being a 'Citizen' in the Smart City: Up and Down the Scaffold of Smart Citizen Participation in Dublin, Ireland." *GeoJournal: Spatially Integrated Social Sciences and Humanities* 84 (1): 1–13. https://doi.org/10.1007/s10708-018-9845-8

Castelnovo, W., G. Misuraca, and A. Savoldelli. 2016. "Smart Cities Governance: The Need for a Holistic Approach to Assessing Urban Participatory Policy Making." *Social Science Computer Review* 34 (6): 724–39. https://doi.org/10.1177/0894439315611103

David, N. 2017. "Democratizing Government: What We Know about e-Government and Civic Engagement." In *International e-Government Development: Policy, Implementation and Best Practices*, ed. M. Pedro, R. Bolívar, and L. Alcaide-Muñoz, 73–96. Berlin: Springer.

David, N., and T. Benson. 2021. "Citizen-Centrism in Smart Cities: Reality or Rhetoric?" In *Holistic Approach for Decision Making towards Designing Smart Cities*, ed. G.C. Lazaroiu, M. Roscia, and V. Dancu, 245–66. Berlin: Springer.

David, N., and A. Buchanan. 2015. "Factors Affecting Youth Participation in Local Government Planning Efforts." Miami: Urban Affairs Association.

David, N., J. Justice, and J. McNutt. 2015. "Smart Cities Are Transparent Cities: The Role of Fiscal Transparency in Smart City Governance." In *Transforming City Governments for Successful Smart Cities*, ed. M. Pedro and R. Bolívar, 69–86. Berlin: Springer.

Davidoff, P. 1965. "Advocacy and Pluralism in Planning." *Journal of the American Institute of Planners* 31 (4): 331–8. https://doi.org/10.1080/01944366508978187

Fainstein, S.S. 2014. "The Just City." *International Journal of Urban Sciences* 18 (1): 1–18. https://doi.org/10.1080/12265934.2013.834643

Forester, J. 1982. "Planning in the Face of Power." *Journal of the American Planning Association* 48 (1): 67–80. https://doi.org/10.1080/01944368208976167

Green, B. 2019. *The Smart Enough City: Putting Technology in Its Place to Reclaim Our Urban Future*. Cambridge, MA: MIT Press.

Healey, P. 1992. "Planning through Debate: The Communicative Turn in Planning Theory." *Town Planning Review* 63 (2): 143–62. https://doi.org/10.3828/tpr.63.2.422x602303814821

Innes, J.E. 1996. "Planning through Consensus Building: A New View of the Comprehensive Planning Ideal." *Journal of the American Planning Association* 62: 460–72. https://doi.org/10.1080/01944369608975712

Innes, J.E., and D.E. Booher. 2004. "Reframing Public Participation: Strategies for the 21st Century." *Planning Theory and Practice* 5 (4): 419–36. https://doi.org/10.1080/1464935042000293170

Johnson, B.J., and G.R. Halegoua. 2014. "Potential and Challenges for Social Media in the Neighborhood Context." *Journal of Urban Technology* 21 (4): 51–75. https://doi.org/10.1080/10630732.2014.971528

Jordan, G., and W.A. Maloney. 1996. "How Bumble-bees Fly: Accounting for Public Interest Participation." *Political Studies* 44 (4): 668–85. https://doi.org/10.1111/j.1467-9248.1996.tb01748.x

Joss, S., M. Cook, and Y. Dayot. 2017. "Smart Cities: Towards a New Citizenship Regime? A Discourse Analysis of the British Smart City Standard." *Journal of Urban Technology* 24 (4): 29–49. https://doi.org/10.1080/10630732.2017.1336027

Kim, S., and J. Lee. 2019. "Citizen Participation, Process, and Transparency in Local Government: An Exploratory Study." *Policy Studies Journal* 47 (4): 1026–47. https://doi.org/10.1111/psj.12236

Kontokosta, C.E., and B. Hong. 2021. "Bias in Smart City Governance: How Socio-Spatial Disparities in 311 Complaint Behavior Impact the Fairness of Data-Driven Decisions." Sustainable Cities and Society 64 (January): 102503. https://doi.org/10.1016/j.scs.2020.102503

Lakewood, WA. 2020. *Comprehensive Plan*. Online at https://cityoflakewood.us/wp-content/uploads/2020/09/0920-LAKEWOOD-COMPREHENSIVE-PLAN.pdf

Lane, M.B. 2005. "Public Participation in Planning: An Intellectual History." *Australian Geographer* 36 (3): 283–99. https://doi.org/10.1080/00049180500325694

Lee, J.-Y., O. Woods, and L. Kong. 2020. "Towards More Inclusive Smart Cities: Reconciling the Divergent Realities of Data and Discourse at the Margins." *Geography Compass* 14 (9). https://doi.org/10.1111/gec3.12504

Levenda, A.M., N. Keough, M. Rock, and B. Miller. 2020. "Rethinking Public Participation in the Smart City." *Canadian Geographer* 64 (3): 344–58. https://doi.org/10.1111/cag.12601

Lombardi, P., S. Giordano, H. Farouh, and W. Yousef. 2012. "Modelling the Smart City Performance." *Innovation: The European Journal of Social Science Research Innovation* 25 (2): 137–49. https://doi.org/10.1080/13511610.2012.660325

Neighborland. 2017. "Central Waterfront and Dogpatch Public Realm Plan." San Francisco. Online at https://handbook.neighborland.com/sfplanning-dogpatch-6589a654fdf5

Oklahoma City. 2015. "Planokc." Online at http://planokc.org/wp-content/uploads/2020/04/01_planokc_final_20201210.pdf

Philadelphia. 2020. City Planning Commission. "Philadelphia 2035." Online at https://www.phila2035.org/plan

Putnam, R.D., R. Leonardi, and R. Nanetti. 1993. *Making Democracy Work: Civic Traditions in Modern Italy*. Princeton, NJ: Princeton University Press.

San Francisco Planning. 2018. *Central Waterfront/Dogpatch Public Realm Plan*. Online at https://default.sfplanning.org/Citywide/Dogpatch_CtrlWaterfront/CWD_Public_Realm_Plan_ADOPTED_Oct2018.pdf

Tomor, Z., A. Meijer, A. Michels, and S. Geertman. 2019. "Smart Governance for Sustainable Cities: Findings from a Systematic Literature Review." *Journal of Urban Technology* 26 (4): 3–27. https://doi.org/10.1080/10630732.2019.1651178

Utz, S., and R. Shipley. 2012. "Making It Count: A Review of the Value and Techniques for Public Consultation." *Journal of Planning Literature* 27 (1): 22–42. https://doi.org/10.1177/0885412211413133

Vanolo, A. 2014. "Smartmentality: The Smart City as Disciplinary Strategy." *Urban Studies* 51 (5): 883–98. https://doi.org/10.1177/0042098013494427

Watson, V. 2002. "Do We Learn from Planning Practice? The Contribution of the Practice Movement to Planning Theory." *Journal of Planning Education and Research* 22 (2): 178–87. https://doi.org/10.1177/0739456X02238446

Wilson, A., M. Tewdwr-Jones, and R. Comber. 2017. "Urban Planning, Public Participation and Digital Technology: App Development as a Method of Generating Citizen Involvement in Local Planning Processes." *Environment and Planning B: Urban Analytics and City Science* 46 (2): 286–302. https://doi.org/10.1177/2399808317712515

Young, I.M. 2000. *Inclusion and Democracy*. Oxford: Oxford University Press.

22 Emerging Inequalities in Citizen-centric Smart City Development: The Perceptible Initiatives in Taipei

I-CHUN CATHERINE CHANG AND MING-KUANG CHUNG

We aim to make Taipei a smart city that is liveable and sustainable, and for all the citizens.
– Taipei mayor Ko Wen-Je, Smart City Mayor Summit, 13 August 2020

Cities across the globe have now widely adopted smart city initiatives to employ real-time, data-driven governance and to pursue a better urban future. The smart city model promises to integrate various urban systems, bring about technological and administrative reforms, and address a wide array of urban issues related to the economy, governance, the environment, mobility, education, and everyday living (Giffinger et al. 2007; Townsend 2013). Early smart city initiatives were often criticized as overly top down and technocratic in their orientation, reinforcing neoliberal logics and serving the interests of corporations and state entrepreneurialism (Greenfield 2013; Kitchin 2014; Rossi 2016; Shwayri 2013; Söderström, Paasche, and Klauser 2014; Wiig 2016). These smart city initiatives often prioritize market-oriented solutions over measures that would better uphold social and political rights and promote the common good for citizenry (Cardullo and Kitchin 2019).

In response to these criticisms, the agenda of smart cities have gradually shifted to citizen-oriented initiatives. For example, the funding arm for smart cities of the European Union, the "European Innovation Partnership for Smart Cities and Communities" program, designates a funding cluster to "citizen-focused" initiatives. Major technology providers and political consultancies for smart cities have also often rebranded their services as focusing on citizens and communities (Cardullo and Kitchin 2019, 2). In one of its recent reports, the McKinsey Global Institute declares that "[a] modern smart city is not just an urban area with a highly developed technological infrastructure, but rather a place where citizens live in a smarter way" (Woetzel and Kuznetsova 2018, 7). Meanwhile, surveys have found many cities adopting smart city initiatives that focus on citizen participation and community engagement in recent years, with the goal of achieving better social inclusion and equality (see Coletta et al. 2019).

Could this new shift towards a "citizen-centric" smart city bring about a different type of technological solution that is less neoliberal and more oriented towards the common good for social inclusion and equality? This chapter sets out to examine recent citizen-centric smart city initiatives in Taipei, the capital of Taiwan. By looking into its "perceptible initiatives" – smart initiatives the city enacted to generate impact that would be widely recognized by citizens – we contextualize the smart development of the city and examine the implications of these citizen-centric smart initiatives. We draw from policy analysis of smart city initiatives, interviews with local residents and community leaders, city government officials, and key industry actors, and participatory observation at the sites of smart projects and related public forums, conferences, and workshops from 2018 to 2020. Based on these extensive data, we argue that the perceptible initiatives were launched both to secure support for the mayor's re-election bid and also to serve as a test bed for market-oriented solutions. In addition to the political success they aimed to achieve, however, we find that these initiatives have led to unequal resource distribution between different areas of the city and across the fault lines of socio-economic status. As a result, smart city development in Taipei has contributed to new forms of inequalities.

Contextualizing Taipei Smart City

Taipei's adoption of smart city development was closely linked to the industrial restructuring of Taiwan. To make up for its eroded comparative advantage in labour-intensive manufacturing, Taiwan in the early 1990s actively pursued new economic niches in knowledge- and capital-intensive information and communications technology (ICT) industries, first by setting up science parks and similar facilities across the island. Then, in 1994, the Taiwanese central government established the National Information Infrastructure Construction Committee to push for the construction of an island-wide cyber network that connects all households to the Internet and to promote Internet-based applications in education, government access, and financial services (Tseng 2007). These initiatives were enacted to create a protected domestic market for Taiwan's nascent ICT industries. Starting in the early 2000s, these infrastructure projects morphed into a national policy campaign of building Taiwan into an entrepreneurial "Silicon Island of East Asia" (akin to the Silicon Valley). These attempts to promote ICT industries reflected a common industrial development strategy of a development-oriented central government. In this process, the central government both directly invested in ICT infrastructure projects and mobilized state-owned enterprises, semi-governmental agencies, and local governments to build and manage Internet-based applications.

Taipei first started its smart city development to complement the development campaign of the central government. Under Nationalist Party (also known as Kuomintang or KMT) mayor Yin-Jeou Ma, Taipei launched the CyberCity project in 1999 to position Taipei as the leader in the national campaign of promoting ICT industries. During Ma's two terms from 1999 to 2006, the CyberCity closely adhered to the plan of the National Information Infrastructure Construction Committee of building ICT infrastructure and promoting Internet-based applications. Taipei partnered with semi-government and government-sponsored ICT corporations to build its citywide broadband and Wi-Fi network; digital education and governmental initiatives were also enacted to promote greater usage of the network.[1] In the late 2000s, Ma's mayoral successor from the same party, Lung-Bin Hau, largely continued with the same smart city development agenda under his Intelligent City initiative. The Hau administration focused on creating universal access to the Internet and e-government services by expanding the ICT infrastructure and integrating governmental websites and databases. The administration also added cloud computing, Big Data, and the Internet of Things as part of its ICT infrastructure construction in the early 2010s (Chung et al. 2011; Huang 2012). The ICT infrastructure and industries were deemed as crucial for the revitalization of Taipei's stagnant economy. The Hau administration proclaimed that the Intelligent City initiative would help transform Taipei into an urban science park that would complement Taiwan's national industrial development strategy and raise the city's economic competitiveness (Taipei City Government 2013).

Taipei's smart city development under the Ma and the Hau administrations can be viewed as a continuation of central-local dynamics under Taiwan's authoritarian developmental state (Hsu 2011). The state relied heavily on the patron-client relationship between the central government, local administrations, and related interest groups (Evans 1995; Wade 1990). The central government directed investment into local development, most often through infrastructure projects, that benefited selective interest groups in exchange for their political support. Taipei's smart city development prior to the early 2010s largely continued with this mode of governance. While many city developments in the global North and West focused on Internet applications and software platforms and involved mostly transnational corporations, Taipei's smart city development under Ma and Hau was chiefly a state-led large-scale infrastructure construction implemented by state and semi-state agencies (Chang, Jou, and Chung 2020).

Taipei's smart urbanism agenda went through a drastic shift starting in 2015 under newly elected mayor Wen-Je Ko. Politically, Ko was the first non-affiliated, and the second non-KMT Taipei mayor, since the end of Taiwan's martial law in 1987. Ko placed smart technology and the smart city agenda at the forefront of his election campaign and later as the anchor of his governing agenda. Rather than passively adhering to the central government's industrial development agenda, Ko presented smart city initiatives as a new platform for making urban governance more transparent and for boosting citizenry participation in policy making. The smart city initiatives, as the Ko administration claims, will lead to a democratic "Open Government" for "ordinary citizens" (Taipei City Government 2017b). Such framing employs a populist narrative that emphasizes "the people" (Mudde and Kaltwasser 2017) vis-à-vis the governing elites and the developmental agenda stipulated by the developmental state.

Using this populist framing, the Ko administration reoriented Taipei's smart city from a top-down infrastructure construction project to city-led experiments in technological governance. Seizing the control of its smart city agenda from the central government, the administration presented Taipei as a citizen-centred "living laboratory" under its "Smart Taipei" campaign. This vision was first enacted in the "5+N" scheme in Ko's first term, aiming to create innovative solutions (the "N") across five major domains of everyday urban living: transportation, public housing, health care, education, and daily monetary transactions (Taipei City Government 2018). Under this scheme, the Ko administration has incentivized ICT and related industries to use Taipei as the experimental site to test various "Proof of Concepts." The administration promised to help commodify and scale up successfully tested smart solutions and adopt them in its city-wide policies.

After Ko was successfully re-elected for a second term, the administration revised its smart city development agenda to a "1+7" scheme in 2019 (Taipei City Government 2020b). Under this new scheme, the core mission of Taipei's smart city is to establish a smart government based on initiatives in seven domains of urban living (building standards, transportation, education, health, the environment, security, and the economy) under four guidelines: citizen participation, open government, open data, and international collaboration. The revision offers a more comprehensive framework for smart city development and strengthens citizen participation in the development agenda. Such revision also prioritizes citizen-centric initiatives and elevates the smart city to be the overarching development agenda for Taipei.

Perceptible Initiatives

In line with Ko's populist platform, the administration presented its citizen-centric initiatives as "perceptible" initiatives. As one of the leading government officials commented in an interview, "smart city is the 'beef' we promised during the election campaign, so we have to find a way to make people see and feel like we kept our promise" (authors' interview with ED01, December 2018). Priority thus has been given to smart initiatives that residents can experience in their daily lives, rather than improving governing functions using invisible algorithms. Our research also suggests that the perceptible initiatives were tailored intentionally to serve certain populations with varying positionalities that together constitute Ko's electoral base. The administration has targeted two specific groups of voters and developed strategic collaborations with related social groups and neighbourhood organizations.

The first targeted group of voters was highly educated working-age adults who feel victimized by Taiwan's developmental state. Similar to other Asian developmental states, Taiwan's export-oriented economy relied on an educated workforce and depressed wages (Wade 1990). While sustaining the island's economy over the past few decades, this mode of development nevertheless produced a large group of well-educated but relatively undercompensated working-age adults. During this same period, housing prices in Taipei increased multifold. The combination of low wages and high prices made it impossible for young working-age adults to purchase a residential unit in Taipei on their own (Li and Hung 2018; Taipei City Government 2017a) and start wealth building. This young cohort often holds grievances against the traditional political parties, which are viewed as responsible for the housing affordability crisis and associated social inequalities. These young people became Ko's most avid supporters in the 2014 mayoral race, and have continued to back his administration since his first election (Chen and Yang 2016).

To court this group, the Ko administration maintained that smart city development would bring about new industries and economic opportunities for young working adults. The administration also presented smart city development as a means to correct the prior urban development trajectory led by traditional governing elites and privileged interest groups under the developmental state regime. Taipei launched several perceptible initiatives aiming to address the housing affordability crisis for young adults, and adopted new technologies in the policy-making process through which this group could feel included.

Smart Public Housing

One of the most prominent perceptible initiatives targeting young working adults is Taipei's smart public housing project (renamed "smart social housing" in 2019). The project was promoted as a tool for advancing housing justice and helping young people stay in the city. The Ko administration pledged during the first election to construct twenty thousand new public housing units in four years, and another thirty thousand units if Ko were elected for a second term. These public housing units were to be "smart." The plan included smart housing features (smart grids and smart metres) and technology-enabled and platform-based service systems that manage services including parking, security, rental management, telecommunications, health care, daycare assistance, cashless retailing, and cloud-based online library services (Smart Taipei n.d.a).

The construction of the public housing units has been slower than originally expected. As of August 2020, when this chapter was written, almost two years into Ko's second term, Taipei had completed 6,233 units, far below even the pledged 20,000 for the first term (Taipei City Government 2020a). The completed units are now accepting applications and welcoming new residents. In order to apply, the applicant's household income has to be below the fiftieth percentile of household income in Taipei ($1,580,000 NTD), and monthly household income per capita has to be no higher than 3.5 times the city's poverty line ($59,518 NTD) (Taipei City Public Housing for Rent n.d.). This application standard is lenient enough that young white-collar workers can usually qualify.

Smart Citizen Participatory Initiatives

To address young working adults' grievances about the developmental state and the politicians who supposedly perpetuated the housing problem, the Ko administration created a series of precipitable initiatives focusing on citizen participation, including Open Government, i-Voting, and Smart Participatory Budgeting. These initiatives were presented as an invitation to ordinary people to join the policy decision-making process, which had been opaque and dominated by partisan politics and interest groups. The new measures aim to increase government transparency and solicit citizens' feedback on urban issues and budget allocation.

In addition, the Ko administration has worked with civic groups organized by young professionals to help implement the smart participatory initiatives. One important partner is g0v.tw, a primarily online community that, according to its website, "pushes information transparency, focusing on developing information platforms and tools for the citizens to participate in society." Technological members of g0v.tw have participated extensively in Taipei's smart city projects. Specifically, they led the open data and data visualization tasks to support i-Voting, Open Government, and Smart Participatory Budgeting, especially during Ko's first term. The Ko administration has advertised its collaboration with g0v.tw on news outlets and social media as part of its commitment to participatory democracy. Members of g0v.tw have supported the administration and its smart city policies on various occasions, especially in online forums. These members often cast the Ko administration as the only local administration with the courage to involve civic communities and upend the corrupt alliance between political and business interests.

Our observations suggest that the Ko administration often prioritizes initiatives that court well-educated young professionals; this was especially the case during Ko's first mayoral term. Close to his re-election bid, however, we observed increasing attention paid to a different flavour of perceptible initiatives: those likely appreciated by citizens with limited digital proficiency, including seniors, retirees, blue-collar workers, and housewives who traditionally support politicians from the major political parties (authors' interviews with LA01 and LA02, January 2019).

In order to determine which initiatives would make this second group of constituents experience the "smartness" that smart city development promises, the city government consulted local community leaders and city council members. The government also hosted workshops and public hearings with residents of designated pilot sites. Based on their input, the city finalized a list of initiatives to push forward, among which some have been scaled up and applied in multiple communities. Most initiatives on this list are related to infrastructure upgrades, waste, pollution and environmental management, and daily mobility concerns. Below is an overview of these initiatives.

Smart-Eco Communities

One of the most prominent precipitable initiatives targeting citizens with limited digital proficiency is the retrofitting of communities to become smart and ecologically sustainable. With intellectual roots in urbanism models such as the eco-city and the garden city, these "smart-eco" communities aim to make infrastructure changes that improve the city's ecological function while opening up spaces in communities for smart experiments. Most of the projects under this initiative are located in public spaces in the city's redevelopment areas. Common infrastructure provisions include sustainable urban landscaping features (such as permeable surfaces, native gardens, storm water management, and watershed management) and smart-technology-enabled daily services (such as smart light poles fitted with data-collection sensors, traffic monitoring, real-time parking tracking and ticketing systems, automatic waste and recycling collection stations, animal tracking systems, and security systems using facial recognition artificial intelligence) (Taipei City Government n.d.).

The Air Box

The Air Box, a portable, low-cost, and user-friendly air quality monitoring sensor, has been widely appraised as the most successful citizen-centric project in Taipei. This initiative was motivated by the need for real-time air quality monitoring as seasonal air pollution, primarily from China, became a serious environmental and health concern for the residents of Taipei. The city's Department of Information Technology formed a partnership with Academia Sinica (Taiwan's premier research institution) and Taiwan-based companies Edimax Technology, Realtek Semiconductor, and Asus to build a prototype sensor capable of measuring temperature, humidity, and the amount of fine particulate matter. The sensor was made in a limited quantity of three hundred, distributed to and installed in elementary schools, public spaces, and interested households (Smart Taipei n.d.b). The sensor collects real-time air quality data to be shared and visualized by the Location Aware Sensor System (LASS) network, an online citizen-scientist community closely related to g0v.tw, for use by the general public. This initiative has also been framed as an effort to prompt environmental education for school children and as a safeguard for public health, making the Air Box highly popular among citizens (authors' interviews with PM 07, PM08, January 2019; TR01, TR02, May 2019. The Taipei city government has widely showcased the Air Box as a successful smart initiative for citizen participation, and also eagerly exported it to eight other cities in Taiwan and several others in South Korea and Malaysia (Smart Taipei n.d.a).

These perceptible initiatives in Taipei bring new dynamics into the discussion on the smart city, social justice, and inequalities. Many smart city initiatives use automation and algorithms to foster more efficient governance decisions and move the decision-making process away from public scrutiny. Critical studies on the smart city have pointed out that such initiatives might depoliticize decision-making and undermine political transparency and accountability (Cugurullo 2020; Kitchin, Cardullo, and Di Feliciantonio 2019), which in turn might reinforce neoliberal logic and widen inequality. The initiatives in Taipei, however, appear to be a different strand of smart city policies, with visibility as their objective. They are designed to (re-)politicize smartness through a populist narrative. Will these perceptible initiatives help achieve more transparent, accountable, and equitable outcomes?

Emerging Inequalities and Exclusions to Being Smart

Politically speaking, Taipei's perceptible initiatives are widely believed to have helped secure a second electoral term for Mayor Ko in 2018. According to several interviewees, voters view these initiatives as evidence that the administration cares about residents' everyday experience. One community leader said: "We residents do not often understand what a high-tech smart city is … but we feel great about those smart initiatives that add convenience to our daily routines … parking, trash, lighting. These are issues that we care about, and we appreciate that the administration cares too" (authors' interview with CU03, August 2018).

Nevertheless, the smart city initiatives are not free of concerns. Many community leaders suggested there was unequal distribution of resources in the smart initiatives and worried about further marginalizing already disadvantaged residents and social groups. We as researchers also observed the increasing influence of private technological providers and a shift towards market-oriented solutions in the perceptible initiatives. Our findings indicate the following concerns about government accountability and smart citizenship.

Unequal Distribution of Resources

Community leaders and residents to whom we talked raised concerns about the city government's prioritization of smart initiatives and their potential to squeeze other budgetary items. As one said, "we appreciate that the administration cares about our lighting and parking issues, but we also noticed that more money has gone to the smart living labs, Internet infrastructure, hardware and software … [W]e might end up with less money to provide for non-smart needs" (authors' interview with LC01, September 2018). This concern is indeed real. The Taipei city government has strategically created a Taipei Smart City Project Management Office (PMO) since 2016 to coordinate smart initiatives across the city and report directly to the mayor's office. Initiatives handled by the PMO are often prioritized over the city's traditional service routines while advertised as "highlights" of the city government's accomplishments.

Meanwhile, community leaders and residents also question the technological modernization assumption underlying the smart city agenda and the expectation of technological literacy in some initiatives, particularly those originally designed to appeal to well-educated young working adults. For some residents, their needs could be fulfilled easily by basic, "non-smart" service provisions. As one resident commented, "I need street lighting and a parking space, but it makes no difference to me if the light is from a smart light pole or the parking space is managed by a real-time system" (authors' interview with TR 05, January 2020). Communities with aging populations are also worried about being left out. In discussing the i-Voting and Participatory Budgeting initiatives, one leader of a local neighbourhood organization told us that most residents in his neighbourhood have little clue about what a smart city is or what projects to propose for i-Voting: "Many of our residents here are elders and can't navigate those online systems. It is difficult to voice our concerns through the smart channels in which the city government now guarantees timely responses" (authors' interview with LC03, December 2019). Meanwhile, the smart public housing project has been criticized for its lenient maximum income threshold and its relatively high rent. Many worry that this would exclude disadvantaged low-income populations from public housing. In all, these concerns about Taipei's smart city initiatives focus on the potential for diverting resources from other much needed investment and the further marginalization of the already disadvantaged.

Tokenization of Citizen Participation

Another frequent concern of citizens who participated in the perceptible initiatives was about how much their input really helped shape smart city development. According to several residents who engaged in the work of g0v.tw and

LASS, the city government largely determined in advance the planning and implementation of these supposedly participatory initiatives; residents were often treated only as application users and data points for feedback and data collection. For example, in August 2017, many Air Box participants observed a record high level of particulate matter in Taipei city, whereas the official observation stations of the national Environmental Protection Administration in the nearby area detected a significantly lower level. This incident triggered a debate on whether the data citizens collected through the Air Box should be considered official and used to send air pollution alerts or inform the city's environmental governing agencies. Several city council members also questioned the city government's decision of introducing the Air Box to the citizens because it was sponsored fully by private firms (Edimax Technology and Realtek Semiconductor) and had limited quality control over accuracy and precision.

In response to the challenges, the city's Department of Information Technology contested that the Air Box's measurements were within the acceptable range of accuracy and precision, and that discrepancy between data sources might be explained by the micro-environment in which the sensitive device is installed (Taipei City Council 2017).[2] Subsequently, the administration announced that the Air Box was only an environmental educational tool and the data collected were not official. The city also cancelled the subscription to the Air Box data visualization app that some citizens had been using. "I felt cheated," one participant in the Air Box initiative said. "Our collective effort put into the Air Box had no actual impact on the city's environmental governance" (authors' interview with TR04, July 2020).

Similar frustration was also expressed by citizens who participated in the i-Voting initiative for the redevelopment plan of Shezi Island, an underdeveloped area with both formal and informal settlements and complicated property ownerships. The Ko administration has been eying Shezi as the site for a large-scale urban regeneration project, including a plan to build a "Smart Shezi Island" as part of the city's smart city development.

Nevertheless, implementing this vision would require overcoming the daunting challenge of relocating at least ten thousand residents. The city government was therefore eager to seek residents' feedback about its redevelopment plans and also to use the opportunity to legitimize the project. The Ko administration hosted an i-Voting session on 27–28 February 2016 over the three redevelopment proposals. The votes did not achieve what the city government had intended. For one thing, the low turnout, at a merely 35.16 per cent, failed to represent the will of the majority of the residents. Despite claims of i-Voting as a ubiquitous, easily accessible voting method (Taipei City Government 2016), many low-income residents of Shezi lacked the digital literacy to participate in the vote. It did not help that the city government treated i-Voting more as a test run of the voting technology than as a channel for seeking residents' input to the government's decision about the development. As the city government stated in several news outlets, this i-Voting provided a learning experience for the Department of Information Technology about the nitty-gritty issues of online voting, but the voting result effectively was cast aside in the remainder of the development process (Yu 2017). Residents and scholars have since criticized i-Voting as only a token for citizen participation (Liao 2016, 2020).

Privatized and Marketized Solutions

Another potential concern is the outsized role of private firms in Taipei's market-oriented approach to smart urban development, including the perceptible initiatives. Budgetary constraints meant that the city could not directly fund much of its smart city development as it would for other routine services. The Ko administration adapted by assuming a "living laboratory" approach, which involves relaxing regulations and opening the city to private firms seeking to test-run their smart living products in the real world. In exchange for services rendered and data produced during the experimental period, the Ko administration promised the firms assistance in collecting data as well as the opportunity to commodify and scale up their products if the city formally adopted them. It is through this public-private partnership model that the majority of precipitable initiatives was launched.

Inherent in this public-private partnership model is the reliance on private firms having readily available solutions and their willingness to provide them for free, at least initially. Typically, the PMO matches private firms' requests for prototype experimentation with communities that have indicated a need for certain types of initiatives (authors' interviews with PM07 and PM08, January 2019). A successful match usually satisfies both the smart development demand of the city and the experimental needs of the firm. Local needs are often left unaddressed, however, if no provider has previously developed the prototype smart solution or if the provider does not find the trade-off in this partnership economical. In the end, without the intent or resources to develop targeted solutions based on the actual needs of the city, Taipei's smart city development is constrained by the conditions and calculations of private providers.

Even for the successfully matched initiatives, their continuation depends on the ability and willingness of the private firms to carry on. Because many of the solution providers experimenting in Taipei are new start-up firms, their operations are highly unstable and their survival rate after two years is low (authors' interviews with ED02, January 2019, December 2019). In one instance, some residents really liked a smart light pole with aesthetically designed streetlight features and various built-in environmental sensors, and requested more of these light poles for the community. The light pole, however, was a one-time prototype and could not be mass produced since the provider had run into operational challenges and no longer considered the project a profitable investment (authors' interviews with LD02 and LD03, January 2019). In other cases with more established firms, some providers have withdrawn from initiatives upon completion of the data collection. Although the perceptible initiatives are presented as citizen-centric, the market-oriented approach in which private firms enjoy much leverage means that citizens are often treated as consumers or data points for experiments.

Conclusion

In this chapter, we have contextualized the smart city development in Taipei and examined the citizen-centric "perceptible" initiatives that the Ko administration has prioritized to secure electoral support. Although these initiatives have helped the administration create a sense of inclusion in the city's development among some groups and residents, they also have led to unequal resource distribution between different areas of the city and across socio-economic groups. Meanwhile, the initiatives still struggle to involve citizen participation meaningfully, as they also reinforce neoliberal, market-oriented logics. Citizens are at risk of being treated only as smart technology consumers and data points, even though citizen participation is instrumental to smart city development. As smart city development has shifted towards a more citizen-centric agenda worldwide, empirical studies prompt us to keep asking: Whom do smart city initiatives serve? And what are the new inequalities and social justice concerns emerging under the name of the "citizens"?

Our findings resonate with many critical smart city studies to date. Although Taipei has developed its smart city in a specific spatiotemporal context and with political rationales that differ from mainstream neoliberal urbanism doctrines, Taipei's smart city still suffers from criticisms similar to those in other smart cities. On the one hand, as the perceptible initiatives specifically target Ko's electoral base, those who are not part of the targeted population have been marginalized from the very beginning. On the other hand, Taipei's living laboratory approach and the perceptible initiatives could be seen as a strategy to attract private firms to deliver public services for free. Such marketized orientation is inheritably incompatible with the level of government accountability and citizen empowerment for which many critical urban scholars and activists have called. Even though the perceptible initiatives seek to promote transparency, citizen participation, and sociopolitical inclusion in the smart city development, the initiatives might divert resources from other, already disadvantaged populations in the city and deepen socio-economic inequalities among social groups.

Although such an outcome might not be as just and equal as many critical scholars might wish, it is still important to acknowledge that the perceptible initiatives successfully (re-)politicized the smart city away from the top-down, developmental state regime's modernization project and towards commitments to the citizens. While we see criticisms on perceptible initiatives arise, we also observe that more and more bottom-up and community-based groups are participating in Taipei's smart city development. As Burns, Fast, and Mackinnon well put in the Introduction to this volume, social justice is a fluid concept and the right to the smart city can be highly contested. Social justice in the smart city involves considerations of procedures, outcomes, equal access, redistribution, the commons, and the public goods. But it is also important to see social justice as allowing the smart city to be a space for alternatives. From this perspective, we see the potential of Taipei's perceptible initiatives in exploring varying forms of smartness. We hope to see more critical studies on these initiatives to help envision alternative urban futures for Taipei.

NOTES

1 Such as providing online life-long education and digital literacy training, setting up websites and databases for all government agencies, creating digitally verifiable personal identity infrastructure for online banking, and promoting a contactless smartcard payment system (EasyCard) for Taipei's metropolitan public transportation services.

2 See also news report, Public Television Services, 17 September 2017, online at https://news.pts.org.tw/article/370645

REFERENCES

Cardullo, P., and R. Kitchin. 2019. "Being a 'Citizen' in the Smart City: Up and Down the Scaffold of Smart Citizen Participation in Dublin, Ireland." *GeoJournal* 84 (1): 1–113. https://doi.org/10.1007/s10708-018-9845-8

Chang, I.C.C., S.C. Jou, and M.K. Chung. 2020. "Provincializing Smart Urbanism in Taipei: The Smart City as Strategy for Urban Regime Transition." *Urban Studies* 58 (3), 559–80. https://doi.org/10.1177/0042098020947908

Chen, M.T., and S.H. Yang. 2016. "Spill-over Effects of the "Ko Wen-je Phenomenon in Taiwan's 2014 Local Elections." *Journal of Electoral Studies* 23 (1): 107–51.

Chuang, S., Y. Lin, Y. Kao, and J. Pan. 2011. "From Governance to Service – Smart City Evaluations in Taiwan." *2011 International Joint Conference on Service Sciences*: 334–7.

Coletta C., L. Evans, L. Heaphy, and R. Kitchin, eds. 2019. *Creating Smart Cities*. New York: Routledge.

Cugurullo, F. 2020. "Urban Artificial Intelligence: From *Automation* to *Autonomy* in the Smart City." *Frontiers in Sustainable Cities* 2: 38. https://doi.org/10.3389/frsc.2020.00038

Evans, P. 1995. *Embedded Autonomy: States and Industrial Transformation*. Princeton, NJ: Princeton University Press.

Giffinger, R., C. Fertner, H. Kramar, R. Kalasek, N. Pichler-Milanovic, and E. Meijers. 2007. *Smart Cities: Ranking of European Medium-sized Cities*. Vienna: Centre of Regional Science. Online at http://www.smart-cities.eu, accessed 25 October 2017.

Greenfield, A. 2013. *Against the Smart City*. New York: Do Projects.

Hsu, J.Y. 2011. "State Transformation and Regional Development in Taiwan: From Developmentalist Strategy to Populist Subsidy." *International Journal of Urban and Regional Research* 35 (3): 600–19. https://doi.org/10.1111/j.1468-2427.2010.00971.x

Huang, W.J. 2012. "ICT-Oriented Urban Planning Strategies: A Case Study of Taipei City, Taiwan." *Journal of Urban Technology* 19 (3): 41–61. https://doi.org/10.1080/10630732.2011.642570

Kitchin, R. 2014. "The Real Time City? Big Data and Smart Urbanism." *GeoJournal* 79 (1): 1–14. https://doi.org/10.1007/s10708-013-9516-8

Kitchin, R., P. Cardullo, and C. Di Feliciantonio. 2019. "Citizenship, Justice, and the Right to the Smart City." In *The Right to the Smart City*, ed. P. Cardullo, C. Di Feliciantonio, and R. Kitchin, 1–24. Bingley, UK: Emerald Publishing.

Li, W.D., and C.Y. Hung. 2018. "Parental Support and Living Arrangements among Young Adults in Taiwan." *Journal of Housing and the Built Environment* 34: 219–33. https://doi.org/10.1007/s10901-018-9620-7

Liao, K.H. 2016. "Op-ed: i-Voting cannot be expected to solve issues related to Shezi Island development (社子島問題不能奢望用假民主的i-Voting解決)." *United Daily News*, 2 February. Online at https://opinion.udn.com/opinion/story/8048/1482785, accessed 20 August 2020.

Liao, K.H. 2020. "Op-ed: Shezi Island development pushed by the Ko administration is the worst lesson for urban planning (柯市府強推的社子島開發計畫是都市計畫最負面示範)." *United Daily News*, 24 April. Online at https://opinion.udn.com/opinion/story/8048/4515732, accessed 20 August 2020

Mudde, C., and C.R. Kaltwasser. 2017. *Populism: A Very Short Introduction*. Oxford: Oxford University Press.

Rossi, U. 2016. "The Variegated Economics and the Potential Politics of the Smart City." *Territory, Politics, Governance* 4 (3): 337–53. https://doi.org/10.1080/21622671.2015.1036913

Shwayri, S.T. 2013. "A Model Korean Ubiquitous Eco-City? The Politics of Making Songdo." *Journal of Urban Technology* 20 (1): 39–55. https://doi.org/10.1080/10630732.2012.735409

Smart Taipei. n.d.a. "Smart Taipei: Government as a Platform, City as a Living Lab." Online at https://smartcity.taipei/uploads/download/download/2/en_1634547169_20211015Smart%20Taipei-EN.pdf, accessed 17 April 2022.

Smart Taipei. n.d.b. "Air Box Makers Education Empirical Project." Online at https://smartcity.taipei/projdetail/50, accessed 20 August 2020.

Söderström, O., T. Paasche, and F. Klauser. 2014. "Smart Cities as Corporate Storytelling." *City* 18 (3), 307–20. https://doi.org/10.1080/13604813.2014.906716

Taipei City Council. 2017. *Gazette* 109 (5). Online at http://tcckm.tcc.gov.tw/tccgazFront/gazatte/readByGaz.jsp?vol=109&no=05&startPage=2043&endPage=2079, accessed 30 August 2020.

Taipei City Government. 2013. "The opening of City Forum could transform Taipei (雲端台北論壇開幕)." Press release, 22 January. Online at https://sec.gov.taipei/News_Content.aspx?n=49B4C3242CB7658C&sms=72544237BBE4C5F6&s=550522B2D863287D, accessed 13 October 2018.

Taipei City Government. 2016. "Final Report on Shezi Island i-Voting on Development Plan (社子島開發方向i-Voting總結報告)." Online at https://www-ws.gov.taipei/Download.ashx?u=LzAwMS9VcGxvYWQvcHVibGljL0F0dGFjaG1lbnQvNjM3MTc1MzU4OTUucGRm&n=NjM3MTc1MzU4OTUucGRm, accessed 20 August 2020.

Taipei City Government. 2017a. Department of Land Administration. "The Aging of Property Owners in Taipei (台北市房屋持有者的高齡化分析)." Online at https://land.gov.taipei/News_Content.aspx?n=0ABE9F8A3E5B75C2&sms=72544237BBE4C5F6&s=0ADBE689FEC847D8, accessed 5 October 2018.

Taipei City Government. 2017b. *Taipei Yearbook 2016*. Taipei: Taipei City Government.

Taipei City Government. 2018. "Smart Taipei Brochure." Taipei: Taipei City Government.

Taipei City Government. 2020a. Department of Urban Development. "Public Housing Project Real-Time Progress Dashboard (社會住宅興建計畫與戰情中心)." Online at https://www.hms.gov.taipei/#!/, accessed 29 August 2020.

Taipei City Government. 2020b. "Taipei 7+1 call for proposals (北市智慧城市1+7領域徵案開跑)." Press release, 30 April. Online at https://doit.gov.taipei/News_Content.aspx?n=4B2B1AB4B23E7EA8&sms=72544237BBE4C5F6&s=D159153F341257D7, accessed 20 August 2020.

Taipei City Government. n.d. Department of Land Administration. "Taipei Smart-Eco Community Demonstration Plan (台北市智慧生態社區示範計畫)." Online at http://www.creatidea.com.tw/html_demo/2017Taipei/doc/%E5%8F%B0%E5%8C%97%E5%B8%82%E6%99%BA%E6%85%A7%E7%94%9F%E6%85%8B%E7%A4%BE%E5%8D%80%E7%A4%BA%E7%AF%84%E8%A8%88%E7%95%AB.pdf, accessed 20 August 2020.

Taipei City Public Housing for Rent. n.d. https://www.public-rental-housing.gov.taipei/, accessed 20 August 2020.

Tseng S.F. 2007. *The Development of Network Society: An Integrated Policy Analysis (網路社會發展政策整合研究)*. Taipei: Research, Development and Evaluation Commission, Executive Yuan.

Townsend, A.M. 2013. *Smart Cities: Big Data, Civic Hackers, and the Quest for a New Utopia*. New York: W.W. Norton.

Wade, R. 1990. *Governing the Market: Economic Theory and the Role of Government in East Asian Industrialization*. Princeton, NJ: Princeton University Press.

Wiig, A. 2016. "The Empty Rhetoric of the Smart City: From Digital Inclusion to Economic Promotion in Philadelphia." *Urban Geography* 37 (4): 535–53. https://doi.org/10.1080/02723638.2015.1065686

Woetzel, J., and E. Kuznetsova. 2018. *Smart City Solutions: What Drives Citizen Adoption around the Globe?* McKinsey Center for Government, McKinsey & Company. Online at https://www.mckinsey.com/~/media/McKinsey/Industries/Public%20and%20Social%20Sector/Our%20Insights/Smart%20city%20solutions%20What%20drives%20citizen%20adoption%20around%20the%20globe/smart-citizen-book-eng.pdf, accessed 28 June 2020.

Yu, C.H. 2017. "Five thousand residents use i-voting to express their thoughts about Shezi Island development." *IT Home*, 16 September. Online at https://www.ithome.com.tw/news/108378, accessed 20 August 2020.

23 Reimagining Smart Citizenship, Reconciling (Im)Partial Truths: POFMA, Digital Data, and Singapore's Smart Nation

ORLANDO WOODS

Data are the building blocks from which decisions are made, ideologies are forged, and power is assembled. Digital data in particular – and the channels through which they are mobilized – have "brought about both a democratisation of communication, representation and participation, and have created a free(r) marketplace in which ideas and knowledge can be exchanged" (Woods 2019, 6). While these apparent freedoms can be seen as emancipatory for citizens, so too can they pose challenges – regulatory, communicatory, and political – to governments. In particular, 'fake news' recently has become a topic of concern due to its potential to "disrupt society by sowing discord and division between groups, influencing political outcomes, legitimizing fringe views, and discrediting establishment journalism" (Chen and Chia 2019, 1). Complicating the matter, and reflecting the spatially promiscuous nature of digital data, is the fact that fake news does not necessarily originate from within a given politico-legal jurisdiction, but can come from other jurisdictions and from non-human actors (i.e., bots) as well. Governments thus are expected to navigate a fluid and opaque terrain of digital (mis)information in the search for an appropriate balance between freedom of speech and mitigating against the potential harm caused by the circulation of untruths. This has become a divisive issue. For example, the Economist (2019a) suggests that among Asia's autocratic governments, mitigation is "usually code for suppressing criticism," with Singapore, in particular, being well known for its "authoritarian government that either silences or co-opts public media" (Lee and Lee 2019, 81). In late 2019 this sentiment was reified with a legislative response to the problem of fake news – the Protection from Online Falsehoods and Manipulation Act (POFMA).

Although the issues surrounding POFMA are many and varied, exploring them through the lens of digital data helps to contextualize them within more wide-ranging discourses. In particular, as a differential construct, data can problematize the terms of *smart* citizenship and what they mean for the digital *right* to the city. Whereas POFMA is based on the principle that data can, as

Singapore's home affairs and law minister K. Shanmugam put it, be weaponized to "attack the infrastructure of fact, destroy trust and attack societies" (cited in the Economist 2020), Singapore's Smart Nation initiative is based on a principle of *open* data that is used to co-opt citizens as stakeholders in the city. By bringing these two divergent perspectives into conversation with each other, I argue that, for smart citizenship to be fully realized in Singapore, it needs to be based on a more *integrative* understanding of digital data – one that necessarily involves a transfer of responsibility and, to a certain degree, of power from the state to citizens. This is an understanding that encourages and enables citizens to *become* smart through data empowerment. In making these arguments, I engage with two interrelated ideas. One concerns the need to "reconfigur[e] the boundaries and modes of citizen participation in line with the ethos of participation and empowerment of smart urbanism" (Ho 2017, 3110); the other concerns the extent to which digital technologies have "brought about a re-balancing of power in and of the media" (Woods 2019, 6). While the first idea speaks of the empowering *potential* of smart urbanism, the second speaks of the empowering *fact* of digital media. The aim of this chapter, then, is to integrate these two ideas and to offer a reimagination of smart citizenship in response.

To be clear from the outset, as a long-term resident of Singapore I am generally supportive of POFMA's aims. That said, I also believe that its ideological underpinnings and implementation could be rethought in alignment with the broader goal of realizing a Smart Nation. These legislative (POFMA) and policy (Smart Nation) regimes have been treated in hitherto discrete terms. Alignment, however, will provide a degree of conceptual clarity that will go a long way to realizing a genuinely *smart* understanding of citizenship. As I explain in more detail later, the Singapore state has a long history of co-opting its citizens into a technologically defined vision of the future. In the same blush, it also wields exact control over what this vision looks like, and what the parameters of inclusion are. While this vision has tended to be rooted pragmatically in skills upgrading and future-proofing the economy, in recent years the Smart Nation initiative has caused this vision to become a more ideological, and thus more contested, construct. POFMA brings these contestations to the fore, and provokes renewed debate about what *smart* citizenship might actually mean. I build my argument through three sections. The first contextualizes the empirical case of Singapore by exploring the compromised *right* to the smart city; the second compares and contrasts the Smart Nation and POFMA; the third concludes by reimagining smart citizenship through the lens of "fourthspace."

A Compromised "Right" to the Smart City?

For a long time, cities have been defined and problematized by the differences that they help to reproduce. While these differences vary, they become more divisive political constructs when they are seen to favour some groups more than others. These divisions foreground concerns about the *right* to the city and, relatedly, the emancipatory politics of urban citizenship (Harvey 1973). These *rights* are typically inspired by Marshall's (1950) threefold delineation of the rights of citizenship – which include civil/legal (including freedom of speech and the right to justice), political, and social rights – and have since been categorically expanded to include symbolic and ideological rights as well. Recently, however, two interrelated processes have complicated the terms of urban citizenship and caused the right to the city to be questioned anew. One is the embedding of neoliberal logics within the urban fabric, the other is the centring of digital data in the practice of urban governance. In the first instance, Cardullo and Kitchin (2018, 817) have argued that neoliberalism "shifts citizenship away from inalienable rights and the common good towards a conception rooted in individual autonomy and freedom of 'choice,' and personal responsibilities and obligations." With this shift, there has been a fragmentation of urban space into individualized compartments that can become increasingly differentiated, and thus removed from the collective. Participation in the public sphere therefore becomes less democratic and more a function of an individual's position in neoliberal urban hierarchies. This is a model of urban citizenship that is rooted in "pragmatic, instrumental and paternalistic discourses and practices" (Cardullo and Kitchin 2018, 814), which, in many respects, has been exacerbated through the rollout of smart city initiatives around the world.

In the second instance, as cities become *smarter*, governance is expected to become a more algorithmically defined construct. Governance decisions become automated and predictive (Kitchin and Dodge 2011), which in turn have been shown to "discourag[e] agonistic spaces of political confrontation as they promote technological solutionism, underpinned by civic paternalism and stewardship, rather than more inclusive forms of citizen engagement and participation" (Cardullo and Kitchin 2018, 818). Simply put, the algorithm replaces the citizen as the arbiter of urban decision-making, which can cause power to become nested among those responsible for implementing and managing the algorithms. As a result, critics have sought to elucidate the ways in which algorithms might foreground new forms of governmentality. Of note is Vanolo's (2014, 883) conceptualization of a *smartmentality* whereby algorithmic governance is interpreted as a "powerful tool for the production of docile subjects and mechanisms of political manipulation." By highlighting the effects of algorithmic governance on the citizens of smart cities – the so-called *smart* citizens – we can begin to see how the *right* to the smart city might be compromised. The rationalization of the city through smart governance can cause pre-existing fissures to become more pronounced and the right of citizenship curtailed. Indeed, by recognizing the role of algorithms in the governance of smart cities, we can begin to see how data can become a divisive construct.

Recognizing and expounding this division, Kitchin, Lauriault, and McArdle 2015) argue that smart city initiatives are often portrayed as "rational interventions designed to improve social, economic and governance systems" (2015, 2), whereas in reality they are inherently politically and ideologically loaded in vision and application, reshaping in particular ways how cities are managed and regulated. Likewise, the *data* within these systems are not neutral and objective in nature, but are situated, contingent, and relational, *framed by the ideas, techniques, technologies, people, and contexts* that conceive, produce, process, manage, analyse, and store them.

What Kitchin, Lauriault, and McArdle (2015) bring to light is the *im*partiality of data and the disjuncture that plays out in different ways between how digital data are perceived (or constructed), and how they are deployed (or leveraged), in the public domain. This disjuncture highlights, in turn, both the justification for data-driven forms of governance, but also its (dis)empowering effects on citizens. Taking this idea further, Gabrys (2014, 30) suggests that smart cities "delimit urban 'citizenship' to a series of actions focused on monitoring and managing data," while Shaw and Graham (2017) more recently have argued that smart citizenship entails an informational *right* to the city. Given that these *rights* are often obstructed, manipulated, or remain partially realized, there have been calls for smart city projects to be guided by principles of data justice to ensure "fairness in the way people are made visible, represented and treated as a result of their production of digital data" (Taylor 2017, 1). The point is that, as cities become more data driven in their management and citizens are discursively constructed *in relation to*, and how they *engage with*, data can be seen to determine the extent to which they can claim a *right* to the smart city. Smart citizenship is premised on the realization of these claims; where they are unrealized, divisions can be reproduced. Echoing this sentiment, Kong and Woods (2018) highlight the fact that, "whilst 'citizenship' is an inclusive form of categorization, *smartness* is potentially or even inherently divisive, meaning the notion of *smart citizenship* has the potential to create social divisions as much as it can overcome them" (693, emphasis in original). This sense of division is clearly observed in Singapore, where there is a marked divergence in the ways digital data are constructed in the public domain. Recognizing and exploring this divergence thus can provide insight into the (un)realized nature of smart citizenship in Singapore.

Paradoxical Parallels, Uneasy Alignments: Singapore's Smart Nation and POFMA

For many years, the digital domain has presented a problem for Singapore's managerialist government. More recently, however, it has presented an opportunity as well. As a problem, it expands public space, causing it to become harder to regulate and control, as "even the endemically marginalised [can] engage with policies" online (Weiss 2014, 96). The risk, then, stems from the idea that the digital domain can serve to weaken, or provide a channel to undermine, the government's grip over society. It foregrounds a redistribution of power that can be used to "critique modes of governance by imagining how it might be possible not to be governed quite so much – or in that way" (Gabrys 2014, 34). In recognition of this risk, the Singapore government has, over the years, made various moves to censor the digital domain. In 1994, for example, it pre-emptively scanned all Singapore e-mail accounts for suspicious content, with the public backlash that ensued revealing the "maladroit steps of a state grappling with a new informational landscape" (Crang 2010, 33). Although these problems and risks endure, in recent years the digital domain has been embraced as a way to ensure Singapore's ongoing urban development and economic competitiveness. In particular, smart urbanism provides a model through which Singapore's future readiness has become enshrined, and thus constitutes a 'neoliberalism-as-development strategy' (Ho 2017). I now expand on these ideas, first by introducing Singapore's Smart Nation initiative, and then by exploring the motivations that led to the tabling and passing of POFMA.

Singapore's "Smart Nation"

Since the launch of the Smart Nation in November 2014, the Singapore government has embraced the potential of the smart city discourse in a comprehensive and centralized way. In many respects, these characteristics are a reflection of, and have been enabled by, Singapore's small size, developed economy, single layer of government, and technologically sophisticated population. While specific projects are wide ranging in scope – covering transportation, health care, government services, and economic alignment – the data that enable them to be realized provide a point of alignment. Central to the vision of the Smart Nation is the "ability to gather data, interpret it, glean insights and then translate those insights into meaningful action" (Foo and Pan 2016, 78). Data are treated as an ideological "good" deployed in the service of urban – and thus national – development. In recognition of the need to offer a more cooperative form of governance that aligns with the principles of the smart city, the Singapore government has started to make selected data available to private app developers, reflecting an ideological shift towards "more horizontal processes of co-creation, rather than the top-down design and implementation of public services" (Kong and Woods 2018, 696). Although the sharing of data might sound like a necessary antecedent to the realization of any smart city, it also involves, to varying degrees, "relinquishing control of the public domain" (696). This poses a significant, and not unproblematic, ideological shift for the government, as relinquishing control in this way assumes that citizens, the state, and other (private sector) stakeholders are ideologically aligned. Data can therefore be reimagined as points of alignment that divert attention from the fact that the Smart Nation is a "top-down narrative that privilege[s] the interests of the country over those of the individual. The assumption is that individual interests are aligned with those of the country, and that the praxis of smart urbanism is justified by the ideology. Where the interests of the individual are misaligned with those of the country, they can, however, undermine the legitimacy and progress of smart urbanism" (690).

The potential for misalignment speaks to some of the fundamental problems underpinning the right to the smart city in Singapore. While citizenship presupposes a diversity and complexity of characteristics, views, and datapoints, being *smart* means to privilege some (the *smart* or aligned) over others (the *not-smart* or mis/non-aligned). Moreover, it also brings to light the role of smart governments in recognizing the need to recalibrate the terms of citizenship in line with the overarching logic of data-driven *smartness*. Thus, although the potential of Singapore's Smart Nation rests on the fact that smart citizens can "exist on terms of relative parity with smart governments" – meaning "power becomes devolved from governments to systems and, as such, governance becomes more participatory" (Kong and Woods 2018, 693) – this vision is selectively realized only within the framework of the Smart Nation. More than that, it brings to light the potential for data to be used to curb civic freedoms and reproduce injustices by reinforcing the power of the state in the name of national cohesion and progress. With these ideas in mind, I now consider how this "potential" has become manifest through POFMA.

The Protection of Online Falsehoods and Misinformation Act

POFMA is a piece of legislation that was several years in the making. In early 2018 the government appointed a Select Committee on Deliberate Online Falsehoods to study in detail the issue of fake news. Based on the committee's findings, the government "robustly defended the need for such legislation, citing the need to preserve Singapore's fragile racial and religious harmony and democratic institutions from the violence and electoral manipulation that [has] plagued other nations" (Lee and Lee 2019, 84). Tabled in parliament in April 2019 and passed later that year, at 62 sections POFMA is "among the world's most far-reaching" (Economist 2019b) pieces of legislation to attempt to curb fake news. Key elements of the act, including its triggers, coverage, and punishments, are:

> Under section 7(1), a person must not do any act in or outside Singapore to communicate a statement knowing or having reason to believe that it is a false statement of fact, and the communication of that statement is likely to, *inter alia*, prejudice national security, public health, public safety, or public tranquillity, or incite feelings of enmity, hatred, or ill-will between different groups of persons …

> Under section 7(2), a person guilty of section 7(1) shall be fined not exceeding $50,000 or be imprisoned for a term not exceeding 5 years, or both. If the person had used an inauthentic online account or a bot to commit the offence, he shall be fined not exceeding $100,000 or be imprisoned for a term not exceeding 10 years, or both.
>
> Under sections 10, 11, and 12, a minister may order a correction direction or a stop communication direction if a false statement of fact has been communicated in Singapore and the minister is of the opinion that it is in the public interest to issue the direction...
>
> Under sections 20, 21, 22, and 23, a minister may order an internet intermediary to make a general correction, make a targeted correction, or disable access if a false statement of fact has been communicated in Singapore and the minister is of the opinion that it is in the public interest to issue the direction. (Chen and Chia 2019, 2–3)

POFMA has attracted considerable criticism and debate from within Singapore and internationally. Debates focus on terminology, and the relative opacity (or, in the terms of this chapter, *im*partiality) concerning what constitutes a "false statement of fact," given that "it is not always easy to distinguish between facts and opinions, characterizations, and misimpressions that result from decontextualization" (Chen and Chia 2019, 5). In recognition of these inherent problems of definition and (mis)interpretation, Chen and Chia (2019, 4) clarify that the focus of POFMA is false, or proven-to-be-incorrect, statements, rather than opinions, criticisms, satires, and parodies (which, it is implied, may be targeted at the government or interpreted as expressions of free speech). They go on to state that, in recognition of these problems, "the primary remedial measure would be corrections, and not removal or takedown of content" and that "criminal offences would apply only to malicious actors seeking to undermine society" (4). Clarifications like these can be seen as attempts to mollify criticisms of how POFMA will be applied. Moreover, contextualizing it globally serves to dilute the severity of its sanctions, with Chen and Chia concluding that "the proportionality of the POFMA sanctions ... would likely be situated in the middle of the spectrum of punishments for online hate speech when compared to current and developing international norms" (2019, 15). While these clarifications and reframings can be seen to desensationalize POFMA, they have not yet managed to appease those critics who focus more explicitly on its symbolic and ideological overtures.

These criticisms do not contextualize POFMA in relation to fake news legislation, but in relation to Singapore's history of media censorship and broader control of the public domain. In this vein, concern focuses on whether the act "unduly restricts the freedom of expression" and could "possibly be abused by the government" (Chen and Chia 2019, 1). International groups such as the International Commission of Jurists, Reporters Without Borders, and Human

Rights Watch have registered their concerns that POFMA "joins a host of other legislation which already keeps critics in check. The country's constitution limits free speech with 'such restrictions as it considers necessary or expedient,'" with Singapore apparently sitting "below Russia, Afghanistan and many of its own neighbours in the latest ranking of press freedom compiled by Reporters Without Borders" (*Economist* 2019b). These criticisms are best interpreted as reminders of Singapore's highly restrictive regulatory landscape when it comes to popular media. More helpful is the targeted concern of Cherian George – a Hong Kong–based academic and renowned critic of governance in Singapore – that POFMA gives "unprecedented discretion to individual ministers to pronounce on what is false or misleading" (cited in the *Economist* 2020). Here we can see a reassertion of centralized power in the face of the datafication of society. Notwithstanding such criticisms, there has been relatively less public opposition to POFMA. Although this might reflect the self-censoring nature of Singapore society and the fact that, after decades of curtailed freedoms, it has been conditioned not to oppose such legislation, it also reveals the easily and intuitively justifiable nature of the act. To this point, Lee and Lee (2019) acknowledge the extent to which the government is now "concerned with establishing *public* legitimacy for [its] governance decisions, including the co-optation of contrarian voices into its public narrative" (82, emphasis in original).

While recognizing the value of these debates, I contend that a more insightful perspective can be developed by shifting the focus from the legislation to the data upon which the legislation is based. The uniqueness of POFMA rests in its focus on *online* falsehoods, which reflects the fact that digital data are distinct from other forms of data. Digital media foreground a data landscape that is saturated with information, meaning "people are offered great opportunities to orient themselves about what news is out there" (Tandoc et al. 2018, 2746). With this in mind, POFMA both recognizes the partiality of data/information/news and offers an *im*partial interpretation of value. Although the incumbent minister of law, K. Shanmugan, stated during parliamentary debates that an "infrastructure of fact" underpins free and responsible public discourse in Singapore, he also revealed the extent to which POFMA "rests on a certain theory of free speech – that an entirely unregulated 'marketplace of ideas' will *not* produce truth and can indeed inhibit the pursuit of the virtue of civic republicanism" (Chng 2019, 2, emphasis in original). Chen and Chia (2019, 7) echo this sentiment, observing how POFMA "presupposes that ... such speech has no inherent value whatsoever, and the best remedy is government intervention." The "virtues" and "remedies" articulated here, however, can be rendered problematic when brought into conversation with the idea of smart citizenship. This problematization proceeds along two axes. One, if *smartness* means the ability to navigate, negotiate and thrive in a data-saturated environment, then this logic arguably should apply to *all* data – irrespective of being true/false, fact/

opinion, quantitative/qualitative – not just those that are used for the purposes of urban governance. Two, if "citizenship" means a sense of inclusion irrespective of difference, then the harm that can emerge from such difference should be self-regulated, not reliant on government intervention.

Fourthspace and the Reimagination of Smart Citizenship

The problem in Singapore is that digital data are non-aligned in their ideological premises. As a result, they are constructed as an uncritical boon for the Smart Nation, but a potential threat when reinterpreted through the lens of POFMA. In this vein, Kitchin, Lauriault, and McArdle's (2015, 3) threefold critique of the technocratic underpinnings of the smart city provides insight into the ways in which the (im)partial treatment of data can block the realization of smart citizenship. One, data are treated in "highly reductionist and functionalist" terms; two, they focus on the "efficient management and manifestations of problems, rather than solving the deep-rooted structural problems underpinning them"; and three, they "centralise power and decision-making into a set of administrative offices, rather than distributing power." The fact is that data themselves hold no inherent value; rather, value (positive, negative, or otherwise lacking) is *attributed* to data. In Singapore, these attributions are government prescribed, and thus highlight the extent to which data are "situated, contingent, relational and framed and used contextually to try and achieve certain aims and goals" (Kitchin, Lauriault, and McArdle 2015, 5). That these framings are so divergent reflects the fact that POFMA hitherto has been treated in terms ideologically distinct from the Smart Nation. Reconciliation, however, could foreground a reimagination of smart citizenship that builds on the notion that "data are a new form of agency, meaning power – and associated 'rights to the city' – rests with those that are able to access and interpret it in ways that can evoke change" (Kong and Woods 2018, 696). These changes are not just about making cities more efficient, manageable, and sustainable, but also about making them more inclusive, participatory, and just.

To realize these ambitions, and to reconcile hitherto (im)partial treatments of data, work needs to be done to recognize and expand the role of "fourthspace" in the smart city. Fourthspace has been conceptualized as the "digitally enabled spaces of urbanism that are co-created, and that contribute to an *expansion* and *diffusion* of social and political responsibility" (Kong and Woods 2018, 679, emphasis added). Through these expansions and diffusions, fourthspace can be seen to realize the "emancipatory potential that comes with digital transformation and the subsequent 'decolonisation' of the city" (Kong and Woods 2018, 694). Data are integral to the formation of new, implicitly more democratic power structures that leverage the equalizing principles of "digital scale" (Woods 2019) to ensure that the (im)partiality of data becomes indexed to the engagement of the masses, rather than the ideologies of elites. In this vision, the responsibility for combatting online falsehoods rests with smart citizens, not the government – with smart citizens expected to minimize their effects through the identification of, and non-engagement with, suspicious content and peer-to-peer and peer-to-government flagging. Likewise, the responsibility for realizing the Smart Nation rests with user-defined initiatives that have the rights and interests of citizens at their core. Notably, these initiatives have already been spearheaded in Singapore by the embrace of hacking collectives, which are being embedded within corporations and government departments in order to "creat[e] opportunities for subversion to become normative" (Kong and Woods 2018, 696). The point, however, is that smart citizenship is based on a symbiotic relationship with smart governments, as each construct becomes ideologically tethered and responsible to, and dependent on, the other.

This reimagination of smart citizenship provides a channel through which more just cities can be realized. Central to the idea of social justice is a set of expectations concerning how people *should* conduct themselves within a given polity. It is an ideal that is realized through recursive processes; it involves shaping, and being shaped by, forces that work towards defining and implementing a public good that is for the benefit of all. This emphasis on individual agency and social responsibility becomes all the more important in the data-saturated context of the smart city, wherein modalities of power become more fragmented. As Taylor (2017, 1) notes, the realization of data justice is predicated on an integrative understanding of how digital data bring about simultaneous processes of (in)visibility, (dis)engagement, and (anti)discrimination, which, in turn, challenge the "growing assumption that being visible through the data we emit is part of the contemporary social contract." Taylor's point is that "just" data are impartial and balance the positive *and* negative effects that emerge when the rights and freedoms attached to digital data are foregrounded. Fourthspace, in this vein, can provide the integrative ether that brings alignment amid divergence and sets the terms by which the impartiality of data is realized. Doing so could bring about an ever-greater rationalization of governance processes, as responsibilities and decision-making become more equitable and achievable in practice. Translating cities into *smart* cities is not a selective process; rather, it is a full-blown articulation of what it means to live in a technology-enabled, data-saturated, urban world.

Taking this idea further, the data through which the smart city is realized are not limited to quantifiable metrics that are non-ideological. Rather, smart cities and the smart citizens therein are implicated in *all* data, whether they be the tracking of mobility patterns or the dissemination of "fake" news. Smart citizenship is, in this view, about enabling individuals to *become* smart through digital empowerment. Although the Smart Nation makes a nod to this effect,

POFMA reflects a paternalistic government that treats its citizens in an anachronistic, non-smart way. Recognizing this fundamental premise provides a first step towards reconciling the (im)partial treatment of data by the Smart Nation and POFMA, and thus reimagining what digital justice and smart citizenship could mean in Singapore and beyond.

REFERENCES

Cardullo, P., and R. Kitchin. 2018. "Smart Urbanism and Smart Citizenship: The Neoliberal Logic of 'Citizen-Focused' Smart Cities in Europe." *Environment and Planning C: Politics and Space* 37 (5): 813–30. https://doi.org/10.31235/osf.io/xugb5

Chen, S., and C.W. Chia. 2019. "Singapore's Latest Efforts at Regulating Online Hate Speech." *Research Collection School of Law*. Online at https://ink.library.smu.edu.sg/sol_research/2921, accessed 1 November 2021.

Chng, K. 2019. "Reflections on Thinking about the POFMA." *Research Collection School of Law*. Online at https://ink.library.smu.edu.sg/sol_research/2986, accessed 1 November 2021.

Crang, M. 2010. "Cyberspace as the New Public Domain." In *Urban Diversity: Space, Culture and Inclusive Pluralism in Cities Worldwide*, ed. C. Kihato, M. Massoumi, B. Ruble, and A. Garland, 99–122. Baltimore: Johns Hopkins University Press.

Economist. 2019a. "Asian governments are trying to curb fake news: Some efforts are more sincere than others." 4 April. Online at https://www.economist.com/asia/2019/04/04/asian-governments-are-trying-to-curb-fake-news, accessed 1 November 2021).

Economist. 2019b. "Singapore strikes its first official blows against fake news." 7 December. Online at https://www.economist.com/asia/2019/12/07/singapore-strikes-its-first-official-blows-against-fake-news, accessed 1 November 2021.

Economist. 2020. "Singaporean ministers can decide what is fake news: And then order websites and social media firms to delete it." 6 February. Online at https://www.economist.com/asia/2020/02/06/singaporean-ministers-can-decide-what-is-fake-news, accessed 1 November 2021.

Foo, S.L., and G. Pan. 2016. "Singapore's Vision of a Smart Nation: Thinking Big, Starting Small and Scaling Fast." *Asian Management Insights* 3: 77–82. https://doi.org/10.1142/10785

Gabrys, J. 2014. "Programming Environments: Environmentality and Citizen Sensing in the Smart City." *Environment and Planning D: Society and Space* 32 (1): 30–48. https://doi.org/10.1068/d16812

Harvey, D. 1973. *Social Justice and the City*. Oxford: Blackwell.

Ho, E. 2017. "Smart Subjects for a Smart Nation? Governing (Smart)Mentalities in Singapore." *Urban Studies* 54 (13): 3101–18. https://doi.org/10.1177/0042098016664305

Kitchin, R., and M. Dodge. 2011. *Code/Space: Software and Everyday Life*. Cambridge, MA: MIT Press.

Kitchin, R., T. Lauriault, and G. McArdle. 2015. "Smart cities and the Politics of Urban Data." In *Smart Urbanism: Utopian Vision or False Dawn?*, ed. S. Marvin, A. Luque-Ayala, and C. McFarlane, 16–33. London: Routledge.
Kong, L., and O. Woods. 2018. "The Ideological Alignment of Smart Urbanism in Singapore: Critical Reflections on a Political Paradox." *Urban Studies* 55 (4): 679–701. https://doi.org/10.1177/0042098017746528
Lee, H., and T. Lee. 2019. "From Contempt of Court to Fake News: Public Legitimisation and Governance in Mediated Singapore." *Media International Australia* 173 (1): 81–92. https://doi.org/10.1177/1329878X19853074
Marshall, T. 1950. *Citizenship and Social Class*. Cambridge: Cambridge University Press.
Shaw, J., and M. Graham. 2017. "An Informational Right to the City? Code, Content, Control and the Urbanization of Information." *Antipode* 49 (4): 907–27. https://doi.org/10.1111/anti.12312
Tandoc, E.C., R. Ling, O. Westlund, A. Duffy, D. Goh, and Z.W. Lim. 2018. "Audiences' Acts of Authentication in the Age of Fake News: A Conceptual Framework." *New Media & Society* 20 (8): 2745–63. https://doi.org/10.1177/1461444817731756
Taylor, L. 2017. "What Is Data Justice? The Case for Connecting Digital Rights and Freedoms Globally." *Big Data & Society* 4 (2): 1–14. https://doi.org/10.1177/2053951717736335
Vanolo, A. 2014. "Smartmentality: The Smart City as Disciplinary Strategy." *Urban Studies* 51 (5): 883–99. https://doi.org/10.1177/0042098013494427
Weiss, M. 2014. "New Media, New Activism: Trends and Trajectories in Malaysia, Singapore and Indonesia." *International Development Planning Review* 36 (1): 91–109. https://doi.org/10.3828/idpr.2014.6
Woods, O. 2019. "Mobilising Dissent in a Digital Age: The Curious Case of Amos Yee." *Geopolitics* 26 (2): 639–60. https://doi.org/10.1080/14650045.2019.1611561

24 From Smart to Sharing Cities: The Promise of Citizen-Led, Place-Based Digitalization

INKA SANTALA AND PAULINE McGUIRK

An increasingly debated dimension of the smart city is the sharing economy: a rapidly emerging peer economy underpinned by online-mediated platforms (see, for example, Burbank 2014; *Economist* 2013). Through novel sharing platforms, citizens have gained practical and convenient access to a pool of material resources, skills, and knowledge (Botsman and Rogers 2010). Not only is the sharing economy promoted as enabling effective use and redistribution of so-called idle assets (Frenken and Schor 2017, 5), but sharing through online platforms is also claimed to save money and support more environmentally friendly practices compared to individual ownership and consumption (Heinrichs 2013; Rifkin 2014).

The sharing economy has sparked a timely debate around social justice in technology-oriented smart cities. Increasingly, scholars have questioned the role of corporate-run sharing-economy platforms in disempowering urban communities and reinforcing profit-oriented organizational norms as part of a broader platformization of the city, embedded in capitalist relations (Barns 2020; Srnicek 2017). By focusing on top-down market solutions, the sharing economy has been criticized for excluding citizens from the creation of urban commons, reinforcing predominantly market-encased citizen behaviours (Martin 2016). Thus, by reinforcing hegemonic urban norms and behaviours, the transactional sharing economy has been implicated in reproducing inequitable power structures and resource distributions (Murillo, Buckland, and Val 2017), restricting collective capacities for change towards more just and sustainable urban futures.

To support smart city development that fosters community empowerment, citizen inclusiveness, and structural equality, an alternative urban narrative has started to emerge in the form of sharing cities (McLaren and Agyeman 2015). Directing attention to citizen-led and cooperatively based sharing platforms, the concept of the sharing city and related grassroots movement (Shareable 2017) is argued to empower urban communities, encouraging more autonomous and collaborative forms of urban agency (Iaione 2016). By supporting bottom-up fabrication of shared urban assets and services, the sharing city includes citizens in urban commoning and co-creation (Labaeye 2019), simultaneously nurturing more proactive and creative citizen subjectivities (Hult and Bradley 2017). Finally, promoting techniques such as place-based mapping, the sharing city visualizes and amplifies alternative forms of sharing agency and diverse sharers' subjectivities that already exist (Sharp 2018), thus building collective capacities to equalize power structures and govern the city as a commons (McLaren and Agyeman 2016).

Presenting more distributed and community-based forms of urban sharing, the sharing city potentially disrupts and redirects the hierarchical and commercially oriented smart city discourse and the transactional sharing economy it sustains (Labaeye 2019). This transformative narrative of the sharing city reframes sharing agency from something marginal and disruptive (Gurran et al. 2018) as a more fundamental part of being, doing, and thinking the city as a shared resource (Thorpe 2018), and challenging contemporary profit-oriented organizational norms. By recognizing sharers not merely as passive users and consumers of services produced by others (Ravenelle 2017), but as active co-creators of the urban commons (Hult and Bradley 2017), the sharing city offers alternatives to predominantly market-encased citizen subjectivities and builds shared capacities to initiate societal change, making it possible to reconfigure currently inequitable power structures (Santala and McGuirk 2019). For McLaren and Agyeman (2015), the sharing city proffers a "truly smart and sustainable city," conceived not as a rigid and structural entity directed by the state and market forces, but as an ongoing and relational process, performed in the here and now.

Despite normative aspirations to foster just and sustainable urban development, to date few empirical studies have unpacked the theoretical or practical contribution of sharing cities and their role in constituting actually existing smart cities. Drawing from participatory action research carried out as part of Inka Santala's PhD project, this chapter uses particular case examples from Sydney, Australia, of a citizen-led sharing platform (ShareWaste), bottom-up fabrication (the Inner West Tool Library), and place-based mapping (The Sharing Map initiative) to analyse their potential to enable and enhance social justice in the city. As the book's Introduction suggests, sometimes teasing out what is unjust helps to reflect on the alternatives and their diverse implications regarding social justice. Thus, the chapter leverages common critiques of the transactional sharing economy as a frame for considering the transformative claims of sharing cities, applying two distinctive iterations of the smart city to explore questions of urban agency, citizens' subjectivities, and collective capacities. Acknowledging the constraints of sharing cities, the chapter provokes discussion of potential supports needed to foster urban social justice in the digital age.

Citizen-Led Platforms and Community Empowerment

Sharing-economy platforms have been regularly critiqued for their commercial orientation. As ideas such as ride sharing and couch surfing have started to gain traction and popularity, corporate entities have recognized the possibility of building efficient and reliable marketplaces while extracting revenue from individual interactions (Srnicek 2017). The expansion of urban sharing platforms has been facilitated by local governments eager to leverage digital platforms and technology innovations to build the "smart city" brand (Barns and Pollio 2019). Thus, ideas that might have originated with a community- and mission-minded ethos have been slowly co-opted by larger and often international corporations for the purpose of creating profit (Martin 2016).

Although this co-option has enabled rapid upscaling and normalization of collaborative consumption (Botsman and Rogers 2010), the focus on direct material benefits downplays the sense of community and caring inherent in the social practice of sharing (Belk 2017). Thus, transactional urban logics that depict sharing as a means to save money or gain convenient access to material assets (Frenken and Schor 2017) can overshadow the fundamental sociality and interdependency of individuals' coming together and agreeing collectively on rules for sharing interactions (Williams 2018). By disregarding the collaborative nature of sharing agency, the transactional sharing economy reproduces contemporary profit-oriented organizational norms, risking the disempowerment of urban communities that are already engaging with more reciprocal and community-based sharing.

To demonstrate the existence of more autonomous and collaborative forms of sharing agency in the smart city, advocates of sharing cities direct attention to the citizen-led platforms currently operating side-by-side with state-facilitated and corporate-run initiatives (see, for example, Gorenflo 2017). These citizen-led initiatives can be anything from online-mediated crop swaps to community-powered platform cooperatives (Scholz 2016), differentiated from many state and corporate-run platforms by their community-owned and -operated basis, which circulates value among those who participate in the sharing practice. These citizen-led platforms seek to empower urban communities to meet their own needs, often beyond state and market (McLaren and Agyeman 2015). Indeed, advocates often frame citizen-led platforms as grassroots solutions to urban challenges that top-down institutions have yet to address (Llewellyn and Gorenflo 2016, 6), as demonstrated by the Australian editor of *Shareable*: "A lot of it is about empowering communities ... given that governments of all persuasions, especially at the national level and international level, seem so unable to respond to climate change and rising inequality."

Citizen-led platforms' potential to empower urban communities is evident in Sydney's ShareWaste initiative, an online database and peer-to-peer application

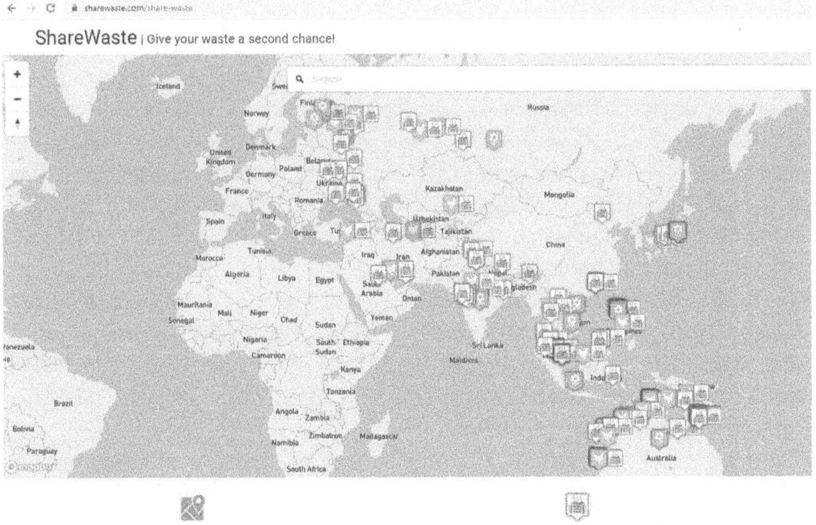

Figure 24.1. Locations of Hosts and Donors in ShareWaste Website
Courtesy of copyright holder.

originating from the city's Inner West. ShareWaste was created by a couple who hoped to connect citizens who generate food scraps with those who compost, helping urban communities to recycle green waste. Since its launch in 2016, the platform has grown quickly to reach a global user base (see Figure 24.1).

Providing a community-based solution to a local problem, ShareWaste complements the lack of formal food-waste recycling in Sydney. By enabling those looking for a sustainable alternative to become engaged and start contributing towards a solution, ShareWaste not only empowers the urban community to co-create a different means of dealing with green waste, but also demonstrates a more "useful doing" in the city compared to previously unsustainable waste practices (Chatterton and Pusey 2020). These benefits were also highlighted by the platform's founder, who saw the purpose of ShareWaste as supporting a sustainable practice while empowering the urban community: "Our motto is that: 'building soil, building the community.' It's 50–50. We're not just a sustainability app, we are a community initiative … Because people are woken up by *War on Waste*,[1] but there is no way how to do those things easily … I think we created a really good solution for people who are half converted but there wasn't a solution for them."

By gathering like-minded people together and connecting them through digital technologies, ShareWaste provides a platform for people to agree on their own rules around food-waste recycling. As there are no formal sharing guidelines, participants need to negotiate around suitable food scraps and the place and means for interaction and taking ownership of sharing practice. According to Thorpe (2018), this form of autonomous and collaborative sharing agency is crucial for people to gain a sense of reshaping the city and taking responsibility for urban commoning. This form of agency was recognized by the founder of ShareWaste, who discussed the different relationship and intimacy of sharing practice compared to the formal recycling process: “The difference between community/council pickup, or just any community drop-off site, is there is the relationship. Once the donor creates some relationship with the host, there’s no reason why they would put in plastic bottles and rotten meat, because they don’t want to have that and they care. And second, there’s also some sort of backup control because the host often is the one who takes the drop-off, so it can be quite intimate.”

As with other citizen-led platforms, however, ShareWaste relies on the founder’s own volunteer time and capabilities, making it vulnerable to shocks such as changing personal circumstances, rising running costs, and the accumulation of responsibilities (Seyfang and Smith 2007). These concerns were also expressed by the platform founder: “It’s like if you’re volunteering for an NGO, but imagine that the whole NGO wasn’t there. And it was just you. Doing everything ... We’re looking into ways how to make it sustainable. But it’s been growing. It keeps growing without any marketing ... But when we grow, it means we have more work, because we’re not just a website, it’s an app. So there’s a lot of client service behind.”

Working with limited time and resources, as the online platform grows it might be at risk of discontinuing or need to look for more commercial means to stay viable within the contemporary profit-oriented and competitive urban landscape (Martin, Upham, and Budd 2015). Although city residents and local government might find the initiative a welcome solution to unsustainable waste practices, taking advantage of the founder’s resourcefulness in the lack of formal procedures and infrastructure undercuts community empowerment in the long term (Avelino et al. 2019). Thus, if citizen-led platforms are to promote social justice via their challenge to contemporary profit-oriented organizational norms and reinstate agency in the urban community, there is a need to prioritize support for initiatives such as ShareWaste.

Bottom-Up Fabrication and Citizen Inclusion

The transactional sharing economy has been critiqued for deploying mainly top-down market solutions that motivate citizens to take part in sharing, not to use and produce shared assets and services collaboratively, but to create financial benefit or gain access to private resources not otherwise available to them (Böcker and Meelen 2017). Although the commodification of private assets has created financial benefits for citizens involved, this relies on atomized and individualized behaviours and consuming practices (Ravenelle 2017), and is premised on a limited understanding of sharers' identities as users and consumers of the services produced by others, largely disregarding the more proactive and creative roles also taking place through the sharing platforms (Richardson 2015; see also Monahan, in this volume). By overlooking these interdependent and socially oriented sharer subjectivities, the sharing economy risks reinforcing predominantly market-encased citizen behaviours, excluding residents from the co-creation of the urban commons.

Advocates of sharing cities have begun to shed light on the bottom-up fabrication that citizens are already involved with across the globe (see, for example, Gorenflo 2017). This fabrication refers to the various means through which citizens are included in urban-sharing initiatives as active makers and co-creators (Hult and Bradley 2017). Providing the space and opportunities for more altruistic behaviours and sustainable practices is often undervalued and taken for granted by top-down institutions (Petrescu et al. 2020). In contrast, bottom-up fabrication includes citizens in urban commoning, rendering them "generous and able to communally self-provide" (Wittmayer et al. 2019). This alters the view of citizen behaviours as necessarily governed, managed, and controlled by competitive market forces and existing political structures (see the dialogue with Rob Kitchin, in this volume).

The potential of bottom-up fabrication to include citizens in co-creation and to nurture related active sharer subjectivities was well recognized by the Australian editor of *Shareable*, who described these active sharer subjectivities as allowing experimentation, cooperation and active urban citizenship: "The new subjectivities are about the courage to experiment and to work with your peers to do so. So, rather than waiting around for change to happen, it is about demonstrating that change is already out there, it's already underway, probably in your own backyard."

Sydney's Inner West Tool Library (IWTL) provides more detailed illustration. The IWTL was initiated as a not-for-profit by a local resident who sought to access expensive machinery without the need to buy it, and came across the idea of a digitally mediated library of things. Based on volunteer labour and donations from community members, the IWTL enables sharing of various material resources, from tools and garden equipment to sports and camping gear (see Figure 24.2). Since its launch in January 2019, the platform has already gained more than 250 donated items with over 1,100 completed loans, all organized by 26 volunteers (Inner West Tool Library 2019).

Enabling citizens to become involved in building the shared tool inventory, IWTL is commoning previously privately held resources. By creating new ways for local residents to contribute by donating their volunteer time or assets, the

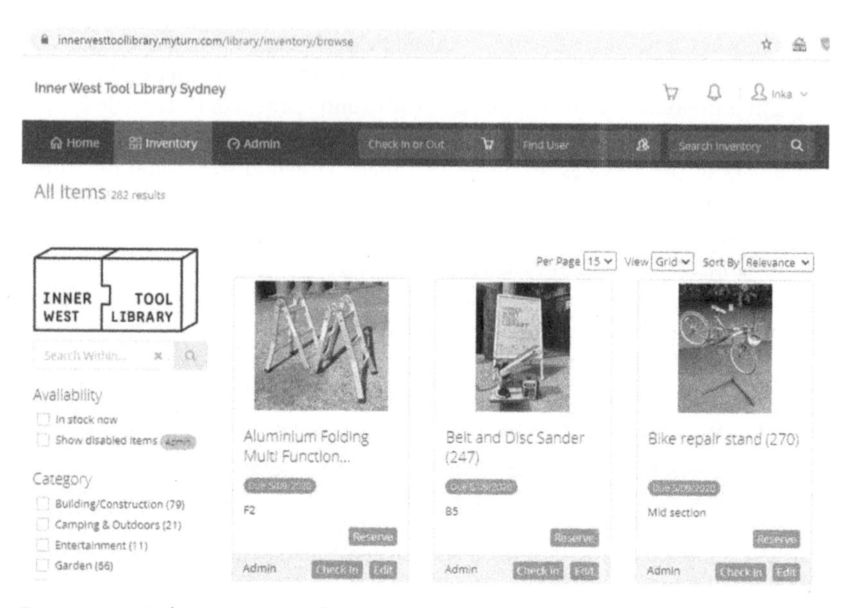

Figure 24.2. Online Inventory of Inner West Tool Library
Courtesy of copyright holder.

IWTL not only includes citizens in the collaborative production of the library, but also acknowledges a different and more proactive way of being in the city (Labaeye and Mieg 2018). This was also recognized by an IWTL volunteer, who explained that joining the tool library had provided him a practical way of contributing to the local community, also enabling the fulfilment of his potential as someone needed, making an impact in the city: "The main thing was doing something … and also try to get something like an impact on the society. And I think this is the perfect thing … it's very easy to see how I can contribute to the Tool Library and to the initiative … and I think that the tool library needs someone like me, so that's super good … Otherwise I would be at home mainly."

By recognizing local residents not only as users and consumers of material resources, but also potential contributors and enablers of sharing practice, the IWTL creates space for citizens to enact interdependent and more socially oriented political subjectivities (see also Williams 2018). Nor are citizens' roles and behaviours limited to lending and borrowing; members can also take on more proactive and creative roles as contributors, librarians, fixers, and advisers, pluralizing the often-narrow depictions of citizenship in data-and-technology-focused smart cities (see Wathne, Karvonen, and Haarstad, in this volume). This diversity of subjectivities and the opportunity to extend them were obvious in the case of the volunteer quoted above, who, despite having no previous experience in community organizing, could now be seen regularly at the library helping with organization, coordinating sharing practice, testing equipment, and advising librarians on safety procedures.

> VOLUNTEER: I'm the librarian coordinator. So my function here is to make sure that all tools are in the best of condition. I also do some tests on the tools. All the power tools need to be tested, so with [the founder] we did a course to certify as testers. And also, trying to find the best way of being, to make things easier to find. I also look after security ...
>
> IS: So is this something you've done previously, organizing and coordinating?"
>
> VOLUNTEER: No, never, never. It's my first experience.

Even though the IWTL enables more proactive and creative means of contributing in the city, sharers' subjectivities are limited by members' embedded habits and identities. As is commonly found with community-based practice (Staeheli 2008), the most active IWTL members are fairly like-minded citizens with similar ideologies and "green" values. While others might be interested in joining the library only for the direct material benefits, as the IWTL founder conceded, "There are people interested in joining, becoming a member, who quite obviously don't care about that – they're just interested in having access. So they're okay with the sharing idea, but they don't really care about why they're sharing."

This suggests the broader question of whether the potential of initiatives such as the IWTL is limited by being convened by a select group of like-minded actors, already oriented to altruistic, green values and identities and to performing proactive and caring roles. If such interests prove to be exclusive of those from different sociopolitical backgrounds and interests (Dillahunt and Malone 2015), then these initiatives' capacities to alter predominantly market-encased citizen's behaviours and to engage citizens as co-creators of the urban commons could be curtailed (see also the dialogue with Alison Powell, in this volume). The task then becomes one of recognizing and nurturing proactive and caring sharer roles and supporting initiatives like IWTL in different local contexts.

Place-Based Mapping and Structural Equality

Acknowledging the risk that corporate-run platforms and top-down market solutions will reinforce profit-oriented organizational norms and individualized behaviours, the sharing economy has been criticized for reproducing inequitable power structures and resource distribution by providing financial and material gains mainly for those who already possess them (Schor 2017) or are trained and skilled to provide quality services (Ravenelle 2017). Transactional sharing-economy benefits accrue in a way that accentuates the existing growing gap between "haves" and "have nots" (Milanovic 2010).

Despite providing new opportunities for gig workers and freelancers, by emphasizing capital-centric urban norms and behaviours the sharing economy restricts collective capacities for broader and lasting social change (Gibson-Graham 2006, 57). This refers to the marginalization of more collaborative forms of sharing agency and proactive sharer subjectivities that are already being enacted through citizen-led platforms and bottom-up fabrication. By undervaluing the social practice of sharing, and thus largely overlooking more altruistic norms and behaviours, the sharing economy limits urban agency and citizens' subjectivities, making it harder to challenge structural inequalities.

To build shared capacities for change, sharing city advocates have been encouraging community members across the world to initiate place-based mapping of local and community-based sharing initiatives (see, for example, Llewellyn 2016). Place-based mapping describes the collaborative production of an online database that lists and visualizes citizen-led platforms and bottom-up fabrication already taking place in the city (Sharp 2018). By visualizing alternative ways of doing and being in the city, place-based mapping encourages the replication and adaptation of sharing initiatives, with a view to unsettling hegemonic urban structures (Labaeye 2017); it also calls for more equal and collaborative forms of urban governance, reframing the city itself as a commons (McLaren and Agyeman 2016). This means revaluing sharing agency as an everyday practice of commoning in the city, recognizing more proactive and creative sharer subjectivities and enabling citizens to act as commoners of the sharing city (Labaeye 2019).

The possibilities of place-based mapping as a tool in redressing structural equalities came to the fore via Sydney's The Sharing Map initiative, an online database of alternative and community-based sharing initiatives operating in the city. The platform was created by a group of community members who wanted to reveal and facilitate access to the city's shared resources and strengthen the local collaborative economy. The ability of place-based mapping to demonstrate alternative economics and urban politics already underway was also highlighted by the Australian editor of *Shareable*: "Rather than look at sharing as a kind of transactional opportunity alone, it is widening that frame of reference to understand how sharing can be leveraged for transformative ends ... Sharing cities are all about building capacity, seeing abundance everywhere, unlocking and appreciating the skills of residents, community groups, everyday people ... it's saying that a new type of economics, a new type of politics, is possible. It's already here, it is just unevenly distributed." In a place-based mapping event in 2018, The Sharing Map gathered Sydney residents together to locate over 250 sharing initiatives, subsequently presented on the project's website (see Figure 24.3).

Figure 24.3. Sharing Initiatives Listed at The Sharing Map
Courtesy of copyright holder.

The project demonstrates already existing collaborative ways of doing and proactive ways of being in the city. By enabling community members to become involved and to identify alternative sharing initiatives, place-based mapping can influence participants' perceptions of the city and its possibilities (Thorpe 2018), in this way promoting more equal and collaborative urban governance. By letting community members self-recognize and contribute to the presentation of sharing in the city, The Sharing Map not only educates the local community, but also creates a different way of relating to the city itself, as a commons (Foster and Iaione 2016). This means understanding the city not as something rigid, controlled by state and market, but as an ongoing process of co-creation (Santala and McGuirk 2019), at the same time reconfiguring the role and nature of sharing platforms to something other-than-normalized by more commercial operators (see Tironi and Albornoz, in this volume). As one of the founders of The Sharing Map described it, the platform is not trying to create anything new, but rather to reframe and amplify what already exists, to build shared capacities for change: "The Sharing Map is trying to show, once again, that things are already happening. It's trying to almost amplify what's already going on as opposed to saying: 'This is how you share, everyone.'"

Although The Sharing Map enables community members to create a different way of relating to the city as a commons, shared capacities for broad and lasting societal change are still limited by the reach and life cycle of the project. As with other place-based mapping projects, The Sharing Map presents a visualization of alternatives at a certain time and location. Its information quickly expires, and learning gathered by the volunteers disappears as they move on to other projects (Seyfang and Smith 2007). As the project loses its core funding and volunteers, there is a risk that the platform will date and slowly cease to exist, as noted by another co-founder of The Sharing Map: "Imagine how much more we could have done if we had had more support, and probably a bit more money. And that wouldn't just suddenly go 'boom,' like it has now, where it stopped. And then all of the things that we've learned and that we now know and the people we've connected with, they just lose all of that … You know, we now have gained all this information and knowledge that's worth something, to them or to the community."

Thus, the project's limited scope and duration risk deteriorating capacities built during its life, leaving little long-term effects. Although participants might have been able to reframe how they perceive not only sharing practices, but also themselves and the city overall, the one-off nature of the project defines the extent to which it can challenge more hegemonic urban structures. If the potential of the sharing city to reconfigure currently inequitable power structures is to be realized, there is a pressing need to enable co-creating, contributing, and relating to the city as a commons by supporting the outreach of place-based mapping such as The Sharing Map.

Conclusion

Drawing on Sydney-based case study examples of ShareWaste, the Inner West Tool Library, and The Sharing Map, this chapter has shed light on actually existing social justice enabled in the sharing city as an alternative iteration of the smart city, one distinct from the transactional sharing economy. We have explored the transformative possibilities of sharing cities focused on citizen-led platforms, bottom-up fabrication, and place-based mapping to enhance community empowerment, citizen inclusion, and structural equality in the city. Acknowledging the constraints on collaborative forms of urban agency, proactive citizens' subjectivities, and collective capacities to transform contemporary capital-centric urban structures, norms and behaviours, we have also pointed to the need for both practical and ideological supports to foster just and sustainable urban development. In doing so, we have sought to demonstrate how the same urban conditions that frame cities as sites of digital (in)justice can be harnessed as political and theoretical tools for sharing cities to generate alternative presentations of (urban and economic) fairness.

The possibilities for, and challenges to, empowerment, inclusion, and equity in the sharing city highlight the need for practical support to counter the smart city's hierarchical and commercially oriented tendencies. First, the resources and time limitations of citizen-led platforms suggests the need to direct material support for initiatives that empower urban communities to bring their own solutions to urban challenges. Second, the limits to bottom-up fabrication by its alignment with a narrow base of user identities suggests the need to enable and facilitate proactive citizenship in different local contexts to expand the base of citizens engaged in urban commoning. And third, the depreciation of knowledge built through place-based mapping initiatives points to the need for support to maintain the resources such initiatives provide and to keep building the collective capacities that enable more just and sustainable urban logics, relations, and practices.

Exploring urban agency, citizens' subjectivities, and collective capacities in the sharing city also demonstrates the need for ideological support to enable the smart city to promote not only technocratic solutions, but also broader justice-oriented societal change. This should begin with shifting focus from direct material outcomes to the nature of social practice and revaluing reciprocal and collaborative forms of urban agency. It requires recognizing the multiple altruistic behaviours and sustainable practices in which urban citizens are already engaging and diversifying the currently limited perception of citizens' subjectivities. And finally, it calls for a progression from replicating and adapting socially innovative initiatives to reconfiguring the neoliberalist logics that have come to dominate urban development, intentionally building collective capacities to achieve urban social justice in the digital age.

NOTE

1 *War on Waste* was an Australian documentary series popular for raising awareness of local waste practices. The series was first presented in 2017 by the Australian Broadcasting Corporation.

REFERENCES

Avelino, F., J.M. Wittmayer, B. Pel, P. Weaver, A. Dumitru, A. Haxeltine, R. Kemp, M.S. Jørgensen, T. Bauler, S. Ruijsink, and T. O'Riordan. 2019. "Transformative Social Innovation and (Dis)Empowerment." *Technological Forecasting and Social Change* 145: 195–206. https://doi.org/10.1016/j.techfore.2017.05.002

Barns, S. 2020. *Platform Urbanism: Negotiating Platform Ecosystems in Connected Cities*. Singapore: Palgrave Macmillan.

Barns, S., and A. Pollio. 2019. "Parramatta Smart City and the Quest to Build Australia's Next Great City." In *Inside Smart Cities: Place, Politics and Urban Innovation*, ed. A. Karvonen, F. Cugurullo, and F. Caprotti, 196–206. Oxford: Routledge.

Belk, R. 2017. "Sharing without Caring." *Cambridge Journal of Regions, Economy and Society* 10 (2): 249–61. https://doi.org/10.1093/cjres/rsw045

Böcker, L., and T. Meelen. 2017. "Sharing for People, Planet or Profit? Analysing Motivations for Intended Sharing Economy Participation." *Environmental Innovation and Societal Transitions* 23: 28–39. https://doi.org/10.1016/j.eist.2016.09.004

Botsman, R., and R. Rogers. 2010. *What's Mine Is Yours: How Collaborative Consumption Is Changing the Way We Live*. New York: HarperBusiness.

Burbank, J. 2014. "The rise of the 'sharing' economy." *HuffPost Australia*, 5 June. Online at https://www.huffpost.com/entry/the-rise-of-the-sharing-e_b_5454710, accessed 31 August 2020.

Chatterton, P., and A. Pusey. 2020. "Beyond Capitalist Enclosure, Commodification and Alienation: Postcapitalist Praxis as Commons, Social Production and Useful Doing." *Progress in Human Geography* 44 (1): 27–48. https://doi.org/10.1177/0309132518821173

Dillahunt, T.R., and A.R. Malone. 2015. "The Promise of the Sharing Economy among Disadvantaged Communities." *Proceedings of the 33rd Annual ACM Conference on Human Factors in Computing Systems*, 2285–94.

Economist. 2013. "The rise of the sharing economy: On the internet, everything is for hire." 9 March. Online at https://www.economist.com/leaders/2013/03/09/the-rise-of-the-sharing-economy, accessed 31 August 2020.

Foster, S.R., and C. Iaione. 2016. "The City as a Commons." *Yale Law & Policy Review* 34: 281–349. https://doi.org/10.2139/ssrn.2653084

Frenken, K., and J. Schor. 2017. "Putting the Sharing Economy into Perspective." *Environmental Innovation and Societal Transitions* 23: 3–10. https://doi.org/10.1016/j.eist.2017.01.003

Gibson-Graham, J.K. 2006. *A Postcapitalist Politics*. Minneapolis: University of Minnesota Press.

Gorenflo, N. 2017. "Introduction." In *Sharing Cities: Activating the Urban Commons*, ed. Shareable, 20–37. Creative Commons Attribution-ShareAlike 4.0 International License (CC BY-SA 4.0): Shareable.

Gurran, N., P.J. Maginn, P. Burton, C. Legacy, C. Curtis, A. Kent, and G. Binder, eds. 2018. *Disruptive Urbanism? Implications of the "Sharing Economy" for Cities, Regions, and Urban Policy*. New York: Routledge.

Heinrichs, H. 2013. "Sharing Economy: A Potential New Pathway to Sustainability." *GAIA* 22 (4): 228–31. https://doi.org/10.14512/gaia.22.4.5

Hult, A., and K. Bradley. 2017. "Planning for Sharing–Providing Infrastructure for Citizens to Be Makers and Sharers." *Planning Theory & Practice* 18 (4): 597–615. https://doi.org/10.1080/14649357.2017.1321776

Iaione, C. 2016. "The CO-City: Sharing, Collaborating, Cooperating, and Commoning in the City." *American Journal of Economics and Sociology* 75 (2): 415–55. https://doi.org/10.1111/ajes.12145

Inner West Tool Library. 2019. *Impact Report 2019.* Online at https://drive.google.com/file/d/1nmLm4VYbYOzo6xxF_DbTvmghSSc-0F9L/view, accessed 13 January 2021.

Labaeye, A. 2017. “Collaboratively Mapping Alternative Economies: Co-producing Transformative Knowledge.” *Networks and Communication Studies, NETCOM* 31 (1–2): 99–128. https://doi.org/10.4000/netcom.2647

Labaeye, A. 2019. “Sharing Cities and Commoning: An Alternative Narrative for Just and Sustainable Cities.” *Sustainability* 11 (16): 43–58. https://doi.org/10.3390/su11164358

Labaeye, A., and H. Mieg. 2018. “Commoning the City, from Digital Data to Physical Space: Evidence from Two Case Studies.” *Journal of Peer Production* 11. Online at http://peerproduction.net/issues/issue-11-city/

Llewellyn, T. 2016. “The Complete Guide to Hosting a #Mapjam in Your City.” Online at https://www.shareable.net/blog/the-complete-guide-to-hosting-a-mapjam-in-your-city, accessed 13 January 2021.

Llewellyn, T., and N. Gorenflo. 2016. *How to: Share, Save Money and Have Fun. A Shareable Guide to Sharing.* San Francisco: Shareable.

Martin, C.J. 2016. “The Sharing Economy: A Pathway to Sustainability or a Nightmarish Form of Neoliberal Capitalism?” *Ecological Economics* 121: 149–59. https://doi.org/10.1016/j.ecolecon.2015.11.027

Martin, C.J., P. Upham, and L. Budd. 2015. “Commercial Orientation in Grassroots Social Innovation: Insights from the Sharing Economy.” *Ecological Economics* 118: 240–51. https://doi.org/10.1016/j.ecolecon.2015.08.001

McLaren, D., and J. Agyeman. 2015. *Sharing Cities: A Case for Truly Smart and Sustainable Cities.* Cambridge, MA: MIT Press.

McLaren, D., and J. Agyeman. 2016. “Sharing Cities: Governing the City as Commons.” In *The City as Commons: A Policy Reader*, ed. J.M. Ramos, 77–9. Melbourne: Transition Coalition.

Milanovic, B. 2010. *The Haves and the Have-Nots: A Brief and Idiosyncratic History of Global Inequality.* New York: Basic Books.

Murillo, D., H. Buckland, and E. Val. 2017. “When the Sharing Economy Becomes Neoliberalism on Steroids: Unravelling the Controversies.” *Technological Forecasting and Social Change* 125: 66–76. https://doi.org/10.1016/j.techfore.2017.05.024

Petrescu, D., C. Petcou, M. Safri, and K. Gibson. 2020. “Calculating the Value of the Commons: Generating Resilient Urban Futures.” *Environmental Policy and Governance* EarlyView: 1–16.

Ravenelle, A.J. 2017. “Sharing Economy Workers: Selling, not Sharing.” *Cambridge Journal of Regions, Economy and Society* 10 (2): 281–95. https://doi.org/10.1093/cjres/rsw043

Richardson, L. 2015. “Performing the Sharing Economy.” *Geoforum* 67: 121–9. https://doi.org/10.1016/j.geoforum.2015.11.004

Rifkin, J. 2014. *The Zero Marginal Cost Society: The Internet of Things, the Collaborative Commons, and the Eclipse of Capitalism.* New York: Palgrave Macmillan.

Santala, I., and P. McGuirk. 2019. "Sharing Cities: Creating Space and Practice for New Urban Agency, Capacities and Subjectivities." *Community Development* 50 (4): 440–59. https://doi.org/10.1080/15575330.2019.1642928

Scholz, T. 2016. *Platform Cooperativism: Challenging the Corporate Sharing Economy*. New York: Rosa Luxemburg Stiftung.

Schor, J.B. 2017. "Does the Sharing Economy Increase Inequality within the Eighty Percent? Findings from a Qualitative Study of Platform Providers." *Cambridge Journal of Regions, Economy and Society* 10 (2): 263–79. https://doi.org/10.1093/cjres/rsw047

Seyfang, G., and A. Smith. 2007. "Grassroots Innovations for Sustainable Development: Towards a New Research and Policy Agenda." *Environmental Politics* 16 (4): 584–603. https://doi.org/10.1080/09644010701419121

Shareable. 2017. *Sharing Cities: Activating the Urban Commons*. San Francisco: Shareable.

Sharp, D. 2018. "Sharing Cities for Urban Transformation: Narrative, Policy and Practice." *Urban Policy and Research* 36 (4): 513–26. https://doi.org/10.1080/08111146.2017.1421533

Srnicek, N. 2017. "The Challenges of Platform Capitalism: Understanding the Logic of a New Business Model." *Juncture* 23 (4): 254–7. https://doi.org/10.1111/newe.12023

Staeheli, L.A. 2008. "Citizenship and the Problem of Community." *Political Geography* 27 (1): 5–21. https://doi.org/10.1016/j.polgeo.2007.09.002

Thorpe, A. 2018. "'This land is yours': Ownership and Agency in the Sharing City." *Journal of Law and Society* 45 (1): 99–115. https://doi.org/10.1111/jols.12081

Williams, M. 2018. "Urban Commons Are More-than-Property." *Geographical Research* 56 (1): 16–25. https://doi.org/10.1111/1745-5871.12262

Wittmayer, J.M., J. Backhaus, F. Avelino, B. Pel, T. Strasser, I. Kunze, and L. Zuijderwijk. 2019. "Narratives of Change: How Social Innovation Initiatives Construct Societal Transformation." *Futures* 112: 1–12. https://doi.org/10.1016/j.futures.2019.06.005

Contributors

Editors

Ryan Burns is Associate Professor in the Department of Geography at the University of Calgary. E-mail: ryan.burns1@ucalgary.ca. At the intersection of digital technologies and human geography, Burns's research advances debates around the social, political, institutional, and economic implications of digital objects, practices, and relations.

Victoria Fast is Associate Professor in the Department of Geography at the University of Calgary. E-mail: victoria.fast@ucalgary.ca. As an urban geographic information scientist, Fast's research is driven by the goal of ensuring all pedestrians have just and sustainable access to the built and digital environment unrestricted by social, economic, and cultural forces such as income, gender, race, and disability.

Debra Mackinnon is Assistant Professor in the Department of Interdisciplinary Studies at Lakehead University, Orillia. E-mail: debra.mackinnon@lakeheadu.ca. Mackinnon's research interests include urban studies, criminology, policing studies, socio-legal theory, surveillance studies, science and technology studies, smart cities, wearables, and qualitative research methodologies.

Dialoguers

Ayona Datta is Professor of Geography, University College London. Datta's work combines interdisciplinary approaches to navigate postcolonial urbanism, marginality, and the complex variegations of smartness.

Stephen Graham is Professor of Cities and Society, Newcastle University. Graham's long-standing work focuses on the transformation of cities through infrastructure, digital media, surveillance, security, and verticality.

Rob Kitchin is Professor and European Research Council Advanced Investigator at Maynooth University. Kitchin's research focuses on the politics and effects of data, software, and related digital technologies.

Vincent Mosco is Professor Emeritus in the Department of Sociology at Queen's University, Kingston, and former Canada Research Chair in Communication and Society. Mosco's work largely orients around the political economy of media and communication technologies.

Alison Powell is Associate Professor of Media and Communication at the London School of Economics and Political Science. Powell's research is a unique examination of the civic use, regulation, and politics of communicate and data technologies.

Authors

Teresa Abbruzzese is Assistant Professor in the Department of Social Science, York University. E-mail: teresa@yorku.ca. Abbruzzese's research interests are interdisciplinary, weaving together critical social, urban, and cultural theory. Abbruzzese's recent empirical investigations focus on questions of care in relation to smart cities and university pedagogy.

Camila Albornoz is Social Anthropologist at the Pontificia Universidad Católica de Chile, Santiago. E-mail: caalbornoz@uc.cl. Albornoz's research interests include digital technologies, urban anthropology, and digital infrastructure/materiality

Matthew Cook is Professor of Innovation as the Open University, Milton Keynes. E-mail: matthew.cook@open.ac.uk. Cook founded and now leads the Future Urban Environments research team. Working on the intersection of urban studies and innovation studies, Cook's research interests are in the development of more sustainable urban and regional environments, with particular reference to mobility and energy infrastructures.

I-Chun Catherine Chang is Associate Professor in the Department of Geography at Macalester College, Saint Paul, Minnesota. E-mail: ichang@macalester.edu. Chang's research interests include global urbanism, urban sustainability, policy mobilities, and Asia.

Ming-Kuang Chung is a postdoctoral fellow at the Institute of Information Science, Academia Sinica, Taipei, Taiwan. E-mail: MKChung843@gmail.com. Chung's research interests include the participatory environmental management and the relationship between geo-spatial technologies and the society.

Federico Cugurullo is Assistant Professor in Smart and Sustainable Urbanism in the Department of Geography at Trinity College Dublin. E-mail: cugurulf@tcd.ie. Cugurullo's research interests include the impact that artificial intelligence is having on the governance, planning, and sustainability of cities.

Joseph Daniels is a PhD candidate in the Department of Geography at the University of British Columbia, Musqueam Territory (Vancouver), and in the School of Geography at The University of Nottingham. E-mail: josephda@student.ubc.ca. Daniels's research interests include alternative economies and economic thinking, money and finance, "platform capitalism," crowds and collectives, and urban governance.

Nina David is Associate Professor in the Joseph R. Biden School of Public Policy and Administration at the University of Delaware. E-mail: npdavid@udel.edu. David's research focuses broadly on land use planning and collaborative governance, and studies plan quality, plan implementation, and the role of public participation in planning processes.

Yonn Dierwechter is Professor in the School of Urban Studies at the University of Washington, Tacoma. E-mail: yonn@uw.edu. Dierwechter's research interests include comparative city-regionalism, urban sustainability/climate action, and spatial planning systems.

Jonathan Gray is Lecturer in Critical Infrastructure Studies at the Department of Digital Humanities, King's College London. E-mail: jonathan.gray@kcl.ac.uk. Gray is co-founder of the Public Data Lab, and Research Associate at the Digital Methods Initiative (University of Amsterdam) and the médialab (Sciences Po).

Håvard Haarstad is Professor in the Department of Geography, and Director of the Centre for Climate and Energy Transformation (CET), University of Bergen. E-mail: havard.haarstad@uib.no. Haarstad's research concerns sustainable transformation of cities, particularly in relation to climate change.

Liam Heaphy is a researcher in the Irish Centre for High-End Computing at National University of Ireland, Galway. E-mail: liam.heaphy@ichec.ie. Heaphy's research interests include environmental and urban science, planning, architecture, and design.

Brandon Hillier is a master's student in the Department of Geography at the University of British Columbia. E-mail: hillier@alumni.ubc.ca. Hillier's research interests include the political economy of central banking, state capitalism, advanced industrial decline in the West, and neoliberal urbanism.

Micah Hilt is a PhD student in the Department of Geography at the University of British Columbia, Musqueam Territory (Vancouver). E-mail: m.hilt@alumni.ubc.ca. Hilt's research interests include urbanization, urban governance and policy, global expertise networks, queer studies, and urban planning.

Andrew Karvonen is Associate Professor in the Department of Urban Planning and Environment at the KTH Royal Institute of Technology, Stockholm. E-mail: apkar@kth.se. Karvonen's research focuses on the social and political implications of technological development in cities.

Maroš Krivý is Associate Professor and Director of Urban Studies in the Faculty of Architecture, Estonian Academy of Arts. E-mail: maros.krivy@artun.ee. Krivý's research interests include the role of urban professionals in neoliberal urbanism and the intersections between urban, digital, and environmental studies.

Karol Kurnicki is a research fellow at the Institute of Advanced Study, University of Warwick, and a senior research associate at the Department of Sociology, Lancaster University. E-mail: kurnicki@gmail.com. Kurnicki's research interests include im|mobilities, digitalization, social practice theory, and critical approaches to urban space.

Carina Listerborn is Professor in Urban Planning at the Department of Urban Studies at Malmö University. E-mail: carina.listerborn@mau.se. Listerborn's research interests include feminist urban perspectives on neoliberal planning, critical housing studies, and smart housing.

Noortje Marres is Professor in the Centre for Interdisciplinary Methodologies at the University of Warwick. E-mail: n.marres@warwick.ac.uk. Marres's work investigates issues at the intersection of innovation, everyday environments, and public life: the role of mundane objects in environmental engagement, intelligent technology testing in society, and changing relations between social life and social science in a computational age.

Michele Masucci is Vice President for Research and Professor of Geography and Urban Studies at Temple University. E-mail: masucci@temple.edu. Masucci's current work examines how barriers to accessing information and communication technologies are interrelated with community development, environmental quality, and access to health, education, and opportunities for civic engagement.

Pauline McGuirk is Professor and Director of the Australian Centre for Culture, Environment, Society and Space at the University of Wollongong, and a Fellow of the Academy of Social Sciences Australia. E-mail: pmcguirk@uow.edu.au. McGuirk's research focuses on critical studies of urban governance, with a current focus on urban governance innovation.

Lorena Melgaço is Associate Senior Lecturer in the Department of Human Geography at Lund University. E-mail: lorena.melgaco@keg.lu.se. Melgaço's research interests include the micropolitics of socio-spatial and technological peripheralization in the postcolony; the intersections of technological dependency, capitalist production of space and the socio-environmental crisis, and the challenges of planning education and practice.

Lígia Milagres is an independent urban researcher who explores the intersection of academic research, socio-ecological practices, and visual arts. She is based in Berlin. E-mail: ligia.milagres@gmail.com. Milagres's research interests include the politics of the collective production of urban commons, socio-spatial pedagogy, and socio-environmental learning practices.

Torin Monahan is Professor in the Department of Communication at University of North Carolina at Chapel Hill. E-mail: torin.monahan@unc.edu. Monahan's research focuses on surveillance, digital platforms, and social inequalities.

Nathan Olmstead is a doctoral candidate in the Department of Political Science at the University of Toronto. E-mail: nathan.olmstead@mail.utoronto.ca. Olmstead's research interests include urban technology, local government, and political theory.

Hamil Pearsall is Associate Professor in the Geography and Urban Studies Department at Temple University. E-mail: hamil.pearsall@temple.edu. Pearsall's research draws on theories and concepts from sustainability science, environmental justice, and urban geography to examine patterns and processes in urban environments using mixed methods.

Helen Roby is Assistant Professor in the Centre for Business in Society at Coventry University. E-mail: ac5036@coventry.ac.uk. Her research interests focus on the use of novel and innovative methods and interventions to influence consumption practices, particularly in the Smart Cities environment. Much of Roby's work is applied, working with industry and commercial partners.

Inka Santala is a PhD candidate at the School of Geography and Sustainable Communities, University of Wollongong, and a member of the Community Economies Research Network. E-mail: ias998@uowmail.edu.au. Santala explores the visions and practices associated with Sharing Cities, with an empirical focus on community-based sharing initiatives in Sydney, Australia.

Zachary Spicer is Associate Professor in the School of Public Policy and Administration at York University. E-mail: zspicer@yorku.ca. Spicer's research interests centre on local government, public administration, and innovation policy.

Alan Smart is Professor Emeritus in the Department of Anthropology and Archaeology at the University of Calgary, Canada. E-mail: asmart@ucalgary.ca. Smart's research interests include political economy, housing, urban anthropology, anthropology of law, borders, zoonotic diseases, smart cities, and posthumanism.

Güneş Tavmen is a visiting research fellow in the Department of Digital Humanities, King's College London. E-mail: mail2gunes@gmail.com. Tavmen's research spans critical data studies, infrastructure studies, STS, and app studies in the context of urbanism.

Martin Tironi is Associate Professor at the School of Design, Pontificia Universidad Católica de Chile, Santiago. E-mail: martin.tironi@uc.cl. Tironi's research areas include digital technologies and mobility, urban infrastructures, and cosmopolitical design.

Miguel Valdez is a lecturer in technology management and innovation and a member of the Future Urban Environments research team at the Open University, Milton Keynes. E-mail: miguel.valdez@open.ac.uk. Valdez's research explores how various publics as well as civic, industrial, and governmental bodies use urban experiments to make sense of innovative technologies and collectively negotiate the future of their city.

Alberto Vanolo, PhD in spatial planning at the Polytechnic of Turin, is Professor of Political and Economic Geography at the University of Turin. E-mail: alberto.vanolo@unito.it. Vanolo's main research fields include urban studies and cultural geography.

Marikken Wullf-Wathne is a doctoral student in planning and decision analysis at the KTH Royal Institute of Technology, Stockholm, and at the Norwegian Institute for Urban and Regional Research (NIBR) at Oslo Metropolitan University. E-mail: marikken@oslomet.no. Wullf-Wathne is currently conducting research on the politics and practice of smart city visioning.

Alan Wiig is Associate Professor of Urban Planning and Community Development at the University of Massachusetts. E-mail: alan.wiig@umb.edu. Wiig's research examines global infrastructure, smart urbanization, and the form, function, and politics of urban and economic development agendas.

Orlando Woods is Associate Professor in the School of Social Sciences at Singapore Management University. E-mail: orlandowoods@smu.edu.sg. Woods's research interests include religion and digital technologies in urban environments in Asia.

Elvin Wyly is Professor of Geography and Urban Studies at the University of British Columbia, Musqueam Territory (Vancouver). E-mail: elvin.wyly@geog.ubc.ca. Wyly's research interests include housing, gentrification, inequality, evolution, and quantitative methods.

Index

Note: *f* indicates figure; *n* indicates note

Milton Keynes UK
Ingram Content Group UK Ltd.
UKHW022213040324
438897UK00023B/209